THE

HISTORY OF ANIMALS

of

ARISTOTLE

and his

TREATISE ON PHYSIOGNOMY

Volume VIII of the Works of Aristotle

Translated by
Thomas Taylor

JOVE HONOURS ME, AND FAVOURS MY DESIGNS
POPE'S HOMER IX, 717

Published by
The Prometheus Trust

Volume XXVI
of
The Thomas Taylor Series

The Prometheus Trust

194 The Butts, Frome, Somerset BA11 4AG, UK.

A registered charity,
number 299648.

The History of Animals

and On Physiognomy

(The Works of Aristotle VIII)

This Edition published in 2004

ISBN 1 898910 25 1

Originally published in 1809

British Library Cataloguing-in-Publication Data.
A catalogue record for this book is
available from the British Library.

Printed in England by Antony Rowe, Chippenham, Wiltshire.

The Prometheus Trust was founded in 1986 in order to help reintroduce to educational establishments those true First Principles which have been the basis of all the world's lasting civilisations. These Principles were given their clearest expression in the West by Plato and those great ones who either provided his grounding in Truth or continued and expanded his beautiful exposition of reality. It is for this reason that the Trust is now making available the whole of the published works of Thomas Taylor, the 'English Platonist', whose philosophical understanding of the principles enshrined in the writings of Plato, Proclus, Plotinus and others is unparalleled in recent times.

The Prometheus Trust would like to thank all those who have contributed to the work involved in the Thomas Taylor Series. We are especially indebted to members of **The Platonic Guild** and would recommend this organisation to all those who wish to make a study of this pure philosophy in order to establish the virtuous life.

The Prometheus Trust is a registered educational charity number 299648. All enquiries should be addressed to: Mr T J Addey, Chairman, The Prometheus Trust, 194 The Butts, Frome, Somerset, BA11 4AG. The Platonic Guild may also be contacted care of the same address.

For the latest information on the Trust and its publishing programme, please visit our website at:

www.prometheustrust.co.uk

* * *

For the preface to the Thomas Taylor Series, a short note on Thomas Taylor, and the Prometheus Trust's introduction to Taylor's *Aristotle*, see the first volume of the Works of Aristotle (Thomas Taylor Series volume XIX).

CONTENTS

Changes to the original text

i. Where Taylor had *the one*, *the one itself*, *the good* or *the good itself* this edition gives these names capital initials; this to distinguish them as the highest names the Platonists gave to God. Other principles have been left in lower case.

ii. A few very obvious grammatical errors and archaic spellings have been corrected; wherever there is any doubt as to the validity of possible errors the original has been followed.

iii. We have followed Taylor's explicit method of printing Greek characters without accents or breathings (see his defence of his Greek at the end of his *The Fable of Cupid and Psyche* and also his reference to this in the Introduction of *Proclus' Commentary on Euclid*).

iv. Some references to works quoted are added, and some original references are given more precise indicators.

v. Standard (Bekker) pagination has been added to the text of Aristotle. **Please note: due to the limitations of our typesetting program the normal pagination convention which gives page number, followed by column indicator of 'a' or 'b', has been modified: column a is indicated by the page number underlined; column b is indicated by the page number without underlining.**

The Thomas Taylor Series

Volume I - *Proclus' Elements of Theology*

Volume II - *Select Works of Porphyry*

Abstinence from Animal Food; Auxiliaries to the Perception of Intelligibles; Concerning Homer's Cave of the Nymphs; The Wanderings of Ulysses -.Taylor

Volume III - *Collected Writings of Plotinus*

Twenty-seven treatises being all the writings of Plotinus translated by Taylor.

Volume IV - *Collected Writings on the Gods & the World*

Sallust On the Gods and the World; The Sentences of Demophilus; Ocellus on the Nature of the Universe; Taurus and Proclus on the Eternity of the World; Maternus on the Thema Mundi; The Emperor Julian's Orations to the Mother of Gods and to the Sovereign Sun; Synesius on Providence; Taylor's essays on the Mythology and the Theology of the Greeks.

Volume V - *Hymns and Initiations*

The Hymns of Orpheus together with all the published hymns translated or written by Taylor; Taylor's essay on Orpheus.

Volume VI - *The Dissertations of Maximus Tyrius*

Forty-one treatises from the middle Platonist, and an essay from Taylor - The Triumph of the Wise Man over Fortune.

Volume VII - *Oracles and Mysteries*

A Collection of Chaldean Oracles; Essays on the Eleusinian and Bacchic Mysteries; The History of the Restoration of the Platonic Theology; An essay on A Platonic Demonstration of the Immortality of the Soul.

Volume VIII - *The Theology of Plato*

The six books of Proclus on the Theology of Plato; to which is added a further book (by Taylor), replacing the original seventh book by Proclus, now lost. Extensive introduction and notes are also added.

Volume IX - *The Works of Plato I*

General Introduction, Life of Plato, First Alcibiades, Republic

Volume X - *The Works of Plato II*

Laws, Epinomis, Timæus, Critias.

Volume XI - *The Works of Plato III*

Parmenides, Sophista, Phædrus, Greater Hippias, Banquet.

Volume XII - *The Works of Plato IV*

Theætetus, Politicus, Minos, Apology of Socrates, Crito, Phædo, Gorgias, Philebus, Second Alcibiades.

Volume XIII - *The Works of Plato V*

Euthyphro, Meno, Protagoras, Theages, Laches, Lysis, Charmides, Lesser Hippias, Euthydemus, Hipparchus, Rivals, Menexenus, Clitopho, Io, Cratylus, Epistles.

Volume XIV - *Apuleius' Golden Ass & Other Philosophical Writings*

The Golden Ass (or Metamorphosis), On the Dæmon of Socrates, On the Philosophy of Plato.

Volumes XV & XVI - *Proclus' Commentary on the Timæus of Plato*

The Five Books of this Commentary, with additional notes and short index.

Volume XVII - *Iamblichus on the Mysteries and Life of Pythagoras*

Iamblichus On the Mysteries of the Egyptians, Chaldeans, and Assyrians; Iamblichus' Life of Pythagoras; Fragments of the Ethical Writings of Certain Pythagoreans; Political Fragments of Archytas, Charondas, Zaleucus, and other Ancient Pythagoreans.

Volume XVIII - *Essays and Fragments of Proclus*

Providence, Fate and That Which is Within our Power; Ten Doubts concerning Providence; The Subsistence of Evil; The Life of Proclus; The Fragments of the Lost Writings of Proclus.

Volume XIX - *The Works of Aristotle I*

The Physics, together with much of Simplicius' Commentary.

Volume XX - *The Works of Aristotle II*

The Organon: The Categories (& Porphyry's Introduction); On Interpretation; Prior and Posterior Analytics; The Topics; The Sophistical Elenchus.

Volume XXI - *The Works of Aristotle III*

The Great Ethics; The Eudemian Ethics; The Politics; The Economics.

Volume XXII - *The Works of Aristotle IV*

The Rhetorics; Nichomachean Ethics; The Poetics.

Volume XXIII - *The Works of Aristotle V*

The Metaphysics (with Commentary from Alexander of Aphrodisiensis and Syrianus); Against the Dogmas of Xenophanes, Zeno and Gorgias; Mechanical Problems; On the World; On Virtues and Vices.

Volume XXIV - *The Works of Aristotle VI*

On the Soul (with Commentary from Simplicius); Sense and Sensibles; Memory and Reminiscence; Sleep and Wakefulness; On Dreams; On Divination by Sleep; The Common Motion of Animals; On the Generation of Animals; On the Length and Shortness of Life; On Youth and Old Age; On Life and Death; On Respiration.

Volume XXV - *The Works of Aristotle VII*

On the Heavens (with Commentary from Simplicius); On Generation and Corruption; On Meteors (with Commentary from Olympiodorus).

Volume XXVI - *The Works of Aristotle VIII*

The History of Animals; On Physiognomy.

THE HISTORY OF ANIMALS

INTRODUCTION

Thomas Taylor

Among the numerous and more popular treatises of Aristotle, his *History of Animals* has in all ages been considered as possessing a very distinguished rank, whether we direct our attention to the comprehensive variety of the matter it contains, or to the accuracy and elegance with which it is written. Other writers on this subject, when compared with Aristotle, may be said to have surveyed the animal creation in a vale, while the soaring and penetrating genius of the Stagirite sees, as from some sublime station,

- whatever breathes or creeps on earth.

Odyssey vi

In the halcyon æra of philosophy, Alexander the Great, who with all his imperfections was certainly one of the most heroic monarchs recorded in history, gave to Aristotle, according to Athenæus, 800 talents† for the expenses necessarily incurred in this arduous work. And Pliny informs us, that Aristotle was desired by Alexander to procure from all Asia and Greece several thousands of hunters, fowlers, and fishermen, that no information requisite to the completion of this great undertaking might be wanting.

Aristotle, therefore, in his four books of *Meteors*, having discussed mixed inanimate natures, as well the imperfect as the perfect, in the following books, as far as to the end of his physical treatises, directs his attention to mixed animated natures; and because among these animals are the principal and the most noble, hence animals are copiously discussed by him in various books. Again, because in animals matter must first be considered, in the second place form, and in the third place properties and accidents; hence in the *History of Animals*, and in the treatises on the *Parts*, and on the progressive *Motion of Animals*, he considers whatever pertains to their matter. In his three books *On the*

† The greater Attic talent was somewhat more than 133 pounds of our money; and, therefore, 800 such talents among to upward of £106,400. But the lesser Attic talent was about 99 pounds sterling; and consequently 800 such talents amount to £79,200.

Soul, he discusses whatever pertains to the form of animals; which form is soul, as being that by which they are internally characterized and distinguished. And in the books subsequent to these, he discusses the properties and accidents of animals, which are sense, memory and reminiscence, sleep, wakefulness, dreams and divination through sleep, common motion and generation, length and shortness of life, youth and old age, respiration and death, health and disease.

Aristotle, also, gives the *History of Animals* a precedence to the other treatises, because without previously knowing the mode in which every thing pertaining to animals subsists, we cannot investigate the causes of its subsistence; and these causes he investigates in the subsequent treatises. Because, likewise, whatever pertains to animals may be reduced to their parts, to sense, to the modes of generating, to actions and lives, that is, to the modes of living and searching for nutriment, and lastly, to manners; but from these the genera and differences of animals are derived, hence Aristotle accurately discusses all these particulars.

In the first, second, third, and prior part of the fourth books, therefore, he discusses the parts of animals; but in the remaining part of the fourth book, he considers the senses, the voice, and the sex of animals. In the fifth, sixth, and seventh books, he treats of the generation of animals, and their various modes of generating. In the eighth book, he discusses the modes in which animals live and search after nutriment. And in the ninth book, he directs his attention to the manners of animals. A tenth book, in which the generation of man is particularly discussed, is, also, ascribed by some writers to Aristotle, and is considered by them as forming a part of this treatise; but, as Diogenes Laertius mentions only nine books of the *History of Animals*, and as this tenth book, is not sufficiently conformable either to the style or the doctrine of Aristotle, I have with the best of his editors and interpreters considered it as spurious, and omitted to translate it.

As I am aware that a genuine modern on perusing this work will be disposed to accuse Aristotle of ignorance for believing in spontaneous generation, of inaccuracy in describing the internal parts of the human body, and of credulity and error in many of the particulars which he relates of animals, it will be necessary to defend him from these accusations, and to show that they originate from an ignorance of true philosophy, and a precipitate perusal of this history.

With respect to the first particular, therefore, spontaneous generation, the language of genuine philosophy on this subject, is as follows:† It is universally admitted that man is generated from man, and from every thing its like. After what manner, therefore, are they generated? For it cannot be said that the generation of these is from chance; since neither nature nor divinity makes any thing in vain. As the generation of men, therefore, is not from chance, it is evidently from seed. Let it, then, be admitted that man is from seed; but seed possesses productive powers in capacity, and not in energy. For since it is a body, it is not naturally adapted to possess productive powers impartibly and in energy. For every where a subsistence in energy precedes a subsistence in capacity; since the latter being imperfect, requires the assistance of something else endued with a perfective power. This something else, you will say, is the nature of the mother; for this perfects and fashions the offspring by its productive powers. For the apparent form of the mother does not make the infant, but nature, which is an incorporeal power and the principle of motion. If, therefore, nature changes the productive powers of seed from capacity to a subsistence in energy, nature must herself possess these productive powers in energy. Hence being irrational and without imagination, she is at the same time the cause of physical productive powers. As the nature of man, therefore, contains human productive powers, does not, also, nature in a lion contain those of the lion; as, for instance, the productive powers of the head, the hair, the feet, and the other parts of the lion? Or whence on shedding a tooth does another grow in its place, unless from an inherent power which is able to make the teeth? How, likewise, does it at the same time make bone and flesh, and each of the other parts? For the same thing energizing according to the same would not be able to fashion such a variety of organization.

But does not nature in plants, also, possess productive powers as well as in animals? Or shall we not say that in these, likewise, the order of generation and the lives of the plants evince that they are perfected from orderly causes? It is evident, therefore, from the same reasoning, that the natures of these, also, comprehend the apparent productive powers. Let us then ascend from these to the one nature of the earth, which generates whatever breathes and creeps on its surface, and which, by a much greater priority, contains the productive powers of plants and animals. Hence the generation of things from putrefaction, and the

† See my Introduction to the *Parmenides* of Plato (TTS vol. XI).

growth of different kinds of plants in the same place without human care and attention, originate from the *whole nature* of the earth containing the productive powers of all these in herself. And thus proceeding, we shall find that the nature in each of the elements and celestial spheres comprehends the productive powers of the animals which it contains. If, also, from the celestial spheres we ascend to the nature of the universe itself, we may, also, enquire respecting this, whether it contains forms or not, and we shall be compelled to confess, that in this, also, the productive and motive powers of all things are contained. For whatever is perfected from inferior subsists in a more excellent and perfect manner from more universal natures. The nature of the universe, therefore, being the mother of all things, comprehends the productive powers of all things: for otherwise, it would be absurd that art, imitating natural reasons,† should operate according to productive principles, but that nature herself should energize without reasons and without inward measures. But if nature contains productive principles, it is necessary that there should be another cause prior to nature, which is comprehensive of forms. For nature verging to bodies energizes in them, just as if we should conceive an artist verging to pieces of timber, and inwardly, by various operations, reducing them to a certain form. For thus nature, merged together with and dwelling in corporeal masses, inspires them with her productive powers and with motion; since things which are moved by others require a cause of this kind, a cause which is properly irrational, indeed, that it may not depart from bodies, which cannot subsist without a cause continually residing with them, but containing the productive powers of bodies, that it may be able to preserve all things in their proper boundaries, and move every thing in a convenient manner. Nature, therefore, belongs to other things, being merged in or co-ordinated with bodies. But it is requisite that the most principal and proper cause should be exempt from its productions: for by how much the maker is exempt from the thing made, by so much the more perfectly and purely will he make: and, in short, if nature is irrational it requires a leader. There is, therefore, something prior to nature which contains productive powers, and from which it is requisite that every thing in the world should be suspended. Hence a knowledge of generated natures will subsist in the cause of the world more excellent than the knowledge we possess, so far as this cause not only knows but gives subsistence to all things; but we possess

† Productive powers (λογοι).

knowledge alone. But if the fabricating cause of the universe knows all things, if he beholds them externally, he will be ignorant of himself, and will be subordinate to a partial soul such as ours; but if he beholds them in himself, he will contain in himself all forms both fabricative and gnostic.

I have continued this reasoning somewhat farther than was requisite to the mere demonstration of the existence of spontaneous generation; but, as in all my publications I have been anxious to lead the reader from matter to mind, from effects to their true causes, and from the investigation of particulars to the knowledge of universals, I was unwilling to omit so fair an opportunity of ascending from material forms to their divine originals in the Maker of the world. I trust, also, that the intelligent reader will immediately see how infinitely superior this reasoning is to the hypothesis of the experimentalists,[1] who from a minute and uninterrupted attention to parts know nothing of the mighty *wholes* in the universe, and the incorporeal powers they contain, and who in order to support their dogma, that every animal and plant are immediately produced from seed, are reduced to the necessity of supposing that the air is replete with millions of all kinds of seeds, which are dispersed every where without order, and which become efficacious only when they meet with a proper recipient. For such an hypothesis directly subverts that most important axiom, and which was held in such veneration by the *pious* philosophers of antiquity, that *God and nature do nothing in vain*; since it is evident that millions of these seeds must perish, in consequence of not meeting with a fit recipient. But experimental enquiries necessarily lead to atheism, when they are not made subservient to the intellectual philosophy.

With respect to the second particular, the inaccuracy of Aristotle in describing the internal parts of the human body, I shall only observe that I am persuaded this accusation principally arises from the hasty and random perusal of his works by those who make it. Thus, for instance, the thorax is defined by Aristotle to be all that cavity which is between the neck and the private parts; but by the moderns it is defined to be, that part of the body situated between the neck and the abdomen. Thus, also, Aristotle comprehends under the term *veins* both arteries and veins; but the moderns denominate them separately. And it is evident in these instances that the disagreement between the two is only a dissonance in words, the anatomical phraseology of Aristotle being in many instances different from that of the moderns. Thus, too, Aristotle defines animal seed to be, "*the excrement of the sanguineous nutriment, which is last of all distributed into the members;*"† but according to the moderns it is *a secretion from the blood.* And it is evident that the difference is only in words. For as Aristotle calls *excrement the relics of nutriment*, seed according to him will be *a secretion of the relics of the sanguineous nutriment which is last of all distributed into the members.* Numerous other instances might be adduced in proof of what I have asserted, as he will find to be true who reads Aristotle with that attention which he so well deserves. In short, I am persuaded that what that eminent and learned surgeon Mr. Charles Barnard has said respecting the skill of the ancients in practical surgery (in Dr. Wotton's Reflections on ancient and modern learning) may with the like justice be said of Aristotle's anatomical skill. "If we enquire (says he) into the improvements which have been made by the moderns in surgery, we shall be forced to confess, that we have so little reason to value ourselves beyond the ancients, or to be tempted to condemn them, as the fashion is among those who *know little, and have read nothing*, that we cannot give stronger or more convincing proofs of our own ignorance as well as our pride."

As to the third accusation, the credulity and error of Aristotle in many of the particulars which he relates of animals, it will be necessary as an apology for him in this particular to unfold to the reader, in the first place, the theory of barren and fertile periods in the sublunary region; for this will be found to be one great cause of the difference between animals in former and in the present times.

Plato, therefore, in the eighth book of his *Republic*, speaking of the dissolution of the city, which he has constituted, observes as follows: "Not only with respect to terrestrial plants, but, likewise, in terrestrial animals, a fertility and sterility of soul as well of body takes place, when the revolutions of the heavenly bodies complete the periphery of their respective orbits; which are shorter to the shorter lived, and contrariwise to such as are the contrary." Now the necessity for such a mutation taking place is this, that all the parts of the universe are unable to participate the providence of divinity in a similar manner, but some of its parts enjoy this perpetually, and others only for a time; some in a primary, and others in a secondary degree. For the universe, being a perfect whole, must have a first, a middle, and a last part. But its first part, as having the most excellent subsistence, must always exist according to nature; and its last part must sometimes subsist according

† See page 275 of my translation of his treatise *On the Generation of Animals.*

to and sometimes contrary to nature. Hence the celestial bodies, which are the first parts of the universe, perpetually subsist according to nature, both the whole spheres and the multitude co-ordinate to these wholes;† and the only alteration which they experience is a mutation of figure, and variation of light at different periods; but in the sublunary region, while the spheres of the elements remain, on account of their subsistence as wholes, always according to nature, the parts of these wholes have sometimes a natural, and sometimes an unnatural subsistence; for thus alone can the circle of generation unfold all the variety which it contains.

The different periods in which these mutations happen are called by Plato, with great propriety, periods of *fertility* and *sterility*; for in these periods a fertility or sterility of men, irrational animals, and plants takes place; so that in fertile periods mankind will be both more numerous, and upon the whole superior in mental and bodily endowments, to the men of a barren period. And a similar reasoning must be extended to animals and plants. The so much celebrated heroic age was the result of one of these fertile periods, in which men transcending the herd of mankind both in practical and intellectual virtue abounded on the earth. And a barren period may be considered as having commenced somewhat prior to the Augustan age,[2] the destruction of all the great ancient cities, with all their rites, philosophy, *etc.* being the natural consequence of such a period.

Hence it is by no means wonderful, if in the time of Aristotle, who lived in a better period than the present, though towards the close of it, animals of every kind should have been on the whole more numerous, and in every respect superior to what they are now found to be; and that in consequence of this, certain particulars in his history of them should be dissonant with the narrations of modern naturalists. This, then, is one apology which may be made for whatever has the appearance of error or credulity in Aristotle's *History of Animals*.

A second, and that no inconsiderable apology may be derived from Aristotle himself, in the eighth book of this *History*. For as he there justly observes (605b23) animals differ according to places; so that as in some places certain animals are not at all produced, in others they are produced, indeed, yet are smaller, live a shorter time, and are subject to a greater number of maladies. And this he illustrates in a great number of instances, which are well worth the attentive perusal of the reader.

† See the Introduction to my translation of the *Timæus* of Plato (TTS vol. X).

He farther adds (607a9) that places, likewise, produce a difference in the manners of animals. "Thus, (says he), the manners of animals in mountainous and rude places are different from those in plain and cultivated places. For their aspect is more ferocious, and they are more robust in the former than in the latter situation: as is evident in the swine in mount Athos. For the males of the valleys avoid the females of the mountain." And he farther adds, that places, also, produce a great difference in the effect of the bite of savage animals. In short, as Proclus well observes in his MS. Commentary on the *Parmenides* of Plato, "if we direct our attention to the generation of particulars, we must assign as the causes of these, the order of the universe, the motion of the heavens, partial natures, the characteristic peculiarities of the seasons, climates, and the powers that preside over these as inspective guardians; since through these powers the mutations of the seasons affect individuals. For the cause being moved, in a certain respect moves together with itself that which is generated from it. Another cause arises from the peculiarities of the presiding causes. For different figures, colours, voices, and motions are appropriate to different individuals. The differences of climates, also, contribute to this diversity. For the generations are different in different places; and partial natures not only proceed from the whole of nature [or from nature considered as one of the great wholes of the universe], but they, likewise, receive something from the characteristic property of the seeds, and are fashioned by this in consequence of verging to bodies, to which they become as it were subject, and no longer possess authority from themselves."† And thus much as an apology for Aristotle, in answer to the accusations of those who through a profound ignorance of the intellectual philosophy, and the consequent want of an enlarged view of things, dare to calumniate the writings of the best and wisest of mankind, and whose little soul, in

† Επειδη παν το γιγνομενον εξ αιτιου γιγνεται τινος, και τουτων (i.e. των καθ' εκαστον) αιτιας θετεον, ως μεν εν τι φαναι, την ταξιν του παντος, ως δε και πλειους αιτιας ειπειν, την κινησιν του ουρανου, τας μερικας φυσιες, τας ιδιοτητας των ορων, τα κλιματα, τους τουτων εφορους. δια γαη τουτων, ορων μεταβολαι δρωσιν εις αυτα. το γαρ αιτιον κινηθεν, συγκινει πως εαυτῳ και το εξ αυτου γενομενον, και αιτιων εφορων ιδιοτθτες. αλλα γαρ αλλοις και σχηματα οικεια, και χρωματα, και φωναι, και κινησεις, και αι των κλιματων διαφοραι. αλλοιαι γαρ εν τοις διαφοροις τοποις αι γενεσεις, και αι μερικαι φυσεις, ου μονον εκ της ολης προιουσαι φυσεως, αλλα και εκ της των σπερμαων ιδιοτητος προσλαμβακουσαι τι, και τυπουμεναι κατ' αυτην την τῳ δυνειν κατα των σωματων, και οιον, εκεινων ειναι και ουχ εαυτων.

the language of the Emperor Julian, is indeed acute, yet sees nothing with a vision healthy and sound.

With respect to the translation of this work, suffice it to say, that in making it, I have endeavoured as usual that it should be as faithful as possible, and in the execution of it have availed myself of the Greek and French edition of M. Camus printed at Paris, 1783. In saying this, however, it must be understood, that I nearly solely availed myself of the Greek text of this addition of the *History of Animals*, as more correct than that of any other; for of the French language I am totally ignorant; though in giving appropriate English names to some of the animals mentioned by Aristotle, I was greatly assisted through the medium of a Dictionary, by the translation of M. Camus. The copious notes which accompany the translation of this gentleman, I would recommend to the perusal of the French scholar, from the numerous and learned citations with which they abound, and of which I should have partially availed myself, had I understood the language in which they are written. I say *partially*, for my object in translating the whole of Aristotle's works was not to comment copiously on all that he has written, but to bring to light all the sublime and most important dogmas of his philosophy, this not having been executed by any one since the time of the Emperor Justinian; and, also, to prove that these dogmas are perfectly conformable to those of Plato. This grand object, to which all my labours in this very arduous undertaking are directed, must be my apology for any deficiencies or inaccuracies in other respects which may be found in this, or any of the other volumes. For if it shall appear that I have faithfully unfolded these dogmas from genuine sources, I shall neither envy those who have illustrated Aristotle better in less particulars, nor regard the criticism of those who may censure me as failing where it would have been easy to excel.

From the edition, also, of M. Camus I have extracted the alphabetical list of the names of the animals employed by Aristotle in this *History of Animals*. For a great part of these names, I could find no appropriate terms in the English language; and I am informed by learned men who are much better skilled in natural history than I pretend to be, in consequence of having made it the whole of their study, that such terms are not to be obtained.

In translating the treatise on Physiognomy, which I have arranged immediately after the History of Animals, though I availed myself of all the helps afforded in the excellent edition of this treatise by Franzius, printed at Altenburg, 1780, and have for the most part adopted his emendations of the very corrupt Greek text, yet the learned reader who compares my translation with the original of this edition, will find that I have frequently differed from Franzius, and have, I may presume to say, in not a few places thrown additional light on this curious and elegant treatise.

Of these emendations and alterations of the text, had I been willing to gratify *mere verbal critics*, I might have made a pompous display in notes. But as I have always found these men to be not only egregious triflers, but, also, void of all principle - men who have dared to assert that I translate from the Latin,[3] when I profess to translate from the Greek, in order to blast my labours, and prevent me from diffusing the salutary light of genuine philosophy, I have thought it much better to frustrate than to gratify their wishes. It, also, occurred to me, that even if they should approve of the alterations which I have made, I ought to pay attention to the Pythagorean precept, that "*the praise of fools is a reproach*."

I shall only add, that I would recommend the treatise of Baptista Porta on Physiognomy, as an excellent commentary on that of Aristotle; and that in the next volume I shall present the reader with some mathematical discoveries of my own, which, I trust, he will deem to be of no mean importance. Among the treatises of Aristotle, of which the next volume will consist, there is one on INDIVISIBLE LINES; and the introduction to this treatise will afford me an opportunity of unfolding these discoveries. I shall there most clearly demonstrate that all the propositions in Dr. Wallis's *Arithmetic of Infinites*, relative to the summation of infinite series, are false; that both the arithmetic of infinities and the doctrine of fluxions are founded on erroneous principles, and are singular instances of the possibility of deducing true conclusions from false premises; and that modern mathematicians have been greatly mistaken in supposing that inferences which are the result of induction in an infinite series are true. For it will be shown, that the aggregate of a series of terms when infinite possesses properties very different from the aggregate of any part of it; and, consequently, that we cannot reason by induction from the parts to the whole of such a series. All the conclusions, likewise, of Bernoulli, in his treatise *De Arte Conjectandi*, respecting what he calls the 12th property of figurate numbers, will be demonstrated to be false, when the number of terms is infinite, the first conclusion excepted.

Additional Notes to Taylor's Introduction

1 (See p. 5) Ray, in his treatise on the *Wisdom of God in the Creation*, exhibits a very remarkable proof of the truth of what is here asserted; for he appears to have had no conception whatever of the agency of Nature, considered as an incorporeal power operating in matter, in subordination to the cause of all things; and from the want of such a conception, he entirely rejects spontaneous generation. For in p. 73, *etc.* he thus writes: "Indeed, a spontaneous generation of animals and plants upon due examination will be found to be nothing less than a creation of them. For after the matter was made, and the sea and dry land separated, how is the creation of plants and animals described, but by a commanding, that is, effectually causing the waters and earth to produce their several kinds without any seed? Now creation being the work of omnipotency, and incommunicable to any creature, it must be beyond the power of Nature or natural agents to produce things after that manner. And as for God Almighty, he is said to have rested from his work of creation after the seventh day. But if there be any spontaneous generation, there was nothing done at the creation but what is daily done; for the earth and water produced animals then without seed, and so they do still." He adds, "such a spontaneous generation seems to me to be nothing less than a creation. For, creation being not only a production of a thing out of nothing, but, also, out of indisposed matter, as may be clearly inferred from the scripture, and is agreed by all divines; this spontaneous generation being such a production, wherein does it differ from creation? Or what did God Almighty do at the first creation of animals, and plants, more than what (if this be true) we see every day done? To me, I must confess, it seems almost demonstrable, that whatever agent can introduce a form into indisposed matter, or dispose the matter in an instant, must be superior to any natural one, not to say omnipotent." Prior to this, however, he confesses that there is one instance which militates against his opinion. "No instance against this opinion (says he) doth so much puzzle me as worms bred in the intestines of man and other animals. But seeing the round worms do manifestly generate, and probably the other kinds too; it is likely they come originally from seed, which, how it was brought into the guts, may afterwards possibly be discovered."

The reader who is at all conversant with the intellectual philosophy of Plato and Aristotle, will easily see that what I have just cited is the result of a very profound ignorance of the following most important truths: That the cause of the universe is an *eternally* energizing Being, and that he does not, therefore, operate at one time, and rest from his work at another, nor begin at a certain time to energize, having been prior to this quiescent; for this is to suppose him a being resembling man. That there are many incorporeal powers proceeding from this cause similar to, though not the same with it, and co-operating with it in the fabrication of the universe. That Nature is the last of these mighty incorporeal powers. And that the universe ever has been, is, and will be incessantly passing into existence, through the energy of these powers in conjunction with the energy of their great eternal cause. Poor Ray for want of this knowledge was reduced to the miserable shift of asserting *that it is likely the worms in the intestines of animals came originally from seed, but that how it was brought into the guts may afterwards possibly be discovered.* Certainly *Ray* was on this subject *rayless.*

2 (See p. 7) It appears to me that this period commenced in the time of Sylla, and I found this opinion on the following passage in Plutarch's *Life of the Great Commander*:- Το δε παντων μεγιστον, εξ ανεφελου και διαιθρου του περιεχοντος ηχησε φωνη σαλπιγτος, οξυν αποτεινουσα και θρηνωδη φθογτον, ωστε παντας εκφρονας γενεσθαι, και καταπτηξαι το μεγεθος. Τυρρηνων δε οι λογιοι μεταβολων ετερου γενους απεφαινοντο, και μετακοσμησιν αποσημαινειν το τερας. ειναι μεν γαρ αυτῳ οκτω τα συμπαντα γενη διαφεροντα τοις θιοις και τοις ηθεσι δι αλληλων, εκαστῳ δε αφωρισκαι χρονων αρισθμον, υπο του θεου συμπεραινομενον ενιαντου μεγαλου περισδῳ και οτον αυτη σχη τελος, ετερας ενισταμενης κινεισθαι τι σημειον εκ γης η ουρανου θαυμασιον. *i.e.* "But the greatest of all [the signs prior to the civil wars] was the following: On a cloudless and clear day, the sound of a trumpet was heard, so acute and *mournful*, as to astonish and terrify by its loudness all that heard it. The Tuscan wise men, and soothsayers, therefore, declared that this prodigy signified the mutation into and commencement of another age. For according to them there are eight ages, differing from each other in lives and manners, each of which is limited by divinity to a certain time of duration, and the number of years of which this time consists is bounded by the period of the great year. Hence when one age is finished, and another is about to commence, a certain wonderful sign will present itself, either from the earth or the heavens." The *mournfulness* of this sound of the trumpet was evidently an indication that a barren period was about to commence. For an account of the *great year*, see the note [18, p. 527, TTS vol. XXV] to the treatise *On Meteors.*

3 (See p. 10) I understand that this has recently been asserted by a writer in a scurrilous and fungous publication called the Edinburgh Review, in a criticism on my translation of Plato; this writer at the same time acknowledging that I could have translated from the Greek had I chosen to do so. In answer to this I deem it will be sufficient merely to ask the intelligent reader whether it is likely, that a man who has several folio volumes of Platonic Greek manuscripts in his possession, of which no translation was ever made, and the substance of which he has given in English in the notes to his translation of the works of Plato, would translate Plato from the Latin, at the same time professing to translate him from the Greek? The iniquity of such an assertion is only to be

equalled by its impudence. In consequence of having spent the far greater part of my life in the study of the philosophy of Plato and Aristotle, I profess to have unfolded from genuine sources all the most abstruse and important doctrines of these two great philosophic luminaries; and I defy any one of these *darkling scribblers* to prove that I have not done this, or that any man else has done it since the destruction of the Greeks. But to such puerile opponents it is no arrogance to say, in Homeric language,

"Give me a *man*, that we our strength may try."

THE HISTORY OF ANIMALS

THE HISTORY OF ANIMALS

BOOK I

486

1 Of the parts of animals, some are incomposite, *viz.* such as are divided into similar parts, as flesh into flesh, but others are composite, being such as are divided into dissimilar parts; as the hand is not divided into hands, nor the face into faces. Some things however of this kind, are not only called parts, but also members; and these are such parts, as being wholes, contain in themselves other peculiar parts; such as the head, leg, and hand, the whole arm, and the thorax. For these parts are themselves wholes, and there are also other parts of them. But all dissimilar parts are composed from similar parts; as, for instance, the hand, from flesh, nerves and bones.

Some animals, however, have all the parts the same with, and some have them different from, each other. But some of the parts are with same in species; as the nose and eyes of one man are specifically the same with the nose and eyes of another man, and flesh with flesh, and bone with bone. This is also the case with the parts of horses and other animals, which we assert to be specifically the same. For as one whole is to another, so is each of the parts to each. But some animals have parts of the same genus, yet different according to excess and defect. I speak of those whose genus is the same, as birds and fishes. For each of these has a difference according to genus, and with reference to genus, 486
and there are many species of fishes and birds. Nearly however most of the parts in them differ in the contrarieties of passive qualities, such as colour and figure, because some of them are more passive to these, but others less. Farther still, they also differ in multitude and paucity, in magnitude and parvitude, and, in short, in excess and defect. For some of them are crustaceous, and others testaceous; some have a long beak, as cranes, and others a short beak; and some have many, and others a few, feathers. Nevertheless in these also some of the parts are different from others; for some of them have spurs and others not; and some have, but others are without a tuft of hair on the forehead. As I have said, however, most parts, and from which the whole bulk is composed, are either the same, or differ in contrarieties according to excess and defect. But some animals neither have parts specifically the same, nor the same according to excess and defect, but according to analogy; just as bone is analogous to the spine, a nail to a hoof, a hand to a claw, and the scales of a fish, to the feathers of a bird. For what a feather is to a

bird, that a scale is to a fish. The parts, therefore, which animals severally possess, are after this manner different and the same; and the like also takes place in the position of the parts. For many animals have
487 the same parts but not similarly situated; as, for instance, some have the mammæ in the breast, but others between the thighs.

But of the similar parts, some are soft and moist, but others are dry and solid; *viz.* they are either entirely moist, or are so as long as they remain in their natural condition; as blood, lymph, fat, suet, marrow, the genital seed, bile, milk in those animals that have it, flesh, and substances analogous to these. Again, after another manner, excrements [are soft parts] such as phlegm, the sediments of the belly, and urine. But the solid and dry parts, are such as nerve, skin, vein, hair, bone, cartilage, nail, and horn. For horn is homonymously predicated with respect to genus, when it is predicated of figure and a whole. The parts also analogous to the before mentioned are hard and dry.

But the differences of animals are, according to lives and actions, manners and parts; concerning which we shall in the first place speak generally, and afterwards discuss the genus of each. The differences, therefore, are, according to lives, actions, and manners; but the causes of these differences are, that some animals are aquatic, and others terrestrial. But the aquatic are twofold; for some of them live, and obtain their nutriment in moisture; which they alternately receive and eject; and they cannot live when they are deprived of this, which happens to be the case with many fishes. Others live in moisture, and derive their nutriment from it, yet they do not receive water, but air, and generate externally. But many animals of this kind are also pedestrious, as the otter, the amphibious animal called latax, and the crocodile. Some of these are also winged, as the plungeon and the didapper; and likewise those which are without feet, as the water-snake. But some obtain their nutriment in moisture, and are not able to live out of it, yet neither receive air, nor moisture, as the shell fish called urticæ,[†] and oysters. Of aquatic animals, however, some are marine, others fluviatic, others live in lakes, and others in marshes, as the frog, and the cordylas. And of marine animals, some pertain to the sea, others to shores, and others to rocks.

But of the terrestrial, some receive and eject air, which is called respiration and expiration; as man, and all such terrestrial animals as have lungs. Others again do not receive the air, but live, and have their food in the earth, as wasps, bees, and other insects. But I denominate

† The urtica is a muscle fish, which pricks and stings when touched.

those animals insects, which have incisions in their body, either in the supine parts, or in these, and also in the prone parts. And many terrestrial animals, as we have said, obtain their nutriment in moisture; 487
but no aquatic animal, and which receives the sea water, derives its aliment from the earth. Some animals, however, first live in moisture; but afterwards change into another form, and live out of moisture; as is the case with the gnats in rivers. But from these the horse-fly is produced.

Again, some animals are stable, but others are mutable. And the stable live in moisture: no terrestrial animal, however, is stable; but many that dwell in moisture live fixed to rocks; as many kinds of shell fish. The sponge also appears to have a certain sense; an indication of which is, that it is with greater difficulty divulsed, as they say, unless the motion by which it is torn away is secret. But some adhere to rocks, which they occasionally leave; such as is a certain genus of what is called the urtica. For some of these feed by night liberated from the rock. Many also when they are liberated are without motion, as oysters, and the spongy excrescences called *olothouria*.† But some have a power of swimming, as fishes, the mollia, and those animals that are covered with a soft shell, such, for instance, as locusts. Some likewise are able to walk, as the genus of crabs; for this animals is naturally aquatic, but has the power of walking.

Of terrestrial animals, however, some are winged, as birds, and bees; and these differ after another manner from each other; but others are pedestrious. And of the pedestrious, some are gradient, others are reptiles, and others have a rolling motion. But there is no animals which has alone the power of flying, in the same manner as fishes have alone the power of swimming. For animals with membranaceous wings are gradient; and bats have feet, and the sea-calf, or phoca, has very small feet. Some birds also have infirm feet, and on this account are called *apodes*, or without feet. But this species of birds has large wings. And nearly those birds which resemble these, have large wings, but have infirm feet, as the swallow, and the *martinet*.‡ For all these are similar in their manners, and their flying, and approximate to each other in their external form. The bird likewise called *apous*, or without feet, is seen in every season of the year; but the martinet is only seen in

† A kind of spongy and marine excrescence.

‡ A bird that breeds on watery banks, or a bank-martin.

summer, when it rains. For then it is both seen and caught, but otherwise this bird is rare. Many animals, however, are both gradient, and have the power of swimming.

There are also such like differences according to lives and actions. For some animals are gregarious; but others are solitary, as well those that
488 are pedestrious as those that are winged, and those that swim. But others participate of each of these natures, so that they are sometimes gregarious, and sometimes solitary. Some animals likewise lead a political, but others a wandering life. The gregarious, therefore, among winged animals, are the genus of doves, the rane, and the swan. But no animal that has crooked nails is gregarious. Among animals that swim also, many genera of fishes are gregarious, such as those which they call flying-fish, the tunny, the pelamides, or pilchers, and hamiæ. But man is both gregarious and solitary. Those animals however are political, of all which there is one common work; which is not the case with all gregarious animals. Man, a bee, a wasp, an ant, and a crane, are animals of this kind. And of these, some live under a leader, but others are without a leader. Thus a crane, and the genus of bees, are under a leader; but ants, and ten thousand others, are subject to no dominion. Some animals likewise, both among those that are gregarious, and those that are solitary, do not migrate from one place to another; but others change their habitations. And some indeed are carnivorous, others feed on all kinds of food, and others feed on fruit. Others again, use a peculiar kind of food, as bees and spiders. For bees feed on honey, and certain other sweet substances; but spiders live from the hunting of flies. Some likewise use fish for food. And some indeed hunt their food; but others hoard it. Others again, are neither adapted to hunt, nor to hoard it. And some indeed procure habitations for themselves, as the mole, the mouse, the ant, and the bee; but many insects and quadrupeds are without habitations. Again, with respect to places of abode, some dwell in caverns, as lizards, and serpents; but others dwell upon the earth, as dogs and horses. Some likewise excavate holes for themselves, but others do not. And some wander by night, as the owl and the bat; but others live in the light.

Again, some are tame, but others are savage. And some are always tame, as man and a mule; but others are always savage, as a panther and a wolf. Some also may be rapidly rendered tame, as the elephant. Farther still, animals may be divided after another manner. For all such genera as are tame, are also savage, as horses, oxen, swine, men, sheep, goats, and dogs. And some indeed are vocal, but others are mute, and others utter a sound. And of these, some have a dialect; but others are without letters. Some likewise are loquacious; but others are silent; some sing, and others do not. It is common, however, to all vocal animals, at the time of coition, especially to sing and talk. And some 488 indeed dwell in fields, as the ring-dove; but others in mountains, as the hoopoop; and some live with men, as doves. Some also are prone to venery, as the genus of partridges and cocks; but others are more chaste, as those birds which belong to the class of crows; for coition with these is rare. Of marine animals also, some dwell in the sea, others dwell about shores, and others near rocks. Again, some are capable of repelling an injury, but others are careful to avoid being injured. And those indeed are capable of repelling an injury, which either attack others, or defend themselves when injured. But those are careful to avoid an injury, which in order that they may not suffer any harm, contain in themselves the means of defence.

Animals likewise differ in their manners. For some of them are mild, are not easily excited to anger, and are not pervicacious, and such are oxen; but others are irascible, pervicacious, and indocile, as the wild boar; and others are prudent and timid, as stags, hares and rabbits. But others are illiberal and insidious, as serpents. Some again are liberal, brave, and noble, as lions, but others are generous, wild, and insidious, as wolves. For that is noble which derives its origin from a good genus; but that is generous which does not degenerate from its own nature. Some also are crafty and malevolent, as the fox; but others are irascible, friendly and fawning, as dogs; and others are mild and tractable, as elephants. Some again are bashful and cautious, as geese; but others are envious and lovers of finery, as peacocks. But man is the only animal that is capable of deliberating and consulting. And though many animals participate of memory and discipline, yet none besides man is capable of reminiscence.[†] Concerning each genus of animals, however, their manners, and their lives, we shall speak hereafter, and with greater accuracy.

2 The common parts however of all animals are those which receive food, and into which it is received. But these are the same and different

† It must be carefully observed by the reader, that reminiscence, according to Aristotle, is an investigation in an orderly series of something which has departed from the memory; and of this no animal but man is capable. See the treatise on *Memory and Reminiscence.*

according to the before-mentioned modes; for they are divided either according to species, or according to excess, or according to analogy, or from position. In the next place, other common parts of most animals are obtained from these, and which emit and receive the excrement of food. For this does not pertain to all the parts. But that part which
489 receives the food is called the mouth; and that into which it is received, the intestine. That which remains is polyonimous. Since excrement however is two-fold, those animals which have parts receptive of humid excrement, have also parts receptive of dry food; but not all such as have the latter have also the former. Hence those which have a bladder, have also the intestines; but not all which have intestines, have also a bladder. For the part which is the recipient of humid excrement, is called the bladder; but that which is the recipient of dry excrement, the intestine.

3 Many of the remaining tribes of animals have these parts, and in addition to these, that part which emits the seed. And of these, with respect to the animals which are adapted to the generation of animals, some emit seed into themselves, but others into others. But the animal which emits seed into itself, is called female; and that which emits it into another, male. In some, however, there is not the distinction of male and female. The form also of the parts which pertain to this fabrication is different; for some have a womb; but others that which is analogous to a womb. These therefore are the parts which are necessary to all, or to most animals. But in all animals, one common sense alone, the touch, is inherent; so that the part in which this sense is naturally adapted to be present, is anonymous. For in some animals it is the same in species, but in others it is analogous.

4 Every animal also has moisture, of which when deprived, either by nature or by violence, it is corrupted. That likewise in which this moisture is produced is different. But this in some animals is blood and vein, and in others that which is analogous to these. These, however, are imperfect; and the one is called fibre, but the other lymph. The touch therefore is produced in a part of the body which consists of similar parts; as in flesh, or in something of this kind; and in short in the sanguineous parts, or in those which are analogous to these. But it is evident that this sense is in all the similar parts; and that the producing powers are in the dissimilar parts of the body. Thus the elaboration of the food [*i.e.* mastication] is in the mouth, and of local motion, is in the feet, or in wings, or in certain things analogous to

these. In addition to these particulars also some animals have blood, as men and horses, and all such as being perfect, are without feet, or are bipeds, or quadrupeds. But others are without blood, as bees and wasps, and among marine animals, the cuttle and the carabus [a kind of crab]; and all such animals as have more than four feet.

5 Some animals also are viviparous, others oviparous, and others vermiparous. And the viviparous indeed, are such as man, horse, the phoca, or sea-calf, and other such animals as have hair; and among aquatic animals, those of the cetaceous kind, as the dolphin, and those that are called cartilaginous. But of these some have a fistula, or pipe, but have no arms, as the dolphin, and the whale. The dolphin, however, has the fistula in the back, and the whale in the forehead. But some have arms disclosed, as the thorn-back, the sea-dog, and other cartilaginous fishes. Again, with respect to a perfect fœtus, that is called an egg, from the principal part of which the generated animal is produced; but the rest is nutriment to that which is generated. And a worm is that from the whole of which another whole animal is generated, the fœtus being distinctly formed and increased. Of viviparous animals, therefore, some [previously] produce an egg in themselves, as the *selache*;† but others [immediately] generate an animal in themselves, as man and horse. When the fœtus however is perfect, either an animal proceeds into light or an egg, or a worm. But of eggs, some have a hard outward case, and a two-fold colour, as the eggs of birds; but others have a soft shell, and one colour only, as the eggs of the selache. Of the worms also produced by vermiparous animals, some immediately move themselves, but others are immoveable. Concerning these, however, we shall afterwards speak more accurately in the treatise *On the Generation of Animals*.

Farther still, of animals, some have feet, but others are without feet. and of those which have feet, some have two, as men, and besides men, birds alone; but some have four, as the lizard and the dog; and others have more than four, as the scolopendra, and bees. But all animals that have feet, possess them in an even number.

Of swimming animals, however, which are without feet, some have fins, as fishes; and of these, some have four fins, two in the superior and two in the inferior part, as the gold-fish, and the sea-wolf; but others fishes that are oblong and smooth have two only, as the eel, and the

† The *selache* are a kind of cartilaginous fishes.

conger. Others, however, have not any fins, as the lamprey, but use the sea as serpents use the earth, and in a similar manner swim in moisture. But of the cartilaginous kind, some have no fins, *viz.* such as are broad and have a tail, as the skate and the pastinaca [*i.e.* certain fish like a thornback]; and these swim by their breadth. Those however have fins whose breadth is not attenuated. But such as appear to have feet, as the fish called mollia, these swim with their feet and their fins; and by these means draw their belly along more swiftly. The sepia and loligo [or the calamary] neither use their fins, nor their feet in swimming; but the polypus uses both. Those however that have a hard skin, as the carabus, swim with their tails, and most rapidly, by the aid of fins, which they have in their tails. The *cordulus* also swims with his feet and his tail, the latter of which is similar to that of the *glanis*, if what is small may be compared with what is great.

But of birds, some are winged, as the eagle and the hawk; others have a dry membrane for feathers, as bees and beetles: and others have leathern wings, as the bird called alopex [or the flying fox] and the bat. Those, therefore, that are [simply] winged, and in a similar manner those that have leathern wings, are sanguineous; but those which are exanguious, have a dry membrane, as insects. All those, however, that are [simply] winged, or that have leathern wings, are bipeds, or without feet. For there are said to be certain snakes of this kind about Æthiopia. The winged genus therefore of animals, is called bird; but the remaining two genera are anonymous in one name. And of the winged genus indeed, but exanguious, some have their wings enclosed in a crustaceous sheath, as the melolonthe [or round beetle], and the beetle. But others are without a crustaceous sheath; and of these, some have two wings, but other four. Those indeed have four wings that are of a large size, or that have a sting in the posterior part of the body. But those have two wings, that either are not of a large size, or have a sting in the anterior part of the body. No one however of those whose wings are enclosed in a crustaceous sheath, has a sting. But those that have two wings have the sting in the anterior part, as the common fly, the gad fly, the ox fly, and the gnat. All exanguious animals, however, are of a less size than those that are sanguineous, except a few that are exanguious in the sea, which are of a larger size, as some of the mollia. And these genera of them, indeed, become the largest in warmer places, and more so in the sea and in fresh water than on the land.

All animals, however, that move themselves, are moved with four members subservient to motion, or with more than four. Those indeed that are sanguineous are moved with four only: as man, with two hands and two feet; birds, with two wings and two feet; quadrupeds with four feet; and fishes with four fins. But those which have two wings, or, in short, have not two, as serpents, are nevertheless moved by four members, (for they have four flexions) or by two members with two wings. Animals, however, that being exanguious have many feet, whether they are winged or pedestrious animals, are moved by many members, as the animal called ephemera; for it is moved by four feet, 490 and four wings. For the peculiarity of this animal is not only that it lives but one day, whence it derives its appellation, but also that being a quadruped, it is winged. All quadrupeds however and multipeds are moved in a similar manner; for they are moved according to a diameter. Other animals, therefore, have two feet as the leaders; but the crab is the only animal that has four leading feet.

6 The greatest genera however of animals, into which other animals are divided, are, bird, fish, and whale. All these, therefore, are sanguineous. But another genus is that of animals, covered with a hard shell, which is called ostreon. Another is that of animals covered with a softer shell, which is anonymous by one name, as carabi,[†] and certain genera of crabs and lobsters. Another is that of the mollia, as the cuttle fish, the large and the small calamary. And another is that of insects. All these genera too are exanguious. Such insects likewise as have feet, have many feet; and some of them are winged. But of the rest of animals there are no longer any great genera. For one species does not comprehend many species; but one species is simple, not having any specific difference, as man; and other species have a difference indeed, but they are anonymous. For all the species of quadrupeds, and which are not winged, are sanguineous, but some of them are viviparous, and others oviparous. Such therefore as are viviparous have not all of them hairs; but such as are oviparous have a covering resembling scales. The scales too are similar to bark. The sanguineous pedestrious tribe of animals, however, which are naturally without feet, is the genus of serpents; and these are covered with scales. Other serpents indeed are oviparous; but the viper alone is viviparous. For not all animals that are viviparous have hairs. For of fishes, some are viviparous, *viz.* those that have hairs, since it must be admitted that there is a certain species of

† The carabus is a marine animal of the genus of crabs, but differing in the tail from other crabs.

hairs, and that some are thorny, such as those of the land hedge-hog and the porcupine. For such thorns, as also the bristles of the hedge-hog, perform the office of hairs, and not of feet. Of the genus, however, of four-footed and viviparous animals, there are many species, but they are anonymous. Many species of quadrupeds indeed are indivisible into other species, as man, lion, horse, stag, dog, and others after the same
491 manner; since there is also one certain genus in those that have a tuft of hair on the forehead, and which are called beasts of burden, such as the horse, the ass, the mule, the colt, and the nag, and the Syrian mules, which are called mules from similitude, though they are not simply the same species; for they copulate and generate from each other. Hence it is necessary that we should survey the nature of each of them separately.

These things, however, have been now asserted after this manner, as in a rude delineation, as a specimen of what is to follow; for afterwards we shall speak more accurately about the subjects to be surveyed, that we may, in the first place, understand the difference and accidents inherent in all animals; and, in the next place, endeavour to discover the causes of these. For thus we shall adopt a method conformable to nature, a history being given of every species of animals; since from these things, the subjects about which and from which it is necessary there should be a demonstration, will be evident. In the first place, therefore, the parts of animals from which they consist, must be assumed. For according to these especially and primarily the whole differ, either because some have them, and others not, or they differ in position and order, or in species, excess, analogy, and contrariety of passions, according to the before-mentioned differences. In the first place, however, the parts of man must be assumed. For, as we determine the value of money by [a standard] which is most known to us, the like method must also be adopted in other things. But man is necessarily the most known to us of all animals. The parts of man therefore are not indeed immanifest to sense; but at the same time, for the sake of not omitting what is consequent, and in order that we may have reason in conjunction with sense, the organic parts must be first enumerated, and afterwards those which consist of similar parts.

7 The following, therefore, are the greatest parts into which the whole body is divided, *viz.* the head, the neck, the thorax, the two arms, and the two legs; the thorax being denominated that cavity which extends from the neck to the pudenda. With respect to the parts of the head, therefore, that which is covered with hair is called the cranium, or the

skull; of which the prior part, or sinciput, is produced posterior to the other parts; for it is formed into a concretion the last of all the bones in the body. But the posterior part of the skull is the occiput; and the middle part between the occiput and sinciput is the vertex or crown of the head. The brain therefore is under the sinciput; but the occiput is empty. The whole of the cranium however is a thin bony substance, round, and covered with a fleshless skin. It has also sutures, one of which, in women, is circular; but in the cranium of men there are three, which, for the most part are conjoined in one. The head of a man however has been seen without any suture. But of the cranium, the vertex or crown is called the middle, and the discrimination of the hairs; and this in some men is twofold. For some are born with two crowns, which are not separated by bone, but by the hairs.

8 In man, however, alone, of all other animals, that part which is under the cranium is denominated the face; for in a fish or an ox, this part is not called the face. But of the face, that part which is under the sinciput, and between the eyes, is the forehead. Those in whom this is large, are of a tardy genius; but those in whom it is small, are mutable. Those also in whom it is broad, are furious; and those in whom it is round, are wrathful.

9 Under the forehead are the two eyebrows; of which those that are in a right line signify soft manners; but those that are infected towards the nose denote an austere person; those that are inflected towards the temples indicate a derider and a dissembler; and those that are depressed denote an envious person. Under these are the eyes, which are naturally two; and of these, the parts upward and downward are the eyelids; the hairs at the extremity of which are called the eyelashes. The humour within the eye, through which we see, is called the pupil. That which surrounds this is the black; and that which is external to this is the white of the eye. But the part common to the eyelid upward and downward, is the canthus, or corner of the eye; of which there are two towards the nose and two towards the temples. These angles, or corners, if they are longer than is proper, denote depraved manners; but if they are fleshy, like the groin, towards the nose, they indicate complete depravity. All the other genera of animals, therefore, except those that are covered with shells, or some other imperfect genus, have eyes; and likewise all viviparous animals, except the mole. Some one however may admit that this animal has eyes after a certain manner;

though simply it has not. For, in short, it neither sees, nor has eyes clearly manifest; but the skin being taken away, the place of the eyes will be seen, and the black part of them. The place also naturally adapted to the eyes, will be found in the part external to the black, as if a skin was superinduced when the eyes began to be formed.

10 With respect to the white of the eye, indeed, it is for the most part
492 similar in all animals; but that which is called the black differs. For in some it is black, in others it is of a very azure colour, in others it is grey, and in others again it is of the colour of a goat's eye; which last is an indication of the best manners, and is most powerful with respect to acuteness of vision. Man, however, alone, in the most eminent degree, or more than other animals, has eyes of various colours; but the eyes of other animals have only one species of colour, except horses, whose eyes are sometimes of an azure colour. Of eyes too, some are large, others small, and others of a moderate size, the last of which are the best. Some likewise are very prominent, or very inward, or in a medium between these. But those that are very inward, see most acutely in every animal. And those that have a middle situation are an indication of the best manners. Some eyes also wink very much, others are intent, and others are between these; the last of which indicate the most excellent manners. But of the other two, the one denotes impudence, and the other instability.

11 That part, however, of the head through which we hear, but which is without respiration, is the ear; for the assertion of Alemæon is not true, that goats breath through their ears. With respect however to the parts of the ear, one part is anonymous; but the other [*i.e.* the bottom part of the ear], is called *lobos*. And the whole ear is composed from cartilage and flesh. The interior part naturally subsists similar to the shell of a snail; but the extreme bone is similar to the ear, into which as into an ultimate vessel, sound arrives. From this part, however, there is no passage to the brain. But there is a passage from the extremity of the ear to the cavity of the mouth, or the palate; and a vein extends from this to the brain. The eyes also terminate in the brain, and each is placed upon a small vein. But man alone has his ears immoveable, among those animals that have this part. For of those that have the sense of hearing, some have ears, but others have not; but have an obvious passage, as is the case with those animals that are winged, or that are covered with scales. All viviparous animals, however, have ears,

except the sea-calf and the dolphin, and other cetaceous animals. For those that are cartilaginous are viviparous; but man alone does not move his ears. The sea-calf, therefore, has manifest avenues through which it hears; but the dolphin hears indeed, yet has no ears. All other animals move their ears. But the ears of man are situated in the same circumference with the eyes, and not above it, as is the case with some quadrupeds. Of the ears also, some are smooth, others are hairy, and others are between these; the last of which are the best for the purpose of hearing. Some likewise are great, others are small, and others are of 492
a moderate size; and they are either very much, or not at all, or moderately erect. Those that are moderate indicate the best manners; but those that are large and very erect, denote folly and loquacity. But the part between the eye, the ear, and the crown of the head, is called the temple.

Farther still, that part of the face which affords a passage to the breath, is the nose; for respiration, expiration, and sneezing, are effected through this. But sneezing is the egression of collected spirit, or breath, and of all spirits is the only augural and sacred indication. Moreover, respiration and expiration are produced into the breast, which could not be effected without the nostrils, because respiration and expiration are from the breast, through the throat, and not from any part of the head. It is possible, however, to live without breathing. Smelling also is effected through this part; and this is the sense of odour. But the nostrils are moved with facility; and are not like the ears naturally immoveable. The part which separates the two cavities of the nose is called a diaphragm, and is cartilaginous; for the nostril is bisected. But in elephants the nostril is very extended and strong, and they use it instead of a hand; for with this they collect and receive food, both moist and dry, and introduce it into the mouth; and the only animal that does this is the elephant.

Again, there are two maxillæ, [*i.e.* cheeks or jaws]. Of these, the anterior part is the chin; but the posterior part is the jaw. But all animals move the lower jaw, except the river crocodile; for he moves the upper jaw alone. After the nose, are the two lips, which consist of easily-moveable flesh. That which is contained between the cheeks and the lips is the mouth, the parts of which are the palate and the jaws. But that which is sentient of sapor is the tongue, in the extremity of which is the sentient power; for in the broad part of the tongue, the sense of taste is less powerful. The tongue likewise is sentient of every thing of which the other fleshy parts of the body have a sensation, such

as the hard, the hot, and the cold, in the same manner as it is sentient of sapor. But the tongue is either broad or narrow, or between these, the last of which is the best, and the most adapted to distinct utterance. The tongue likewise is either free, or tied, as in those that stutter. It also consists of a rare and fungous flesh. Of this, that part which is at the root of the tongue is called the epiglottis. The twofold structure of
493 the mouth also is called tonsillæ; and the multifarious conjunction of the tonsillæ forms the gums. These are fleshy, but the teeth within are bony. In the more interior part of the mouth, there is also another part, which has the form of a grape, and is called uvula, or the palate, and columella, or a little pillar. This when it swells in consequence of being inflamed by moisture is called uva, a grape, and causes suffocation.

12 The neck is that part which is between the face and thorax; and of this the anterior part is the throat, but the posterior part is the oesophagus. Of the throat also that part which is cartilaginous and anterior, through which voice and respiration are transmitted, is the aspera arteria; but the fleshy interior part, which is situated before the spine, is the oesophagus; and the posterior part of the neck, is the cervix. These are the parts, therefore, as far as to the thorax. But of the parts of the thorax some are anterior, and others posterior. And, in the first place, after the neck, in the anterior parts is the breast, naturally distinguished by two mamillæ. In females, each of these has a papilla or nipple, through which the milk is percolated. The mamillæ also consist of rare flesh. Milk likewise may be produced in the mamillæ of males; but the flesh in these is dense, while in women it is fungous, and full of pores.

13 After the thorax, in the anterior parts is the belly; and the root of this is the navel. The part under this root, and which is twofold, is the ilia; and the single part which is under the navel is the abdomen. At the extremity of this are the pudenda or privities; but the part above the navel is the hypochondrium. The part common to the hypochondrum and ilia, is called cholas. But in the posterior parts, those parts which are opposite to the hypochondria at the cincture, are called the loins, (οσφυς), whence also the name is derived; for it appears to be, as it were, *ισοφυς*, *i.e.* endued with an equal breadth. The part through which the dry excrement is emitted, is subdivided into two parts, of which the fleshy part, subservient to sitting, is called nates; but the part in which the thigh is turned, is the acetabulum. The part peculiar to the female

is the vulva; and to the male, the penis, which extends outwardly to the end of the thorax. And the extremity of it indeed is fleshy, and, as I may say, equally smooth, and is called glans penis; but the skin which surrounds it is anonymous; which if it is cut off does not regenerate, as neither does the skin of the cheek, nor of the eyelids. To this penis and its glans the prœpuce is common; but the remaining part is cartilaginous, is easily increased, and has an egression and ingression contrary to that of beasts of burden or animals that have a tuft of hair on the forehead. Beneath the penis are the two testicles, which are surrounded with a skin called the scrotum. But the testicles are neither of the same nature with flesh, nor yet very remote from it. With respect to the manner, however, in which they subsist, and in short, concerning all such parts, we shall speak more accurately hereafter.

14 The pudenda, however, of a woman are fashioned contrary to those of a man; for the part under the pubes is hollow, and is not prominent, as it is in the male. The urethra also is external to the womb, affording a passage to the seed of the male. But there is an egression both in the male and female of humid excrement. The part common to the neck and breast is the throat; but to the side, arm, and shoulder, is the arm-pit; and to the thigh, and inward part of the nates, the groin. But the external part of the thigh and nates, is formed by the substance which is under the buttocks. Of the anterior parts of the thorax indeed we have already spoken.

15 Behind the breast, however, is the back, the parts of which are the two scapula, or shoulder-blades, and the spine. Underneath are the loins, which are opposite to the belly. But the ribs, which are on each side eight in number, are common both to the superior and inferior part; for with respect to what is reported of the Lychnians, that they have [only], seven ribs, we have not heard this asserted by any one who is worthy of belief.

Man, however, has the parts of his body distinguished by the superior and inferior, the anterior and posterior, and the right and the left. The right and left parts, therefore, are nearly similar and the same, except that the left parts are more imbecile; but the posterior are dissimilar to the anterior, and the inferior to the superior parts; except that the inferior correspond in a certain proportion to the superior. Thus the lower belly corresponds to the face in corpulency, or gracility, and the legs to the arms. Hence those whose arms are short, have also for the

most part, short thighs; and those whose feet are small have likewise small hands. Of the superior extremity, however, there is a twofold part, the arm; and the arm is divided into the *humerus*, the *lacertus*, the *cubitus*, or the *cubit*, the *ulna*, and the *hand*.[†] But the parts of the hand are the palm and the five fingers. And of the finger indeed, that part which can be inflected is the joint, but that which cannot be inflected is the part between the joints. The great finger also, or the thumb, has only one joint; but the other fingers have two joints. The inflection, however, both of the arm, and all the fingers, is inward; and the arm is inflected towards the gibbous part of the cubitus. The interior part of the hand, the palm, is fleshy, and is distinguished by lines, in those that are long-lived indeed, by one or two lines which extend through the
494 whole of the palm: but in those that are short-lived, by two lines, which do not extend through the whole. But the joint by which the hand is connected with the arm, is the wrist. And the external part of the hand is nervous, and without a name.

Another twofold division of the inferior extremity consists of the legs. The first part of the leg, however, is the thigh, which has a head at each extremity. The second part is the knee-pan; and the third part is the tibia. This part follows after the first flexure of the leg, which consists of two bones. And of this indeed, the anterior part is the *antiknemion*, but the posterior part is the *gastroknemia*, or sura, which is composed of nervous or veiny flesh. Those that have large loins have the sura elevated towards the knee; but it is drawn downward in those that have small loins. The extremity of the antiknemion is the ankle, which is twofold in each leg. And that part of the leg which consists of many bones, is the foot. Of this, however, the posterior part is the heel; but the anterior part of the foot is divided into five toes. The inferior fleshy part of the foot is the sole; but the superior part is nervous and without a name. Again, of the finger, one part is the nail, but the other the inflexion. And in all the fingers the nail is in the extremity only; but all the toes are inflected. With respect to the foot, however, those that

† There are neither any Latin nor English names, which accurately correspond to the Greek names assigned by Aristotle to the division of the arm. They may, however, be described by the names assigned in the text. The beginning of the arm is the humerus. The second part of the arm is the lacertus, which is the part extended from the humerus to the first curvature of the arm. The third part is the cubitus, which is the part of the arm that extends from the first curvature to the extremity of the fingers. The fourth part of the arm is the ulna, which extends from the cubitus to the hand. And the fifth part is the hand, which is divided into the five fingers and the palm.

have the inward part of it thick, and not concave, but so full that they press the ground in walking with the whole of the inward part; these are crafty. But the common part, by which the tibia is connected with the thigh, is the knee.

These therefore, are the common parts of the male and the female. But as to the position of the parts, how they subsist, with respect to the superior and inferior, anterior and posterior, right and left, though the order of the external parts is obvious to sense, yet it must be described by us, for the same reason for which we have made the preceding enumeration, that the discussion may be accomplished in a continued series, and that we may not be ignorant how the parts of men, and other animals, subsist with respect to order and position.

Man, therefore, has the superior and inferior parts of his body arranged, according to natural places, in a greater degree than all other animals. For his superior and inferior parts are arranged, according to the superior and inferior parts of the universe; and after the same manner his anterior and posterior, his right and left parts, have a natural position. But with respect to other animals, some have and others have not the parts thus arranged, but they are more confusedly disposed. The head, therefore, indeed, in all animals is upward with respect to the other part of the body; but man alone, as we have just observed, being fashioned conformably to the order of the universe, has this part corresponding to the heavens. After the head is the neck, then the breast and the back, the former having an anterior, but the latter a posterior position; and to these succeed the belly and the loins, the pudendum, and the hip; afterwards the thigh and the tibia; and, in the last place, the feet. With respect to the legs, however, they are inflected towards the anterior part, towards which part also they are moved; and the more motive part of the foot and its flexure, are also anterior. But the heel has a posterior position. And each of the ankles has the same position in the tibia, as the ears have in the head. The arms hang from the right and left side of the body, and have their flexure inward; and are so inflected, especially in man, that the incurvation of the legs and the arms have a position opposite to each other. With respect to the senses, however, and the sensoria, *viz.* the eyes, the nostrils, and the tongue, man has them situated in the same, *i.e.* in the anterior part: but he has the hearing, and its sensorium, the ear, in a lateral position indeed, but in the same circumference with the eyes. Man likewise has his eyes proportionally less distant than other animals. But of the senses, the most accurate in him is the touch, the next to this in accuracy is the taste, but in the other senses he is excelled by many animals.

16 Thus, therefore, the parts in the external superficies of the human body are arranged, and, as we have said, they are especially denominated and known from custom. The contrary however is true of the inward parts; for of all the parts belonging to man they are the most unknown, so that it is necessary to survey them by referring them to the parts of other animals to which they are similar. The brain therefore occupies the first situation in the head, and is posited in the anterior part of it; and this similarly takes place in other animals that have this part. But all sanguineous animals have it, and besides these the mollia. Man, however, in proportion to the magnitude of his body, has the largest and the moistest brain. Two membranes likewise contain it; one about the bone, which is the stronger of the two; but the other about the brain itself, which is less strong. The brain however in all animals is twofold, and to the posterior part of it, that which is called the cerebellum, is joined, which has a different form† from the brain, both according to the touch and the sight. But the posterior part of the head in all animals
495 is [partially at least] empty and hollow, according as the head is larger or smaller. For in some animals that have a larger head, the part of the face which is subjected to the head is small; as is the case with those animals that have a round face. Other animals, on the contrary, have a small head but long cheeks, such as all the genus of labouring cattle, or animals that have a tuft of hair on the forehead. The brain however in all animals is exanguious, has no veins, and is naturally cold to the touch; and in the middle of it, in most animals, there is a small cavity; but the membrane which surrounds it is veiny, and of the nature of skin. Above the brain also, there is a bone which is the thinnest and weakest of all the bones of the head, and is called the sinciput. But their avenues, or ducts, tend from the eyes to the brain; the largest indeed, and the middle, tend to the cerebellum; but the smallest tends to the brain itself. And the smallest of these ducts, is that which especially verges to the nose. The greatest ducts, therefore, are parallel to each other, and never coincide; but the middle ducts coalesce. This however is especially evident in fishes; and these middle are nearer to the brain

† The word used here by Aristotle is μορφη, which, as we have frequently observed in the other volumes of this translation, pertains to the colour, figure, and magnitude of superficies.

than the large ducts. But the two smallest ducts are more distant from each other than the rest, and do not coincide.

Within the neck are what is called the oesophagus, which has its denomination from its narrowness and length, and the aspera arteria. The aspera arteria, however, is prior in position to the oesophagus in all animals that possess it; and all animals have it that have lungs. But this artery is indeed naturally cartilaginous, has but little blood, and is surrounded by many thin veins. It is also placed in the superior part towards the mouth, opposite to the ducts which extend from the nostrils to the mouth. Hence it happens, that if when we are drinking, the liquor enters into the aspera arteria, and is thrown upward, it departs through the nostrils. The head of the aspera arteria, in the superior part in which it extends to the mouth, has, what is called the epiglottis, which is in the middle of the two avenues that reach from the nostrils to the mouth. The aspera arteria, therefore, in the superior part is conjoined with the tongue, but with the other extremity it descends into the middle of the lungs; afterwards from this it is divided into each part of the lungs. For the lungs are bipartite in all the animals that have them. In viviparous animals, however, the separation is not similarly apparent, and it is in the smallest degree apparent in man. But the lungs 495
of man are not multifidious, as they are in some viviparous animals, nor are they smooth, but anomalous. In oviparous animals, however, as in birds, and such quadrupeds as are oviparous, the two parts of the lungs are very much separated from each other, so that they seem to have two lungs. Since the lungs therefore are bipartite, the aspera arteria which is one, when it arrives at the lungs is divided into two canals, as it were, and insinuates itself into each part of the lungs; it is also connected with the great vein, [or vena cava] and with another vein which is called the aorta. When in respiring, however, we inflate the aspera arteria, then the breath enters into the hollow parts of the lungs. But the cavities of the lungs have certain little pipes inserted in them which are always narrower and narrower, till at length they end in a sharp point; and from these pipes certain cavities are deduced, which so extend through the whole substance of the lungs as to be always less and less. The heart also is conjoined with the aspera arteria by fat, cartilaginous, and fibrous bonds, and in that part in which it depends from the aspera arteria, it is hollow [and sinuous]. But though in some smaller animals it is not manifest, whether when in respiring they inflate the aspera arteria, the breath not only enters the lungs, but likewise the heart, yet in larger animals it is evident that the breath insinuates itself through the aspera

arteria into the heart likewise. The aspera arteria therefore subsists after this manner, and its office consists in alone receiving and emitting the breath, but it does not receive any dry or liquid nutriment. Indeed, if food, either liquid or dry, happens to fall into the aspera arteria, the animal is in pain till it is ejected.

The stomach however, [or oesophagus] depends in the superior part of it from the mouth, adheres to the artery, and is conjoined both with the spine and the artery by membranaceous ligaments. But it passes through the transverse septum, [by which the breast is divided from the belly] and ends in the ventricle. It has also a fleshy nature, and is adapted to be extended both according to length and according to breadth. The human, however, is similar to the canine ventricle; for it is not much larger than the intestine, but resembles, as it were, a broader intestine. Under the ventricle the intestines are situated, one of which is simple and winding, but the other is broad; and the lower cavity, [or the alvus, which contains the intestines] is similar to that of a sow; for it is broad, and the part under the alvus, which is extended to the seat, is short and thick. From the middle of the alvus the omentum depends; which is naturally a fat membrane, in the same manner as in all other animals which have only one ventricle, and teeth on each side of the mouth. Above the intestines is the mesentery, which is membranaceous, broad and fat. But from the great vein and the aorta many veins are deduced,
496 which are diffused through the mesentery in that part by which it is connected with the intestines, and beginning from the summit of the mesentery, they are diffused as far as to its lowest parts.

17 The oesophagus, therefore, the aspera arteria, and the belly, subsist after this manner. But the heart has indeed three cavities, and is situated towards the superior[†] part of the lungs, and about that part where the aspera arteria is bisected. It has also a fat and thick membrane, through which it is conjoined both to the great vein and the aorta. But the cuspis or point of the heart in all animals that have a breast, is similarly directed towards the breast; and in those animals that are without a breast, the cuspis of the heart verges to the anterior part. This however is frequently latent, because during the dissection of the animal, the heart is removed from its natural situation. But the convex and curved part

† The original is *κειται δ᾽ανωτερω του πνευμονος*, *i.e.* "the heart is situated above the lungs;" but as this is obviously false, I conclude with Sealiger, that the text is corrupt, and have accordingly adopted his emendation.

of the heart is upward; the cuspis of it is for the most part fleshy and dense: and nerves are inherent in the cavities themselves. In other animals, however, that have a breast, the heart is situated in the middle of the breast, but in man it verges to the left side, a little inclining from the division of the mammæ, in the upper part of the breast. Nor is it large, and its whole form is not oblong, but has something of rotundity, yet so to end in a point. It has also, as we have said, three cavities, the greatest indeed in the right, but the least in the left parts, and the middle in magnitude has an intermediate situation. All these cavities, likewise, through certain avenues are pervious to the lungs, which however is only manifest in one of them. Through the greatest of these cavities the heart is joined to the great vein, with which also the mesentery is conjoined; but through the middle cavity it is joined to the aorta.

Avenues also tend to the lungs from the heart, which are divided after the same manner as the aspera arteria, and follow the canals of the artery through the whole of the lungs. These avenues however are in the superior part of the heart; nor do they accord in any common avenue, but they are severally conjoined to the lungs, and from this conjunction receive hot air, and transmit it to the heart. For one of the avenues leads to the right, but the other to the left cavity of the heart. Concerning the great vein, however, and the aorta, we shall separately speak in common hereafter. But in animals that have lungs and are viviparous, so that they not only produce, but also conceive an animal in themselves, there is no member which contains so much blood as the 496
lungs. For the whole of them indeed are spongy, and receive in their several canals blood from the great vein. Those however who think that the lungs are empty are deceived, in consequence of surveying the lungs taken from dissected animals, which in that state are wholly destitute of blood. But of the other viscera the heart alone contains blood; for the lungs contain blood, not in their own essence but in the veins; but the heart contains blood in itself. For in each of its cavities it has blood; but the most attenuated blood is in its middle cavity.

Under the lungs is the transverse septum, through which the breast, or the superior part of the thorax, [in which the heart and lungs are contained], is separated from the belly [in which the liver, spleen, *etc.* are contained]. The septum transversum is also called the diaphragm, præcordia, and cinctura: the diaphragm is connected with the spine of the back, and also with the ribs and hypochondria. But the middle part of the diaphragm is thin and membranaceous; and [many] veins are extended through it. The veins of a man however are thick, considered with reference to the magnitude of the human body. Under the transverse septum in the right-hand parts, the liver is situated, but the spleen is in the left-hand parts; and the liver and spleen are similarly disposed in all animals that have these parts naturally, and not monstrously situated; for in some quadrupeds these parts have been seen in an inverse order. Both the liver and the spleen, however, are in the lower belly, near to the omentum [by which the intestines are covered]. The spleen however of a man is narrow and long, and resembles that of a hog. But the liver for the most part in most animals is without the gall, but in some animals it is connected with the gall. The liver of a man also is round, and resembles the liver of an ox. It happens however in some victims that there is a privation of the gall. Thus, for instance, in a certain place of Chalcis in Eubœa, the cattle have no gall; but in Naxus nearly all the quadrupeds have so large a gall as to astonish the strangers that sacrifice there, who fancy that this circumstance is prognostic, and that it is not natural to these animals. The liver also is connected with the great vein, but has no communication with the aorta. For a vein which proceeds from the great vein passes through the liver, in that part of it where what are called the gates of the liver are situated. The spleen likewise is conjoined with the great vein alone; for a vein extends from the great vein to the spleen.

After these, the reins or kidneys are situated near the spine of the back [under the liver and spleen] which [in men] are of a nature similar to the
497 reins of oxen. But in all animals that have reins, the right is higher than the left, but it is less fat and more dry than the left. This likewise similarly takes place in all other animals. Avenues, however, or ducts, lead to the reins from the great vein and the aorta, except that they do not reach to the cavity [which is in the middle of the reins]. For the reins have a cavity in the middle, some a large and others a less cavity, except the reins of the phoca or sea-calf; and these resembling the reins of an ox, are the most solid of all. The avenues, however, which extend to them, are consumed into the body of the reins. But that these avenues do not terminate in the cavity of the reins, is indicated by their having no blood, either fluid or congealed. They have therefore, as we have said, a small cavity; but two solid avenues lead to the bladder from the cavity of the reins, and other strong and continued avenues tend also from the aorta. From the middle, however, of each of the reins, a hollow and nervous vein depends, and which extends by the spine of the back through narrow avenues. These veins when they arrive at the hip no longer appear, but afterwards become manifest, and at length end at

the bladder. In this last part, therefore, in which the aforesaid veins are terminated, the bladder is situated. It is also bound to the veins, and depends from them through those two solid avenues which are extended from the cavity of the reins to the neck of the bladder, which neck of the bladder tends to the urinary passage. The bladder likewise is nearly every where circularly involved in thin and fibrous membranes, and which after a manner are similar to the diaphragm of the thorax. The human bladder however is sufficiently large.

The pudendum likewise is fastened to the neck of the bladder, the orifice of which is most external; and the orifice of the urethra tends to the same point, but is a little inferior. Of these orifices, the one tends to the testicles, but the other which is nervous and cartilaginous, to the bladder. From this the testicles depend in males, concerning which, we shall afterwards show how they subsist, when we discuss the parts in common. All these too are naturally adapted to subsist after the same manner in the female. For the female differs in no respect from the male in the internal part, except in the womb; the form of which must be surveyed from anatomical descriptions. The situation of it, however, is in the intestines; but the bladder is above the womb. We shall however speak of all wombs in common in what follows; for they are neither similar in all animals, nor subsist similarly. These therefore, and 497
such as these, are the internal and external parts of man, and after this manner they subsist.

BOOK II

1 With respect, however, to the parts of other animals, some are common to all animals, as we have before observed, but others are peculiar to certain genera. But they are the same with or different from each other, in the way we have frequently mentioned. For nearly all animals which are of a different genus, have most of their parts specifically different. And some indeed have their parts only analogously the same, but generically different; but others have them generically the same, and specifically different; and there are many parts which some animals have, but others have not.

Viviparous quadrupeds therefore [accord with man, and with each other, in this, that they] have a head and neck, and all the parts in the head. But the several parts differ from each other in their form. And the neck of the lion indeed, consists of one bone, but has no vertebræ. All the interior parts however of a lion are similar to the interior parts of a dog. Viviparous quadrupeds also have the forelegs instead of arms; but those forelegs which have multifidous feet, are especially analogous to hands; for they use them for many purposes instead of hands. The left legs, however, of quadrupeds, except those of the elephant, are less agile and expedite than the left hands of men. But the elephant has the division of the five toes in the foot, indistinct, and has the fore much larger than the hind legs. In his hind legs also, he has small mallets. He has however a nose of such a kind and so large, that it serves him instead of hands. For he both eats and drinks extending his nose [or proboscis] to his mouth. He also extends this to his keeper when riding on his back, tears up trees with it, and respires by its assistance when he walks through the water. The extremity likewise of the proboscis may be bent, but cannot be folded; for it is cartilaginous, [though the other parts of it may also be folded.] Man however is the only animal that is ambidexter, or has the use of both his hands.

With respect to the breast, however, all animals have this part analogous, but not similar to the breast of man; for the human breast is broad, but that of other animals narrow. But no other animal besides man has mammæ in the anterior part; for the elephant has two mammæ
498 indeed, yet they are not in the breast, but near [*i.e.* a little below] the breast.

The legs however of other animals, as well the fore as the hind legs, have flexions contrary to each other, and to the flexions of the legs and arms of man, the elephant being excepted. For viviparous quadrupeds have their forelegs bent towards the fore part, but their hind legs bent towards the hind part, and they have the cavity of one flexion opposite to the cavity of the other. What is asserted of the elephant, however, by some, is not true [*i.e.* that he cannot bend his legs, nor sit] for he can do both, except that he cannot on account of his weight at one and the same time bend each foreleg, and recline on each side, but he can alone bend one leg, either the right or the left, and alone recline on one side, and in this manner he sleeps, [leaning against some wall or tree.] But he bends his hind legs in the same manner as men.

Oviparous quadrupeds, however, such as the crocodile, the lizard, and others of this kind, have both their fore and hind legs bent so as to be hollow towards the fore part, and with a flexion a little inclining to the oblique. The like also takes place in other animals that have many feet; except that it is always ambiguous to what part the middle flexions of the last legs verge, which are rather obliquely inflected.

But man has contrary flexions of his arms and legs. For he bends his arms so as that they are hollow in the anterior part, the cavity a little inclining to the oblique; but he bends his legs so as to form a cavity in the posterior part. There is no animal, however, that bends its legs so as that the cavity of the fore and hind legs is posterior. But the flexion of the shoulder in men and quadrupeds is contrary to that of the cubit in men, and the forelegs in quadrupeds; and the flexion of the parts through which the hind legs are joined to the thorax, is contrary to that of the thigh. Hence, since man bends some of his members in a way contrary to that of many other animals, if there are certain animals which in such members imitate human flexions, they also bend such members in a way contrary to that of many other animals.

Birds also have inflexions similar to those of quadrupeds. For, being bipeds, they bend their legs towards the posterior part; but instead of arms and forelegs, they have wings, the flexion of which is towards the anterior part.

The sea-calf however is, as it were, a mutilated quadruped; for immediately after the shoulder-blade, it has [two fore] feet similar to hands, such as are the [fore]feet of bears. For the [fore] feet of the sea-calf have five toes, each of which has three flexions, and a nail which is
not large. But the [two] hind feet, have indeed five toes, and flexions 498
and nails similar to the fore feet, but in their form they resemble the tails of fishes.

The motions however both of quadrupeds, and animals that have many feet, are according to a diameter, and in this manner also they stand. But the principle of motion in all animals is from the right hand parts.

The lion also, and both the Bactrian and Arabian camel, walk foot by foot. But to walk foot by foot, is when the right foot does not go before, but follows the left.

Such parts, however, as in man are anterior, are inferior in quadrupeds, and turned towards the earth; and the parts which in man are posterior, are superior in quadrupeds, and directed towards the heavens.[†] Farther still, most animals have a tail; for the sea-calf also has a small tail, similar to that of a stag. But concerning apes we shall speak hereafter.

All such animals likewise as are quadrupeds and viviparous are, as I may say, hairy, and in this respect they do not resemble man, who, except in his head, has but few and short hairs, but in his head is the most hairy of all animals. In other animals however that have hairs, the prone parts are more hairy, but the supine parts are either entirely smooth, or less hairy. But the contrary takes place in man. Man also has hairs on both his eyelids, in his armpits, and about his privities. No other animal however has any one of these, nor have they hairs on the inferior eyelid, but some of them under the inferior eyelid have a few hairs.

With respect however to quadrupeds that have hairs, of some indeed the whole body is hairy, as the hog, the bear and the dog; but in others the neck is every where similarly hairy, as is the case with such as have a mane, as the lion. Others again have longer hairs in the superior part alone of the neck, as far as to the shoulders, as the horse, the mule, and of savage and horned animals, the bonasus.[‡] The hippelaphus[§] also, and the wild beast called hippardium,[◇] have two thin manes, one in the summit of the back, the other in the summit of the neck. It is likewise
499 peculiar to the hippelaphus to have a beard depending from the throat. Both the hippelaphus and the hippardium however are horned and bisulcous; but the female hippelaphus is without horns. The magnitude also of this animal is similar to that of a stag. But the hippelaphi are natives of Arachosia [in India,] in which place also there are wild bulls.

† Thus the breast and the belly in man are anterior, but in quadrupeds they are inferior, so as to be directed towards the earth; but the superior part of quadrupeds corresponds to the shoulder-blade, and the other posterior parts of man.

‡ This animal has the head of a bull, but the body and mane of a horse.

§ So called because it partakes of the nature of a horse and a stag.

◇ This animal receives its appellation from partaking of the nature of a horse and a leopard.

Wild, however, differ from tame bulls, in the same manner as wild boars differ from domestic swine; for wild bulls are black, very strong, and have a nose nearly aquiline; but the curvature of their horns is reversed. The horns however of the hippelaphi are similar to those of a roe-buck. But the elephant is the least hairy of all quadrupeds. The tails also of quadrupeds, when they have any magnitude, follow the condition of the body, with respect to abounding with or being deprived of hairs; for some have so little hair that it can scarcely be perceived.

Camels, however, differ from other quadrupeds in this, that they have what is called a tuber or swelling on the back. But the Bactrian differ from the Arabian camels. For the Bactrian camel has two swellings on the back, but the Arabian has only one. Each camel also has another swelling in the inferior similar to that in the superior part, by which the rest of the body is supported when they bend their knees. The camel likewise has four mamillæ, in the same manner as cows, a tail similar to that of an ass, and the pudendum behind. It also has but one knee in each leg, and not many flexions as some say; but through the form of the belly [which is very small and drawn up] it appears to have many knees or flexions. And it has indeed an ankle-bone similar to that of an ox, but a small thigh in proportion to the magnitude of its body. It is also bisulcous, and has no teeth in the upper jaw. But it is bisulcous in such a way, that in the inferior part of the foot it is bipartite, as far as to the second flexion of the toes, but in the superior part it is quadripartite, as far as to the first joints of the toes. The intervals of the fissures also are covered with a certain membrane, similar, as it were, to that with which the intervals of the fissures in geese are covered. The sole too of the foot is fleshy, like that of a bear; and on this account when they lead camels to battle, to prevent their feeling pain in this part, they shoe them with that kind of shoes which are called carbatinæ.[†]

All quadrupeds, however, have bony, nervous and fleshless legs; and, in short, this is the case with all other animals that have feet, except man. They are also without hips, and especially birds. On the contrary, man has hips that are very fleshy with respect to the other parts of his body, and his thighs and tibiæ are very fleshy; for that part of the tibia which is called sura, is very fleshy.

But of sanguineous and viviparous quadrupeds, some are multifidous, as the hands and feet of man; for some have feet with many fissures, as

† A coarse kind of country shoe.

the lion, the dog, and the panther; but others have bifidated feet, and instead of nails have hoofs, as sheep, goats, stags, and the hippopotamus. Others have their feet without fissures, such as those that have solid hoofs, as the horse and the mule. The genus of swine, however, is ambiguous; for in Illyria, Pœonia, and other places, the swine have solid hoofs. Bisulcous animals, therefore, have two fissures behind and before; but animals that have solid hoofs have feet without fissures.

Some animals likewise have horns, but others are without horns. Most of those therefore that have horns are naturally bisulcous, as the ox, the stag, and the goat; but we have seen no animal that has solid hoofs, and is bicornous. There are however a few animals that have one horn and solid hoofs, as the Indian ass. But the oryx[†] has one horn, and is bisulcous. The Indian ass also is the only animal that has an astragalus[‡] (αστραγαλος) of those that have solid hoofs. For the hog, as we before observed, is ambiguous. Hence its astragalus is not well formed. But of bisulcous animals, men have an astragalus. Of multifidous quadrupeds, however, no one has been seen with this *astragalus*, and man also is without it; but the lynx indeed has that which is similar to a semi-astragalus, and the lion in the place of the astragalus has a certain winding bone like a labyrinth. All animals, however, that have the astragalus, have it in their hind legs, so that it is erect in the flexure by which the foot is conjoined to the leg. The astragalus also in its prone part [which because it is a little convex is called the back] is external; but in its supine part [which because it is hollow is called the belly] is external. The parts likewise of the astragalus which are called veneres, are inwardly turned towards each other; but the parts of it which are called calus, have an external direction; and that part of the astragalus which is turned upward, is called antenna. The position, therefore, of the astragalus in all animals that have it, subsists after this manner. There are however some animals that are bisulcous, have manes, and are bicornous, as the bonasus, which is produced in Pœonia and Media. But all such animals as are horned are quadrupeds, unless some animal should be metaphorically said to have a horn, as the Egyptians assert of the Theban serpents, as if something resembling a horn projected from their

† A kind of goat produced in Africa.

‡ By the αστραγαλος, Aristotle means that bone which is in the joint of the foot of some animals, and has many faces. This bone was collected by ancient shepherds after the slaughter of the victims, and was used by them as an instrument of play; whence the game of dice originated.

head. Of animals however that have horns, the stag alone has a horn wholly solid; but the horns of other animals are to a certain extent hollow, and at the extremity are solid. The interior and concave part of the horn therefore seems to be naturally formed from skin; but the exterior and solid part is formed from bone, and appropriately surrounds the interior part; and such are the horns of oxen. But the stag is the only animal that casts off its horns annually, beginning to do so at the age of two years; and other horns grow in their stead. Other animals, however, never lose their horns unless they are deprived of them by violence.

Farther still, quadrupeds differ from each other, and also from man, both in the mammæ, and in the organs subservient to generation. For some animals have the mammæ in the anterior part of the body, *i.e.* in the breast, or a little under the breast; and they have also two mammæ and two papillæ or nipples, as man, and the elephant, as we have before observed. For the elephant has two mammæ under the shoulders; but the female elephant has perfectly small mammæ, and not in proportion to the magnitude of the body, so that obliquely they can scarcely be seen. The male also, as well as the female elephant, has very small mammæ. The bear however has four mammæ. Again, other animals have two mammæ, and two papillæ between their thighs, as the sheep; but others have four papillæ, as the cow. Others again have the mammæ, neither in the breast, nor between the thighs, but in the belly, as the dog, and the swine, yet they are not all equal. Other animals, therefore, [as the dog, the swine, and many other quadrupeds,] have more than four mammæ in the belly; but the panther has four, and the lioness has two, in the belly. The camel also has two mammæ, and four papillæ, in the same manner as the cow. The males, however, of those animals that have solid hoofs are without mammæ, except such offspring of these as resemble their mother, which sometimes happens to be the case in horses.

But with respect to the pudenda of males, some of them have this member externally, as man and horse, and many other animals; but others have it internally, as the dolphin. And of those that have it externally, some indeed have it in the anterior part of the body, as the above-mentioned animals. Of these again, some have both the penis and the testes hanging loose, as man; but others have them adhering to the belly; and in some animals they are more, but in others less loose; for this part is not similarly detached in a boar and a horse. The elephant also has a penis similar indeed to that of a horse, but small, and not in proportion to the magnitude of his body; but his testicles are not externally apparent, but are inward and near the reins. Hence in coition he is rapidly liberated. The female elephant, however, has the pudenda in the place in which the mammæ of sheep are situated; but when she ardently desires coition, she draws this part upward, and turns it externally, so that copulation may be easy to the male. The opening also is proportionally very large. In most animals, therefore, the pudenda subsist after this manner. But some of them discharge the urine backward, as the lynx, the lion, the camel, and the hare. The males, therefore, of these animals differ from each other in the manner we have shown, but all the females discharge the urine backward; for the female elephant has the pudenda in the same manner as other animals, between the thighs. There is, however, a great difference in the pudenda of animals. For some have the penis cartilaginous and fleshy, as man; the fleshy part of which cannot be inflated, but the cartilaginous part admits of increase; other animals have it nervous, as the camel and the elephant; and in others it is bony, as in the fox, the wolf, the polecat, and the weasel; for the weasel also has a bony penis.

In addition to this also, man when he has acquired his proper magnitude, has his superior less than his inferior parts; but in other sanguineous animals, the contrary takes place. We call however the superior part that which extends from the head as far as to the part in which there is an egress of the excrements; and we denominate the inferior part, that which remains of the body from the places through which the excrements are emitted. In those animals, therefore, that have feet, the hind legs determine the magnitude of the inferior part; but in those that have no feet, the magnitude of the inferior part is determined by the tail, and whatever is instead of a tail. Such, therefore, are animals when they have arrived at perfection. For man in his infancy has the superior larger than the inferior parts; but the contrary takes place as he
501 increases through age. Hence man alone has not the same progressive motion when he is an infant, and when he has arrived at perfection; but at first he creeps on the ground with his hands and feet, like a quadruped. Some animals also have an analogous increase of the superior and inferior parts, as the dog. But in others, the superior parts are at first less, and the inferior larger, but as they advance to perfection, the superior parts become larger, as animals which have a tuft of hair on their forehead. For these receive no increase in the inferior part which extends from the hoof to the hip.

There is however much difference in the teeth of other animals, both with respect to each other, and those of man. For all such quadrupeds as are sanguineous and viviparous have teeth; but in the first place, some have teeth on both sides, and others have not. For such animals as horses have not teeth on both sides; since they have not the fore teeth of the upper jaw. But there are some animals which neither have teeth on both sides, nor horns, as the camel. And some animals indeed have teeth projecting beyond the mouth, as male swine, but in others they do not project. Again, some animals have serrated teeth, as the lion, the panther, and the dog; but others have them in a continued series, as the horse and the ox. For those animals have serrated teeth, which have at the same time the order of the sharp teeth changed, [so that the space intercepted between the two inferior teeth is occupied by the superior tooth, and the space intercepted between the two superior teeth is occupied by the inferior tooth.] No animal, however, has at one and the same time projecting teeth and a horn; nor is there any animal which has serrated teeth, and is horned, or has projecting teeth. But most animals have the fore teeth sharp, and the inward teeth broad. The sea-calf, however, has all its teeth serrated, as being an animal allied to fishes; for nearly all fishes have serrated teeth.

None also of these genera of animals have double rows of teeth; though if we may believe Ctesias, there is a certain animal which has. For he says, that in India there is a savage animal, the name of which is *Martichora*, that has a triple row of teeth in each jaw. He adds, that this animal is equal in magnitude to a lion, that it is similarly hairy with, and has feet resembling those of a lion, but that its face and ears are like those of a man, and that its eyes are azure, and its colour resembling that of cinnabar. It has also a tail armed with a sting, and resembling the tail of the land scorpion. It is likewise furnished with certain native darts, which it hurls forth; and it utters a sound resembling that of a pipe and a trumpet. But it runs with no less swiftness than the stag, and is savage and feeds on human flesh.

Man, therefore, sheds his teeth, and also other animals, such as the horse, the mule, and the ass. Man, however, sheds his fore teeth, but no animal sheds the maxillaries or grinders. And the swine, in short, does not shed any of its teeth.

2 It is dubious, however, with respect to dogs whether they shed their teeth or not. And some indeed are of opinion that they do not shed any of their teeth; others assert that they only shed their canine teeth; and

others think that they shed their teeth in the same manner as men, but that this is latent, because they do not shed them before other teeth equal to them grow inwardly in their stead. It is reasonable also to suppose that the like takes place in other savage animals, since they are said to shed only their canine teeth. The age of dogs, however, is known from their teeth; for those that are young have white and sharp teeth; but those that are elderly have black and obtuse teeth.

3 The contrary, however, takes place with respect to teeth in other animals and in horses; for other animals as they advance in age have blacker teeth, but the horse as he grows old has whiter teeth. But the teeth which are called canine separate the sharp from the broad teeth, participating of the form of both; for beneath indeed they are broad, but above they are sharp. The males also have more teeth than the females, in men, in sheep, in goats, and in swine; but in other animals this has not yet been seen. For the most part, however, those animals that have more teeth live longer; but those that have fewer and thinner teeth are for the most part shorter lived.

4 In the human species, however, the last maxillary teeth, which they call genuini, [or the innermost cheek teeth] grow about the twentieth year, both in men and women. But in some women, and also in some men, the maxillary teeth have grown when they have been only eighteen years of age, but in rising have been attended with pain. This, however, has happened to such as have not had the innermost cheek teeth in their youth.

5 & 6 But the elephant has indeed four teeth on each side, by which he elaborates his food, and grinds it into the appearance of flower. Besides these also, he has two other large teeth. The male elephant, therefore, has these large, and turned upward; but the female has them
502 small, and in a position contrary to the males; for the teeth of these are turned downward. The elephant also has teeth as soon as he is born, those however that are large, being at first immanifest. But he has a tongue very small, and so inward that it is difficult to see it.

7 Animals also differ in the magnitude of the mouth. For the opening of the mouth of some is very wide, as in the dog and the lion, and all those that have serrated teeth. But other animals have a small mouth, as men; and others have a mouth of a moderate size, as the genus of

swine. The Egyptian hippopotams, however, has a mane, in the same manner as a horse; but is bisulcous, in the same manner as an ox; and is flat-nosed to the view. He has also the astragalus, in the same manner as bisulcous animals; teeth somewhat prominent, a tail resembling the tail of a hog, and a voice like that of a horse. But the magnitude of his body is that of the ass. His skin likewise is so thick that hunting shields are formed from it; but his inward parts resemble those of a horse and an ass.

8 & 9 Some animals however have an ambiguous nature, because they partly imitate man and partly quadrupeds, such as apes, the cœbi [a kind of apes] and the cynocephali. But the cœbus has the tail of an ape; and the cynocephali have the same form with apes, except that they are larger and stronger, and have a more canine face. Their manners also are more savage, and they have teeth more canine and strong.

Apes, however, as being quadrupeds, are hairy in the prone part of their body; and in a similar manner, in their supine parts, as having in these the human form. For hairiness, as we have before observed, subsists contrarily in men and quadrupeds; except that apes have very thick and dense hairs, both in the prone and in the supine part. But the face of the ape has many similitudes to that of man; for it has similar nostrils and ears; and both its fore and maxillary teeth resemble those of man. And farther still, since other animals have not hairs on both eye-lids, the ape has, though they are very thin, especially on the inferior eye-lid, and very small; for other quadrupeds are deprived of these. The ape also has two papillæ of small mammæ in the breast. It has likewise arms in the same manner as man, except that they are hairy: and they bend these and their legs in the same way as man, *viz.* so that the curvatures of the arms are contrary to the curvatures of the legs. Besides this also, apes have hands, fingers and nails, similar to those of the human species, except that all these are of a more brutal form. The have however feet peculiar to themselves; for they are as it were large hands; and with respect to their toes, in the same manner as in the fingers of the hand, the middle is longer than the rest; and the sole of the foot resembles the palm of the hand, except that it extends itself as far as to the extremity, or heel. But the last part of their foot is callous, and rudely and obscurely imitates the heel of the human foot. They likewise use their feet, both as hands and feet, and bend their [fore] feet in the same manner as hands. The cubit however and thigh of apes are short in proportion to the other part of the arm and the tibia. They likewise have not a prominent navel, but a certain hard substance in the place of the navel. The superior part of the body also of these animals is much larger than the inferior part, in the same manner as in quadrupeds; for in quadrupeds the length of the superior is to the length of the inferior part, nearly as five to three. They degenerate therefore from men on this account, and also because they have feet similar to hands, and which are as it were composed from the foot and the hand; from the foot indeed, according to the extremity of the heel; but from the hand according to the other parts; since the toes have what is called the palm of the hand. They likewise live for a longer time prone after the manner of quadrupeds, than erect [after the manner of men]. And they neither have hips, as being quadrupeds, nor a tail as being a biped; for what they have as a tail is very small, as if it were designed merely as a mark. But the private parts of the female ape are similar to those of a woman; and those of the male more resemble the privities of a dog than of a man. The *cœbi* also, as we have before observed, have a tail. The whole genus of apes, however, in their interior parts resemble men. The external parts therefore of viviparous animals subsist after this manner.

10 All quadrupeds however which are oviparous and sanguineous, (and no terrestrial and sanguineous animal is oviparous, which is not a quadruped or without feet) have a head, neck, back, and the prone and supine parts of the body. They have also fore and hind legs, and that which is analogous to a breast, in the same manner as viviparous quadrupeds; and most of them have a long tail, but a few [as tortoises] have a short tail. All of them however have feet, consisting of many toes, and divided into many parts. Besides this likewise, all of them
503 have sensoria and a tongue, except the Egyptian crocodile. But this animal resembles certain fishes. For, in short, as fishes have a thorny tongue, and which adheres to the jaw bone, thus crocodiles have the place destined to the tongue so smooth, that no vestige of a tongue is apparent in them, unless the lips are very much drawn up. All oviparous quadrupeds likewise have no ears, but have only a certain avenue subservient to hearing; nor have they either mammæ, or a penis, or testicles, externally apparent, but they are inwardly latent. They are also without hairs, but all of them are covered with scales. All of them likewise have serrated teeth.

River crocodiles, however, have eyes similar to those of swine, but their teeth are large and prominent, their nails are strong, and their skin consists of impenetrable scales. They also do not see well in the water, but out of it they see most acutely. For the most part of the day, therefore, they live on the land, but at night in the water: for the water is then warmer than the air.†

11 The chamæleon has the whole figure of its body resembling a lizard, but its sides tend downward, and are conjoined under the belly, in the same manner as in fishes, and the back bone rises above the other parts in a manner similar to that of fishes. The face also of the chamæleon is most similar to that kind of ape which has the face of a swine, but it has a very long tail which ends in a point, and is convolved like a thong of leather. It is more erect however from the earth than the lizard: but the flexions of its legs subsist in the same manner as in the lizard. Each of its feet too is bipartite, and the two parts into which each is divided have a position with respect to each other, similar to that which the thumb has to the rest of the hand. Each of these parts of the foot of the chamæleon also is divided as it were into certain fingers. The first part of the foot therefore corresponds to a thumb, but the second part, to the rest of the hand, and is therefore divided into many other fingers. The second part however of the fore feet, is not divided in the same way as the second part of the hind feet, but the second part of the fore feet is divided inwardly, *i.e.* towards the sole, into three parts, and as it were fingers, but outwardly it is divided into two parts only. But the second part of the hind feet, is divided towards the sole into two fingers as were, and outwardly, *i.e.* in the superior part, it is divided into three parts resembling fingers. These parts also are covered with crooked nails. The whole body however of the chamæleon is rough, in the same manner as that of the crocodile. But its eyes are situated in a cavity, are very large and round, and are covered with a skin similar to that of the rest of the body; and in the middle of the skin that covers the eye there is a small avenue, through which it sees, which avenue is never covered with any skin. The motion also of its eye is circular, and 503
it transfers its sight to all places, and thus sees what it wishes to see. But the mutation of its colour takes place when it is inflated. It has also a black colour, not very remote from that of the crocodile, and a pale colour like lizards, variegated with black spots like the panther. A mutation likewise of this kind takes place through the whole of its body;

† The air in the day is hotter than the water; but in Egypt the water at night is warmer than the serene air, because the water preserves longer than the air the heat produced from the sun in the day.

for its eyes and its tail are changed in a manner similar to the rest of the body. But its motion is very slow like that of the tortoise. When it is dying also it is of a pale colour, and it has a colour of this kind when it is dead. But the parts about the stomach, and the *aspera arteria*, are situated similarly to those in the lizard. It has, however, no flesh, except in the head, the jaws, and the extremity of the tail, and these parts have a little flesh. It likewise has blood only about the heart, the eyes, the parts which are above the heart, and such veins as extend from these parts. But in these there is very little blood. Its brain also is placed a little above the eyes, but in continuation with them. And if the skin which covers the eyes is taken away, the eyes will be found to contain a shining substance resembling a thin brazen ring. Many strong membranes likewise are extended nearly through the whole of its body, and which are much superior in strength to those which cover the rest of its body. And when it is wholly dissected, it breathes for a long time, a small motion strenuously yet remaining in it about the heart. It is likewise capable of contracting all the parts of the body, and especially the parts about the sides. It has not however any where the spleen apparent; and it conceals itself in caves in the same manner as lizards.

12 In like manner birds also have certain parts in common with the above-mentioned animals. For all of them have a head, neck, back, the supine parts of the body, and that which is analogous to the breast. They differ likewise from other animals in this, that they are bipeds in the same manner as man, except that their legs are bent like those of quadrupeds, in a way contrary to those of man, as we have before observed. They have, however, neither hands nor fore-feet, but [instead of them] they have wings, which are peculiar to birds. Farther still, they
504 have a hip similar to a thigh, which hip is long, and extends as far as to the middle of the belly, to which it is affixed, so as to appear when divulsed to be a thigh. But the thigh in the other part is joined to the tibia. Birds, however, that have crooked nails have the largest thighs, and their breast is more robust than that of other birds. But all birds have feet armed with many nails, and all of them are after a manner multifidous; for the feet of the greater part of birds have toes. Those birds also which seek their food by swimming, though they have broad feet, and which are covered with a skin, yet they have perfect and distinct toes. All such birds, however, as fly high have four toes; for they have three for the most part in the anterior part of the foot, and one in the posterior part, in the place of the heel. But some few have

two toes indeed in the anterior part, and two in the posterior part, as is the case with the bird called *jynx*. This bird, however, is a little larger than the chaffinch, and its plumage is various. But it has something peculiar in the disposition of its toes, and its tongue is similar to that of serpents; for it can extend this to the length of four fingers out of the mouth, and again draw it within its beak. Farther still, it can also turn its neck round to the hind part of it, the rest of the body being quiescent, in the same manner as serpents. It has also large nails, and which are naturally similar to those of a jack-daw; and it emits a shrill voice.

Birds, however, have indeed a mouth, but one peculiar to themselves; for they have neither lips nor teeth, but a beak; nor have they ears or nostrils, but they have the avenues of these senses; of the nostrils indeed, they have the avenue in the beak, but of the hearing in the head. All of them, however, have two eyes, in the same manner as other animals, but without eye-lids. Heavy birds also [*i.e.* those which do not fly high] wink, that is, shut and open the eyes with the inferior eye-lid; and they likewise cover the eyes with a certain membrane emitted from the corner of the eye. Nocturnal birds, however, wink with the superior eye-lid. Animals also that are covered with scales do the very same thing, such as lizards, and others that are of a kindred nature to these animals. For all of them wink with the inferior eye-lid, though all of them do not cover the eyes with a membrane emitted from the corner of the eye.

Again, birds have neither scales, nor hairs, but feathers. But all feathers have a quill. And they have not indeed a tail like that of quadrupeds, but their tail consists of feathers. Those birds also which have long legs and broad feet, have a short tail, but those which have short legs and not a broad foot have a longer tail. Those birds too, which have a longer tail, when they fly have their legs contracted to their belly; but those that have a short tail, when they fly have their legs extended to their tail. All birds likewise have a tongue, but this is dissimilar; for some have a long, others a short, others a broad, and others a narrow tongue. Some genera of birds also, beyond all animals after man, utter the sound of our letters; and this is especially the case with those birds whose tongue is broad. No oviparous animal, however, has the epiglottis on the *aspera arteria*, but they are able to contract and dilate the avenue of the *aspera arteria*, so as to prevent any weight from falling on the lungs. Some genera of birds also have a spur; but there is no bird which has at the same time crooked nails and a spur. Some birds, however, that are armed with crooked nails fly high; but those birds that have spurs are heavy [and fly near the earth]. Again, some birds have a comb, or crest, consisting of raised feathers; but the cock alone has a peculiar crest, which is neither flesh, nor very remote from the nature of flesh.

13 Among aquatic animals, however, the genus of fishes alone is divided into many species. For they have a head, and the prone[†] and supine parts; and in the supine part of the body their belly and intestines are contained. In the last part of the body also they have a tail, which is in continuity with, and is not divided from, the rest of the body; and this tail is not similar in all fishes. But no fish has a neck, or hands, or feet, or in short testicles, either internally, or externally; nor have they any mammæ. To be deprived of mammæ, however, is not peculiar to fishes, but is common to them with many other animals, since all animals that are not viviparous are without mammæ, and of viviparous animals, those only have them which do not first conceive an egg in themselves, but an animal. For the dolphin is viviparous; and hence it has two mammæ, yet not in the superior, but in the inferior part, and near the organs of generation. It has not, however, like quadrupeds, apparent papillæ, but in the place of these it has as it were two orifices, one on the right, and the other on the left side, from which the milk flows. From these it is sucked by its young that follow it; and this by certain persons has been evidently seen.

Fishes, therefore, as we have said, have neither mammæ, nor any exterior and apparent genital member. They have, however, certain peculiar members, *viz.* gills and fins, by which they return the water which they receive through the mouth. Most fishes likewise are furnished with four fins, yet there are some that have only two; for certain long fishes, such as eels, have only two, near the gills. In a similar manner also, the mullets have only two fins, which are produced in the lake in Siphæ [in Tuscany]; and this is likewise the case with the fish called *tœnia*. But some of the long fishes, such as lampreys, have neither fins nor gills entire and perfect like other fishes. Of those,
505 however, that have gills, some have them concealed and others apparent; but all the cartilaginous kind have them apparent. And all those, indeed, that have their gills concealed, have them obliquely situated. But of

† The prone parts in fishes have a direction towards the heavens, and the supine parts towards the earth.

cartilaginous fishes, those that are broad, as the torpedo and the ray, have their gills in the inferior and supine part; but those that are long have them in their side, as is the case with all the genera of lampreys. The frog fish, however, has them obliquely and concealed, yet not with a thorny covering, like those fishes that are not cartilaginous, but with a skinny covering. Farther still, of some fishes the gills are single, but of others two-fold; but that part of the two-fold gills which is turned towards the body of the fish is not two-fold but single. And some fishes indeed have few gills, but others many. In either case, however, they have as many on one side as on the other. But those that have the fewest gills, have one on each side; and this is two-fold, as in the fish called the boar. Others again, have two on each side, the one simple, but the other double, as the conger and the *scarus*; but others have four simple gills on each side, as the ellops or sturgeon, the *synagris* or denticus, the lamprey, and the eel. Others have four gills indeed, but they are in a two-fold order, except the last, as the sea-thrush, the perch, the *glanis*, and the carp. All those, however, that are of the nature of lampreys have two-fold gills, and five on each side; but the sword-fish has eight two-fold gills. And thus much concerning the multitude of gills in fishes.

Farther still, fishes differ from other animals, not only in their gills, but also in the covering of their body. For they neither have hairs in the same manner as terrestrial viviparous animals; nor have they scales[†] like some oviparous quadrupeds; nor are they covered with feathers, as the genus of birds: but most of them, indeed, have thin scales, a few of them are rough, and a very few of them are smooth. Of the cartilaginous kind, therefore, some are rough, but others are smooth; but the conger, the eel, and the tunny, are among the number of the smooth.

All fishes, likewise, have serrated teeth, except the scarus. All of them too have their teeth sharp, and some of them have many rows of teeth, and inserted in the tongue. They have also a hard and thorny tongue, and so adherent and fixed, that sometimes they seem not to have a tongue. With respect to the mouth, in some fishes it is very open, in the same manner as in some viviparous quadrupeds. But with respect to the sensoria or organs of sensation, they neither have the sensoria of hearing or smelling, nor the avenues of these apparent; but all of them have eyes without eye-lids, though their eyes are not hard. The whole 505

† *i.e.* Scales resembling those of a serpent.

genus of fishes, therefore, is sanguineous; but some of them are oviparous, and some of them are viviparous. All those indeed that are covered with thin scales are oviparous; but all the cartilaginous kind are viviparous, except the frog-fish.

14 Of sanguineous animals, however, there remains the genus of serpents. But they partake of the nature both of terrestrial and aquatic animals. For most of them are terrestrial, and but a few of them are aquatic, and which live in potable water. There are also marine serpents similar in form to the terrestrial genus, except that their head more resembles that of the conger. There are, however, many genera of marine serpents, and they have an all-various colour; but they are not generated in very deep places. Serpents, also, are without feet, in the same manner as the genus of fishes.

There are, likewise, marine scolopendra, resembling the terrestrial in their form; but their magnitude is somewhat less. These marine scolopendra are generated in rocky places [of the sea]; but their colour is more red, they have a greater number of feet, and their legs are thinner, than those of terrestrial scolopendra. These, likewise, are not generated, as neither are serpents, in very deep places of the sea.

There is, also, a certain little fish found on rocks, which some call *echeneida*, or the sea lamprey, and some use it as a philter, and for the purpose of obtaining victory in judicial processes; but it is useless for food. Some also say that it has feet, but it has not. It appears, however, to have them, because it has fins resembling feet. And thus we have shown what and how many the external parts of animals are, and what their differences are with respect to each other.

15 In the first place, therefore, we must [next] show how the internal parts subsist in sanguineous animals. For the greatest genera of animals differ from the rest in this, that the one are sanguineous, but the other exanguious. Sanguineous animals, however, are man and viviparous quadrupeds; and besides these, oviparous quadrupeds, birds, fish, cetaceous animals, and others which are not contained in these greatest genera, but are signified by the name of species, as serpent and crocodile.

All such animals, therefore, as are quadrupeds and viviparous, have a stomach and *aspera arteria*, and these are situated in the same manner as in men. This is also the case with such quadrupeds as are oviparous, and
506 with birds; but they differ in the form of these parts. In short, all such animals as by receiving the air respire and expire have lungs, the *aspera*

arteria, a stomach; and the position of the stomach and *aspera arteria* is similar, though these parts differ in species from those of other animals. On the contrary, the lungs are not only specifically different in different animals, but have also a different position.

Farther still, all sanguineous animals have a heart, and a transverse septum or diaphragm; but in small animals, the blood is not similarly apparent, on account of its tenuity and paucity, except in the heart. It is also peculiar to a certain genus of oxen, but not to all oxen, to have a bone in the heart; and the heart of horses, likewise, has a bone. All sanguineous animals, however, have not lungs. Thus, for instance, fishes, and all animals that have gills, are without lungs. All such animals, too, as have blood have a liver; yet not all, but the greater part of them, have a spleen. Many animals, likewise, that are not viviparous, but oviparous, have a spleen so small that it can scarcely be perceived, as is evident in most birds; as in the dove, the kite, the hawk, and the owl. But the bird called *ægocephalus*, [*i.e.* having the head of a goat] is entirely without a spleen. The like also takes place in oviparous quadrupeds; for these have a very small spleen, as is evident in the tortoise, the rubeta or land toad, the lizard, the crocodile, and the frog.

Again, of those animals that have a liver, some have the gall annexed to the liver, but others have not. Of viviparous quadrupeds, indeed, neither the stag, nor the fallow-deer, nor the mule, nor the ass, nor the sea-calf, has the gall; and this is also the case with some rats. With respect to stags, however, those which are called *achaini*, seem to have the gall in their tail; but they have, indeed, a colour in their tail resembling gall, yet it is not fluid like gall, and in the interior part is similar to the spleen. All these stags, too, have in their head living worms; and these are generated under the concavity of the tongue, about the vertebra, by which the head is connected with the neck. They are also not less in magnitude than the largest worms. But they are generated at once, and in continuity, and they are about twenty in number. Stags, therefore, as we have said, are without a gall, but their intestine is so bitter, that the dogs are not willing to eat it, unless the stag himself is very fat. The liver of the elephant also is without a gall, 506
yet if it is cut in that part to which the gall usually adheres, a liquor flows from it which is more or less bilious.

But of animals which receive the sea water, and have lungs, the dolphin alone has no gall. All birds, and other fishes, however, have a gall, and all oviparous quadrupeds, in short, have it more or less; but some fishes have the gall in the liver, as the sea-dog, the glanis, the skate, the smooth ray, and the torpedo; and among long fishes, the eel, the acus or needle fish, and the *zagœna* or libella. The fish also which is called *callionymus* has the gall in the liver, and larger than in other fishes, in proportion to the magnitude of its body. Other fishes have the gall, not in the liver, but in a certain intestine depending from the liver, which intestine has very slender avenues. The fish, therefore, called hamia has the gall in a certain intestine, which is extended in an equal length, and which is frequently redoubled; but others have the gall at the intestines, some more remote from, and others nearer [to the liver], as the frog-fish, the ellops or sturgeon, the *synagris*, or dentex, the lamprey, and the sword-fish. It frequently happens, however, that the same genus of fishes is seen to have the gall in both situations, as the conger; and some indeed have it at the liver, but others at the intestines. The like also takes place in birds. For some of these have the gall joined to the belly, but others to the intestines, as the dove, the crow, the quail, the swallow, and the sparrow. Some birds, likewise, have the gall at the same time, both at the liver and at the belly, as the *ægocephalus*; but others have it, at the same time, at the liver and the intestines, as the hawk and the kite.

16 All viviparous quadrupeds, likewise, have the kidneys and the bladder; but among oviparous animals, both birds and fishes are without kidneys and a bladder. And of oviparous quadrupeds, the marine tortoise alone has kidneys proportional in magnitude to the rest of the body, but they are similar in proportion to the kidneys of oxen. The kidneys, however, of the ox are composed as it were from certain small kidneys. The *bonasus* also has all the interior parts similar to those of an ox. Such animals, likewise, as have these parts, have them in a similar situation.

17 The heart, too, is in the middle of the breast in all animals except
507 man; for he has it rather in the left side, as we have before observed. But the point of the heart in all animals is towards the anterior part, except in fishes, in which it is not turned towards the breast, but to the head and mouth. The extremity of the heart also, in fishes, is suspended from that which conjoins the right and left gills with each other. But there are other tubes, or avenues, which tend from the heart of fishes to the several gills, which are greater or less, according to the greater or less magnitude of their bodies. In large fishes, however, the tube which has its egress from the point of the heart is very thick and white. Very few

fishes also have a stomach; such as the conger and the eel, and in these it is very small.

Those animals, likewise, that have a liver, which is not divided but continued, have the whole of it in the right side; but those that have it divided from the beginning, have not the whole liver, but the greater part of it in the right side. For some animals have the two parts of the liver not conjoined, but divided from the beginning; and of this kind, among fishes, is the sea-dog, and a certain genus of hares, which are found elsewhere, and about the lake Bolba, in what is called Sycina. These hares have the liver so divided, that some one might think they had two livers, because the avenues, [and fibres] of the two parts of the liver are conjoined at a distance from each other, in the same manner as in the lungs of birds.

The spleen, also, is naturally in all animals in the left side. A certain quadruped, however, on dissection, has been found to have the spleen in the right side, and the liver in the left; but things of this kind must be considered as prodigies.

The *aspera arteria* also in all animals extends to the lungs, but in what manner we shall afterwards explain. But the stomach, in such animals as have it, extends to the belly, through the diaphragm; for fishes, for the most part, as we have already observed, have no stomach, but the belly in these animals is immediately connected with the mouth. Hence it frequently happens, that while some of the larger fishes pursue the less the belly falls into the mouth.

All the above-mentioned animals, however, have a belly, and similarly situated. For it is directly situated under the transverse septum, and the intestine adheres to the belly, and ends at the egress of the food; which part is called the anus. They have, however, bellies of a various form. For, in the first place, such viviparous and horned quadrupeds as have not teeth in both jaws, have four such avenues, or ducts; and these 507
quadrupeds are said to ruminate. For the stomach, beginning from the mouth, extends downward to the lungs, from the diaphragm, and ends in the [first, which is called] the great belly, the internal parts of which are rough and wrinkled. To this first belly, in that part in which it is joined to the stomach, a second cavity and belly are united, which form a similitude to the fillets of women they call reticula; for this fillet is externally similar to the first belly, but internally resembles folded fillets. This reticulum also is much less than the first belly. After the second sinus, or belly, follows the third, which is called omasum; which in the interior part is rough, and distinguished by certain cells after the manner

of a sponge, and is similar in magnitude to the second belly, or reticulum. After this follows the fourth belly, which is called abomasum; which is of a more ample magnitude than the omasum, but its figure is more oblong; and inwardly it contains many large and smooth cells. This also is followed by the intestine. Horned quadrupeds, therefore, and which have not teeth in both jaws, have such a belly as we have explained. They differ, however, from each other in the figure and magnitude of the belly, and also because the stomach in some extends to the middle belly, but in others to the side of the belly.

On the contrary, animals which have teeth in both jaws, as man, the dog, the lion, the wolf, the swine, and the bear, have but one belly [to which the intestine is joined]. But the *thos* [an animal generated from a hind and a wolf], has all the internal parts similar to those of a wolf. All animals, therefore, that have teeth on both sides, have one belly, and after this the intestine; but some have a larger belly, as the swine and the bear, the belly of the former of which has a few smooth cells; and others have a much less belly, and which is not much larger than the intestine, as the dog, the lion, and man. But other animals differ in the form of the belly; for some have it similar to that of the swine, and others to that of the dog. These bellies, also, differ according to magnitude and figure, thickness and thinness, and also in the position in which the stomach is connected with the belly.

The nature of the intestines also differs in each of the above-mentioned animals (*viz.* in those which have not teeth on both sides, and in those that have), in magnitude, thickness, and reduplications. In all animals, however, that have teeth only on one side, the intestines are larger; for these animals also are larger; since but few animals that have teeth only on one side are small; for they are horned, and no horned animal is entirely small. Some animals, also, that have teeth only on one side,
508 have certain appendices of intestines. But the intestine of an elephant is so sinuous, that it seems to contain four bellies. In this, also, the food is received; for separate from this it has no other receptacle. The elephant likewise has viscera, similar to the viscera of swine, except that its liver resembles that of an ox, but it is quadruply larger; just as the other parts of the elephant are much larger than the parts of the ox. The spleen, however, of the elephant is much less than its proportion with the rest of the body seems to require.

Oviparous quadrupeds, also, *viz.* both the land and sea tortoise, the lizard, the river crocodile, and the crocodile which is not fluviatic, and in short all such animals, have the belly and intestines after the same manner as quadrupeds which have teeth on both sides. For they have one simple belly, and in some it resembles the belly of swine, but in others that of dogs.

The genus of serpents, however, resembles, and has nearly every part similar to, lizards, which are terrestrial and oviparous animals, if we give length to the lizard, and deprive it of feet. For serpents are covered with scales, and have both the prone and supine parts similar to those of the lizard; except that they have no testicles: instead of which they have two avenues in the same manner as fishes, which concur in one place, and a small and bifidated vulva. But their other internal parts are the same as those of lizards, except that all their viscera, on account of the length and narrowness of their bodies, are long and narrow, and can scarcely be distinguished from those of the lizard. For they have the *aspera arteria* very long, but the stomach in them is even longer than the artery. But the beginning of the artery is so near to the mouth, that the tongue seems to be under it. The artery, however, appears to be above the tongue, because the tongue is contracted and subsides, and does not remain in the same manner as in other animals. But the tongue of serpents is thin, long, and black, and when it is drawn out may be extended farther. This, however, is peculiar to serpents and lizards, that the extremity of their tongue is bifidated; and this is especially the case with the tongue of serpents; for the extremities of their tongues are as thin as hairs. The tongue of the sea-calf also is bifidated.

The serpent also has a belly, which is as it were a more broad intestine, resembling the belly of the dog; and after the belly, they have a long and thin intestine, which is uniform to the end. But a small and long heart, and having the form of the kidneys, adheres to the pharynx; on which account it may sometimes seem that the point of the heart is not towards the breast. In the next place the lungs follow, which are simple, fibrous, replete with pipes or little tubes, very long, and very distant from the heart. The liver also of serpents is long and simple; but their spleen is small and round, in the same manner as the spleen of the lizard. Their gall, however, resembles that of fishes; for aquatic serpents, indeed, have the gall at the liver; but other serpents have it for the most part at the intestines. All serpents, however, have serrated teeth. But they have as many sides as the month has days; for they have thirty. They also say that a circumstance happens to serpents, which happens to the young of swallows; for they report that if their eyes are dug out, they will be re-produced. The tails also of serpents and lizards when cut off are re-produced.

Fishes also accord with serpents as to the belly and intestines; for they have one simple belly, but the figure of it is different in different fishes. For in some the form of it is entirely different from that of other fishes, as in the fish called scarus, which is the only fish that appears to ruminate. The magnitude also of its intestine is simple, and has a reduplication, which is analyzed into one thing. It is, however, peculiar to fishes, and the greater part of birds, to have certain appendices, [and as it were branches of intestines]; though birds have only a few of these, and in the inferior part of the intestines: but fishes have many appendices, and in the supreme part of the intestines, near the belly. Some fishes also have more of these than others, as the gudgeon, the sea-dog, the perch, the scorpion, the saw-fish, the barbel, and the *sparus*; but the mullet in one part of the belly has many appendices, but in the other part has one only. Some fishes, however, have but few appendices, as the hepatus, or liver fish, the *glaucus*, and the gold-fish. Fishes, also, of the same genus differ from each other, in the number of these appendices; since among gold-fishes some have many, and others have but few, of them: and, in short, there are some fishes which have no appendices, as most of the cartilaginous kind; but, of the rest, some have a few, and others very many. All fishes, however, have these appendices near the belly.

Birds, however, differ from each other, and from other animals, as to their internal parts; for some of them have the craw before the belly, as the cock, the ring-dove, the pigeon, and the partridge. But the craw is a hollow and large skin, in which the food is first received before it is concocted. The skin, also, in the superior part, in which it is joined to the stomach, is more narrow; but afterwards it is broader, and in the inferior part, in which it is joined to the belly, it is again narrower. Most birds likewise have the belly fleshy, and compact or callous; and under the fleshy part there is a strong skin, which may easily be separated from it. Other birds, on the contrary, have no craw, but instead of it have a wide and broad stomach, which extends either
509 through the whole, [*i.e.* which is either wholly so] so in that part in which it is joined to the belly; as the jackdaw, the raven, and the crow. The quail, also, has the inferior part of the stomach broad; and the ægocephalus, and the owl, have this part a little broader. But the duck, the goose, the sea-gull, the diver, and the bustard, have the whole of the stomach wide and broad; and in a similar manner many other birds. Some birds, however, as the kestrel, have a belly similar to the craw. There are, likewise, some birds which neither have a stomach nor a wide

craw, but have a long belly; and this is the case with such small birds as the swallow and the sparrow. A few, also, have neither a craw, nor a wide but a very long stomach, and these are such as have a long neck, as the bird which is called porphyrion.† Nearly, however, all these emit moister excrements than other birds. But the quail has this peculiarity with respect to other birds, that it both has a craw, and before the belly a wide and broad stomach. The craw also of this bird, according to the proportion of its magnitude, is very distant from that of the stomach, which is near to the belly.

Most birds, likewise, have a thin intestine, and which when unfolded is simple. Birds, however, as we have before observed, have but few appendices, [and as it were branches of intestines], and these not in the superior part as fishes, but in the inferior part, near the end of the intestine. Yet not all but most birds have these appendices, as the cock, the partridge, the duck, the night-raven, the stork, the ascalaphus, the goose, the swan, the bustard, and the owl. Some little birds also, as the sparrow, have these appendices, but they are very small.

† An aquatic bird, so called from the purple colour of its beak and feet.

BOOK III

1 As we have discussed, therefore, the other interior parts of animals, and have described their number, qualities, and differences, it remains that we should speak concerning the parts which are subservient to generation. For these in all females are internal, but in males they have many differences. For, in short, of sanguineous animals some have no testicles, but others have, yet have them internally. And of those that
509 have them internally, some have them near the loins, about the region of the kidneys, but others have them near the belly; and others externally. Some animals also have the penis adhering to the belly, but in others it is detached in the same manner as the testicles. The penis, however, does not adhere to the belly, in the same manner, in animals that discharge the urine in the anterior, as in those that discharge it in the posterior, part of the body.

Among fishes, therefore, none have testicles, nor any other animal that may have gills, nor the whole genus of serpents, nor, in short, any animal without feet, unless they are such as generate an animal within themselves.† But birds have indeed testicles, yet they have them internally, near the loins. And such quadrupeds as are oviparous have them situated in the same manner; as the lizard, the tortoise, the crocodile; and, among viviparous quadrupeds, the hedgehog. Of animals, also, which are without feet, the dolphin has testicles, not near the loins, but near the belly; and among viviparous quadrupeds, the elephant: but other animals have the testicles apparent. We have, however, before shown what the difference is, between those animals which have the testicles at the belly, and those which have them in the region near to the belly. For in some animals, as in the genus of swine, they adhere to the posterior part of the belly, and are not detached; but in others, as in men, they are detached.

Fishes, therefore, as we have before observed, and serpents have no testicles, but they have two avenues or ducts, which, being suspended from the transverse septum or diaphragm, pass through each side of the spine, and unite in the superior part; in the inferior part, also, they unite in that place in which there is an egress of the excrement. But I call the part which is at the spine the superior part. These avenues at the time of coition are full of genital seed, and, from the mutual attrition of the male and female, a white seed proceeds from them. The difference,

† Such as the dolphin; which, though it is without feet, and has gills, yet, because it generates an animal within itself, is not without testicles.

however, of these avenues, with respect to each other, ought to be surveyed from dissections, and will be afterwards more accurately explained by us in the peculiar discussions about each part. But all oviparous animals, whether they are bipeds or quadrupeds, have testicles near the loins, under the transverse septum, some being whiter, and others more pale, and surrounded with very thin veins. From each of the testicles, also, an avenue extends; and these avenues, or ducts, coalesce in one above the egress of the excrement, in the same manner as in fishes. This, however, is the genital member, which in small oviparous animals is immanifest, but in those that are larger, as in geese and other birds of a similar magnitude, becomes more manifest, especially after recent coition. But these seminal avenues, as well in fishes and serpents, as in quadrupeds and bipeds, adhere to the loins under the belly and intestines, between the great vein, from which avenues extend to each of the testicles. As in fishes, however, at the 510
time of coition these avenues are very conspicuous, and are found to be full of seed, but when the time of coition is past, the avenues also are sometimes immanifest: thus, with respect to the testicles of birds, before they copulate some indeed are small, but others are perfectly immanifest; but when they copulate they have them very large. This, however, happens most manifestly in ring-doves and partridges, so that some persons in winter fancy that they have no testicles.

But of animals that have the testicles in the anterior part, some have them internally, near the belly, as the dolphin; but others have them externally, near the end of the belly. And those that have the testicles within, or external to the belly, agree in other things, but differ in this; that those that have the testicles within have them not covered with any skin, but those that have them external to the belly have them covered with a skin, which is called the scrotum.

The testicles, however, in all pedestrious and viviparous animals, subsist in the following manner: Veiny avenues or ducts extend from the aorta as far as to the head [or superior part] of each of the testicles; and two other avenues extend from the kidneys [to the pudendum]. And these latter avenues indeed contain blood; but those which extend from the aorta are exanguious. From the head, also, an avenue extends to the testicles, which is more dense and nervous than the testicles, and which in its progress is again inflected in each to the head of the testicles. But from the head both the avenues again coalesce, and by their interior part enter the pudendum. Though the testicles, however, have as well avenues resident in as proceeding to the head of the testicles, yet the

difference of such avenues is with difficulty distinguished; because such avenues are covered with the same membrane; so that unless the membrane is disjoined, they appear to be one and the same avenue. The avenue, therefore, which is resident in the testicles themselves, contains a certain sanguineous liquid, but less than that which is contained in the avenues extending from the aorta. The avenues, however, which are not resident in the testicles, but extend to the head of them, contain a white liquid. An avenue, also, extends from the bladder to the neck of it; and this urinary duct is surrounded as it were with a shell, which is called the pudendum. What has been said, however, may be surveyed from the following description.† A is the beginning of the two avenues which originate from the aorta, and descend to BB, *i.e.* the head or superior part of each testicle. CC are the avenues resident in the testicles themselves. DD are the avenues which, proceeding through the testicles, are reflected, and return to the head of the testicles. In these avenues a white liquor is contained. E is the pudendum; F the bladder; and GG
510 are the testicles. When the testicles, however, are cut off, or taken away, the avenues [or little veins through which the seed descends] are drawn upward. But [shepherds] are accustomed to castrate animals in two ways; for they either break the testicles while the animals are yet young; or cut them off when the animals are adult. It has also happened that a bull, when castrated, has immediately copulated and generated.

The particulars, therefore, respecting the testicles of animals subsist after this manner. Wombs, however, neither subsist after the same manner in those animals that have wombs, nor are similar in all animals, but differ from each other in viviparous and oviparous animals. All wombs, therefore, that are placed near the genital member have two cavities, one in the right, but the other in the left side. These cavities, however, have but one beginning, and one mouth, [or neck of the womb], which in most and the largest animals is very fleshy and cartilaginous. The interior part, also, is principally called the womb, and *delphys*, whence all brothers are called *adelphi*; but the neck and mouth of the womb are denominated the matrix.

Such animals, therefore, as are viviparous, and bipeds or quadrupeds, as man, the dog, the swine, the horse, and the ox, have the womb beneath the transverse septum or diaphragm; and this is, likewise, the case with all horned animals. But the extremities of most wombs, which are called the horns, have certain convolutions [or wrinkles]. Those

† See the figure in the Commentary of Sealiger on this place.

animals, however, which bring forth eggs, have not all of them wombs similarly situated, but birds have their wombs near the transverse septum, and fishes have them beneath it, in the same manner as viviparous bipeds and quadrupeds; except that the wombs of the latter are thin, membranous, and long. Hence, in very small fishes, each cavity of the womb appears to be one egg, as if these fishes had two eggs; but this egg is friable; for it is not one but many eggs, and on this account it may be dissolved into many. The womb of birds, however, has a fleshy and callous neck in the inferior part, but the part near the transverse septum is membranous, and very thin, so that the eggs appear to be out of the womb. In larger birds, therefore, this membrane is more manifest, and if it is inflated through the neck of the womb, it is elevated, and becomes sinuous; but in small birds these things are more immanifest. The womb, also, subsists after the same manner in oviparous quadrupeds, such as the tortoise, the lizard, the frog, and other 511
such like animals. For the neck of the womb in these is fleshy, and in the inferior part; but the division of it into two cavities is in the superior part, near the transverse septum, and in this division the eggs are contained.

With respect to such animals, also, as are without feet, and which bring forth an animal externally, but conceive an egg in themselves, such as the sea-dog, and other animals that are called cartilaginous; (and any animal that is without feet, and having gills, is viviparous, is denominated cartilaginous), these have a two-fold womb; and in a similar manner with the womb of birds, it extends to the transverse septum, so that in the superior part it is narrower, and afterwards becomes broader. The eggs, therefore, of such animals are generated, and consist in the superior and middle part of the womb, but afterwards falling into the inferior and broader part, they exclude the animal. The difference, however, of these wombs from each other, and from those of other fishes, may be more accurately surveyed in anatomical figures.

The genera of serpents, likewise, differ as to wombs from each other, and from other animals. They differ indeed from each other; for all other genera of serpents are oviparous, but the viper alone is viviparous, first conceiving an egg in herself. Hence she has the particulars respecting the womb similar to those of cartilaginous animals. The womb, however, of serpents is long, in the same manner as their body, and it extends downward, beginning from one continued avenue, and afterwards stretches by a two-fold passage from each side of the spine, as far as to the transverse septum. In this womb, likewise, eggs are generated in a regular order, and are not brought forth one by one, but in a continued series. Such animals, however, as conceive an animal in themselves, and bring it forth externally, have the womb situated above the abdomen;[†] but in all such animals as conceive and bring forth an egg, the womb is situated beneath, near the loins. And in such animals as bring forth an animal externally, but conceive an egg in themselves, the situation of the womb is ambiguous. For that part of it which contains the eggs is beneath, near the loins; but the other part from which the eggs make their egress, is situated above the intestines. Farther still, there is also this difference in wombs, with respect to each other. For some horned animals, and which have not teeth on both sides, have acetabula[‡] joined to the womb, when it contains the embryo; and among animals that have teeth on both sides, the hare, the mouse, and the bat, have these acetabula. All other viviparous animals, however, that have feet and teeth on both sides, have the womb smooth, and in these the embryo is suspended from the womb, and not from the acetabulum. After this manner, therefore, the dissimilar parts of animals, as well the exterior as the interior, subsist.

511 2 Of the similar parts, however, the blood is the most common in all sanguineous animals, and the part in which the blood is naturally adapted to be ingenerated; and this part is called a vein. Afterwards that which is analogous to these follows, *viz.* lymph and fibre. That, also, which is especially the body of animals is flesh, and, whatever is analogous to this; and likewise bone, and that which is analogous to bone, as the spine and cartilage. Skin, also, membrane, nerve, hair, nail, and whatever is analogous to these, may be considered as principally pertaining to the body of animals. And besides these, fat, suet, and the excrements; but the last of these are dung, phlegm, and yellow and black bile.

Since, however, the nature of the blood, and of the veins, seems to be the principle [of the other similar parts], we must first consider these, especially since some of the ancients do not speak truly concerning them. But the cause of their ignorance arises from the difficulty of

† *i.e.* Between the intestine called the alvus, in which the dry excrement is contained, and the bladder.

‡ The acetabula are certain little vessels, with which the navel of the fœtus is conjoined, so that through them it may derive nutriment from the veins of the maternal womb.

surveying these. For, in dead animals, the nature of the most principal veins is immanifest, because these especially immediately subside and fall on the egress of the blood; since the blood is collectively poured out from these, as from a vessel; for all the blood is contained in the veins, and not in the other parts of the animal, except a little which is in the heart. And in the animal while living, it is impossible to survey how the veins and blood subsist, since the nature of them is within the body. Hence it happens, that those who survey the veins and blood in dead and dissected animals, are unable to detect the greatest principles of them; but those who have surveyed these in very lean men while living, have formed their opinion of the principles of the veins, from circumstances which have been then externally apparent.

Syennesis, indeed, a Cyprian physician, thus writes on this subject: "The thicker veins originate as follows. Two veins extend from the eyes, near the eyebrow, and proceed through the back to the lungs, under the mammæ; one indeed, from the right side to the left, but the other from the left side to the right. The vein, therefore, which extends from the left side proceeds through the liver to the kidneys and the right testicle; but the vein which extends from the right side proceeds to the spleen, the kidneys, and the left testicle; and from hence to the pudendum."

Diogenes of Apollonia, however, says as follows. "In man there are two greatest veins; and these extend through the belly, near the spine of the back; one of them, indeed, to the right, but the other to the left; each of them also extends to the foot, through the leg proximate to itself; and each of them ascends to the head, through the shoulders and neck. From these two veins other smaller veins are diffused through the 512
whole body, from the right to the right, and from the left to the left. Two of the largest of these smaller veins, indeed, extend to the heart, near the spine of the back; but two others, one of which is called *splenitis*, from pertaining to the spleen, but the other *hepatitis*, from pertaining to the liver, extend a little higher through the breast, under the arm-pit near to them, to the arms, and when they arrive at the hands, they are so divided in their extremities, that one part extends to the thumb, but another part to the palm of the hand; and from these, through slender branches as it were, they are diffused into the whole hand, and all the fingers.

"Other thinner veins, also, extend from the first greatest veins; some of which extend from the right to the liver, but others from the left to the spleen and kidneys. Those veins, however, which extend to the legs are divided at the thigh, in that part in which the thigh is joined to the buttock, and thus extend through the whole thigh; but the greatest of them proceeds through the posterior part of the thigh, and there becomes sensibly thick; and the other proceeds through the anterior part of the thigh, and is somewhat less thick than the other. Afterwards these veins extend through the knee to the legs and feet, in the same manner as the veins which extend through the arms are diffused into the hands and fingers; then they pervade to the soles of the feet, and from thence extend to the fingers. From the greatest veins, also, many other more slender veins are diffused into the belly and sides.

"But the veins which extend through the throat to the head appear to be large in the neck. From these many ramifications are derived, which are diffused through the whole head, some of them proceeding from the right to the left, but others from the left to the right; but each of them terminates near the ear. Besides the larger vein, there is also in the neck another vein a little less, into which, on both sides, [*i.e.* both on the right and left side], many of the veins of the head extend, and afterwards are derived into the shoulder-blade, arms, and hands. Other veins, also, a little less, appear near the veins called *splenitis* and *hepatitis*, which they cut when there is an abscess under the skin. But in diseases of the ventricle, it is requisite to cut the *splenitis* and *hepatitis*.

"Other veins, also, extend from these to the mammæ. And there are other slender veins, which extend from each of these through the marrow of the back to the testicles. Others, again, extend under the skin, and through the flesh to the kidneys; in men terminate in the testicles, but in women in the womb. The veins, however, when they first proceed from the ventricle are broader, but afterwards become more thin, while they change their situation, so that the right veins proceed to the left side, and the left to the right side. These veins, likewise, which, proceeding from the right to the left, and from the left to the right side, are gradually attenuated, are called seminal veins. And with respect to the blood, the thickest part of it is imbibed by the flesh; but that part of it which arrives at the seminal and more subtle veins becomes thin, hot and frothy."

3 Such, then, are the assertions of Syennesis and Diogenes. But Polybus thus writes, [concerning the origin and propagation of the veins]. "There are four conjugations of the veins. The first conjugation, indeed, contains two veins, which originate from the posterior part of the head, and passing through the exterior part of the neck, and the sides

of the spine, descend to the hip and legs, and afterwards arrive through the tibia, and exterior ankle bones, at the feet. Hence physicians bleed those who have pains about the back and hip, by opening the veins of the ankle and the hind part of the knee.

"But the second conjugation contains two veins, which originate from the head, near the ears. These veins are called jugular, and descending inwardly through the back, near the spine, arrive at the testicles, and afterwards through the loins, the hind part of the knee, and the tibia, tend to the interior ankle bones, and the soles of the feet. Hence, in pains of the loins and testicles, physicians bleed those who are so afflicted from such veins of the hind part of the knee and ankles.

"The third conjugation contains two veins, which originate from the temples, and through the neck descend under the shoulders to the lungs. One of these, also, proceeding from the right to the left part, tends under the left mamma to the spleen, and the left kidneys; but the other, proceeding from the left to the right part under the mamma, tends from the lungs to the liver, and afterwards to the right kidneys; and both of them at last terminate in the testicles.

"The fourth conjugation, likewise, contains two veins, which originate from the anterior part of the head, and proceed from the eyes and under the neck to the clavicles of the shoulders. Afterwards through the arm, cubit, and elbow, they descend to the palm of the hand and the fingers. From such veins under the arm-pit a branch originates, which passes through the ribs, till one of the veins arrives at the spleen, but the other at the liver; and afterwards each, through the superficies of the belly, terminates in the genital parts."

Such nearly are the assertions of others, [concerning the origin and progress of the veins]. There are, also, certain physiologists, who, though they have not written on the veins with so much accuracy as physicians, yet all of them similarly ascribe the origin of the veins to the head and the brain; which, nevertheless, is false. Since, however, as we have said, the survey of the veins is difficult, they can alone be sufficiently inspected by those who pay attention to things of this kind, in animals which have first wasted away by leanness, and are afterwards strangled.

The nature of the veins, however, subsists in the following manner. There are two veins in the thorax, situated near the spine, of which the greater is in the anterior part, but the less is in the posterior part. And the greater, indeed, declines more to the right side, but the less to the left side. Some, likewise, call the less vein the aorta, because the nervous part of it is also seen in dead bodies. These veins, however, originate from the heart. For the whole of them extend through the other viscera, preserving their own nature as veins; but the heart is, as it were, a part of them, and is more manifestly the principle of the larger vein, which is in the anterior part; so that one of the two principal veins is above, but the other beneath, and the heart is in the middle of them.

All hearts, however, contain in themselves a [triple] cavity; but in very small animals, the greatest cavity of the heart can scarcely be seen. But in animals of a moderate size, the first cavity of the heart, which is the greatest, is not only visible, but also the second; and in the largest animals all the three are visible. The point, however, of the heart is directed to the anterior part, as we have before observed; the greatest cavity being in the right, and highest part of the heart, the least cavity in the left part, and that which is of a moderate magnitude being in the middle of both the others. But the other two are much less than the greatest. All these cavities, likewise, are connected by perforations with
513 the lungs; but, except one, they are immanifest through the smallness of the cavities.

The great vein, therefore, originates from the greatest cavity of the heart, which is constituted in the highest and right part of the heart. Afterwards passing through the middle cavity, it again receives in it the form of a vein, since the middle cavity of the heart is a part of the greater vein in which the blood stagnates. But the aorta originates from the middle cavity, yet not after the same manner as the great vein originates from the greatest cavity, but it communicates with it by a much narrower canal. And the great vein, indeed, extends through the heart, but the aorta from the heart. The great vein, also, consists of membrane and skin; but the aorta is narrower than this vein, and is very nervous. Being far extended, also, from the heart to the head, and the inferior members, [*i.e.* to the feet], it becomes more narrow, and perfectly nervous.

One part, therefore, of the great vein extends in the first place to the lungs, and in that place, in which it is conjoined with the aorta, is divided into two branches, one of which proceeds to the lungs, but the other to the spine of the back, and the last vertebræ of the neck. The vein, therefore, which extends to the lungs, because the lungs are bipartite, is also first divided into two branches, corresponding to the two parts of the lungs, and afterwards is divided into many other branches, through which it enters the several pipes and avenues of the lungs; the larger, indeed, through more ample veins, but the less through narrower veins, and in such a manner that there is no part of the lungs in which there is not an avenue and a little vein. For the last avenues and parts of the small veins are immanifest, on account of their smallness. The whole lungs, however, are seen to be full of blood; and under the little veins there are pipes, which are derived from the artery. But the other branch of the great vein, which, as we have said, tends to the spine, and the vertebræ of the neck, is again extended through the spine; which vein Homer speaks of when he sings:

> Antilochus, as Thoon turn'd him round,
> Transpierc'd his back with a dishonest wound:
> The hollow vein that to the neck extends,
> Along the chine, his eager javelin rends.
>
> *Iliad*, XIII, 546. Trans. Pope

Small veins extend from this vein to each of the sides and vertebræ. This same vein, also, is divided into two other branches, near the vertebra, which is situated above the kidneys.

After this manner, therefore, these parts of the great vein are divided. The vein, however, which extends from the heart through the spine of the back, and the vertebræ of the neck, not only sends various ramifications of itself to the ribs and vertebræ, but, in its superior part, the whole of it is again divided into two places. For some of these ramifications tend to the sides and shoulders, and afterwards are diffused through the arm-pits; in men into the arms, in quadrupeds into the 514
anterior legs, in birds into the wings, and in fishes into the superior fins. These veins, however, when they first begin in the throat to be divided into many branches, towards the throat and shoulders, are called jugular veins; but after the great vein, being divided in the throat, has ascended through the neck to the head, two of its branches proceed near that artery of the lungs which is called the aspera arteria. And if any one externally compresses these two veins, which proceed from the neck to the head, those in whom they are compressed will fall down, without suffocation, indeed, yet they will become senseless,† and will close their eye-lids. These veins, therefore, extend through the neck to the head, so that the artery comes between them, and they proceed as far as to the ears, *i.e.* to that part in which the jaws are conjoined to the head. Again, however, from the interval of the ears, they are divided into four

† Because by this compression of the veins the animal spirits are intercepted, so that they cannot ascend from the heart to the head.

veins [or branches], one of which, passing through the neck and shoulders, is conjoined in the curvature of the arm with the other branch of the greater vein, which, as was before observed, proceeds through the arm-pits to the arms. But the second vein, or branch, descends from the interval of the ears as far as to the hand and fingers, and in them is terminated. And the third branch tends from each seat of the ears to the brain, and to the membranes which cover the brain, and which are called by the Greeks the *meninges*, and is then divided into many thin branches. [Though the membranes of the brain, however, are full of many little veins], yet the brain itself is entirely exanguious, and neither a small nor a large vein terminates into it. But of the remaining ramifications which proceed from this vein, some circularly surround the head, but others end in the sensoria and the teeth, in small and very thin veins.

4 After the same manner, also, the parts of the less vein, which is called the aorta, are diffused, since the branches of the aorta are the attendants of the branches of the greater vein; except that the avenues of the aorta are less than the avenues of the greater vein.

The veins, therefore, from the heart, are propagated after this manner to the superior parts of the body. But the part of the great vein which is under the heart, extends directly through the transverse septum or diaphragm, by which the thorax is separated from the belly, and adheres both to the aorta and spine by membranous and pliant avenues. This vein, also, having passed through the transverse septum, is divided into two branches, one of which tends to the liver, which is in the right, but the other to the spleen, which is in the left part. The branch, likewise, which extends to the liver, is short but broad; from which many thin veins, extending to the liver, are obliterated. But two larger branches proceed from the vein of the liver, of which one ends in the transverse septum, and what are called the præcordia, but the other returning
514 upward, and descending through the arm-pit to the right arm, concurs in it with other veins, near the interior curvature of the arm. Hence this vein of the arm is cut by physicians, that, by taking blood from it, certain diseases of the liver may be cured. Another branch, likewise, of the great vein, which is also short but broad, tends to the spleen, which is in the left part, and being diffused in the spleen into small veins, becomes no longer apparent. But another branch of the great vein ascends to the left arm. Yet though the vein which passes through the liver ascends to the right arm, yet the vein which passes through the

spleen does not ascend to the left arm, but the vein which ascends to the left arm is different from the vein which passes through the spleen. Again, from the greater vein, which proceeds from the inferior part of the heart, two other veins are derived, one of which tends to the omentum or caul, but the other to what is called the pancreum, [which is so denominated because the whole consists of a certain glandulous flesh]. But from the vein which tends to the pancreum many veins are scattered through the mysentery; and all these end at length in one large vein, which is extended through the whole intestine and belly as far as to the stomach. From this vein, also, many branches are diffused about the parts through which it is extended.

Both the aorta and the great vein, therefore, proceed as far as to the kidneys, and here they rather adhere to the spine. Both of them, likewise, are divided into two parts, after the manner of the letter V, and the great vein verges to the posterior part of the body, somewhat more than the aorta. The aorta, however, especially about the heart, adheres to the spine by nervous and small veins. But the aorta in its first egress from the heart is very ample and capacious; but in its progress it becomes narrower and more nervous. Veins, also, extend from the aorta to the mysentery, in the same manner as the branches from the great vein, except that they are much less; for they are narrow and fibrous; since they terminate in thin, hollow, and fibrous little veins. No vein, however, tends from the aorta to the liver and the spleen. But those branches into which each vein is divided extend to the hip, and both of them insert themselves into the bone of it. Veins, also, tend to the kidneys, from the great vein and the aorta, except that they do not insinuate themselves into the cavity of the kidneys, but into the body of them. From the aorta, therefore, two other strong and continued avenues tend to the bladder, and others proceed from the cavity of the kidneys, but have no communication with the great vein. From the middle, also, of each of the kidneys a hollow and nervous vein is suspended, and which extends, near the spine itself, through the [ramifications of] veins; then, in the first place, they enter each hip, and 515 are no longer apparent; but afterwards being extended to the hip, they become again visible. The extremities of them, however, in males are connected with the bladder and the penis, but in females with the vulva. And from the great vein, indeed, no vein extends to the womb, but many and large veins tend to the womb from the aorta. Other less branches, likewise, are derived from the branches of the great vein and the aorta; of which, in the first place, some that are large and capacious tend to the groin, and afterwards through the legs, and in the feet and fingers. Others, again, proceeding through the groin and thigh, change their course, so that the right proceed to the left part, and the left to the right, and conjoin themselves with other veins at the concave part of the knee.

After what manner, therefore, the veins subsist, and whence they originate, is from hence evident. All sanguineous animals, however, accord in the principles of the veins, and in the greatest veins;[†] but they do not accord in the multitude of the other veins. For the same number, and the same mode of the subsistence of the veins, is not in all animals, or at least is not apparent in all; but their progress is especially apparent in very sanguineous and the largest animals. For in small animals, and which are not very sanguineous, either naturally, or from the fatness of their body, the secondary veins and their mode of subsistence can scarcely be perceived. For, on account of the fat, the avenues of the veins are confused, and resemble rivulets which contain a great quantity of mud. But very small animals have fibres instead of veins, [and therefore they can scarcely be distinguished]. The great vein, however, in all animals, and even in those that are small, is very conspicuous.

5 The nerves, however, subsist in animals after the following manner: The principle of these, indeed, is from the heart; for the heart contains nerves in its greatest cavity; and what is called the aorta is a nervous vein, and its extremities are perfectly nervous, since they are not hollow [like the veins], but have an extension like nerves, and end in the flexions of the bones. The nature of the nerves, however, is not derived from one principle [*i.e.* from one principal nerve] with which it is in continuity, in the same manner as the veins. For the veins are distributed through the whole body, so as to resemble the painted streams of a fountain, and hence, in emaciated bodies, all the parts are seen to be full of veins. For the same place in those that are emaciated is veiny; but in those that are corpulent is fleshy. But the nerves are divulsed about the limbs and the flexions of the bones. If, however, the nature of them was continued, the continuity of all of them would become apparent in emaciated bodies.

† For all of them, according to Aristotle, have the cavities of the heart as the principles of the veins; and they have, likewise, the great vein, and the aorta, or artery.

But the principal nerves are about that part of the body to which the province of leaping is assigned; and this part is called the ham. In the next place, there is a certain double nerve, which is called a tendon. In the third place, there is a nerve which is an auxiliary to the strength of the body, and which in Greek is denominated *epitonos*, and *omoiaia*, or pertaining to the shoulders. But the other nerves, which are about the flexion of the bones, are anonymous. For all the bones, whether they touch each other, or so subsist that the one is inserted in the other, are bound together by nerves; and there is a multitude of nerves about all the bones, except the bone of the head, in which there is no nerve, but the head is connected by the sutures of its bones. The nature of nerve, however, is divisible according to length, but is indivisible according to breadth, and possesses much tension. A mucous moisture, likewise, white and glutinous, is produced about the nerves, by which they are nourished, and from which they appear to be generated. A vein, therefore, may be burnt [without being destroyed]; but every nerve is destroyed when burnt; and if it is cut off it will not be reproduced; nor does any part of the body experience a numbness where there is no nerve. Most of the nerves, likewise, are about the feet and the hands, the shoulders, neck, and arms. All sanguineous animals, too, have nerves; but those animals that are without flexions or curvatures of the members, and are deprived of feet and hands, have thin and immanifest nerves. Hence, in fishes, the nerves near the fins are especially conspicuous.

6 The fibres, however, have a nature between nerve and vein; but some of them have the moisture of lymph, and tend from the nerves to the veins, and from them to the nerves.† There is, also, another genus of fibres, which subsists, indeed, in blood, yet not in the blood of every animal; and the blood cannot be congealed if these fibres are taken from it, but if they are not taken from it, it can be congealed. Fibres, therefore, are inherent in the blood of most animals; but there are no fibres in the blood of the stag, the deer, the buffalo, and certain other animals. Hence, their blood is not congealed in a manner similar to that 516
of other animals; but the blood of stags is congealed similar to that of hares. The congelation, however, of the blood of both these is not of a firm consistence, like that of other animals, but is fluid, so as to

† *i.e.* They differ from the veins so far as they participate of the nerves; and they differ from the nerves so far as they participate of the veins.

resemble milk without the coagulum. But the blood of the buffalo is congealed in a greater degree; for it is similar, or not much less congealed than the blood of sheep. After this manner, therefore, vein, nerve, and fibre, subsist.

7 All the bones, likewise, in animals, depend from one bone, and are in continuity with each other, in the same manner as the veins; but there is no bone which subsists itself by itself. The spine of the back, however, is the principal in all animals that have bones. But the spine is composed from vertebræ, and extends from the head as far as to the hips. All the vertebræ, therefore, are perforated. But the bone of the head, which is called the cranium, is in continuity with the last vertebræ; and the place in which one part of the cracium is inserted in the other is called the suture. The cranium, however, does not subsist similarly in all animals; for in some, as in the dog, the cranium consists of only one bone [without any sutures]; but in others, as in man, it is composed from many bones: and the female, indeed, in the human species, has the parts of the cranium conjoined by a circular cuture; but the male has three sutures, which are conjoined triangularly at the vertex. The head of a man, however, has been seen without sutures. But the head is composed, not from four, but from six bones; and two of these, which are small when compared with the rest, are about the ears. The bones of the jaw, however, are extended from the head; and in all other animals the lower jaw is moved; but the river crocodile is the only animal that moves the upper jaw. The genus of teeth, also, is inherent in the jaws; and consists of bones, which are partly solid and partly perforated, and the bone of the teeth is the only bone which it is impossible to carve.

From the spine, the fibula, the clavicles, and the ribs, originate. The breast, also, is placed in the ribs; and of these, some are conjoined, but others are unconjoined. For there is no animal which has a bone about the belly. Farther still, there are also the bones in the shoulders, and what are called the scapulæ, or shoulder-blades, and the bones of the hands and arms, which are annexed to them. These bones, however, subsist after the same manner in animals which have fore legs, as in man. But beneath, where the spine of the back ends, and after the hip, is the cotyloid cavity, and the bones of the legs, from which the thighs and
516 tabiæ consist, and of which the ankle bones are a part, and of these what are called spurs in animals that have an ankle bone. In continuity with these, also, are the bones of the feet.

The bones, therefore, of such sanguineous and pedestrious animals as are viviparous are not very different, but rather differ analogously in hardness, softness, and magnitude. Again, of the bones in the same animal, some have marrow, but others have not. Some animals, however, do not in short appear to have any marrow in their bones, as is the case with the lion, because it has but very little and thin marrow, and but in few of its bones; for it alone has it in its thighs and fore feet. But the lion, beyond all other animals, has its bones solid; for they are so hard, that from the attrition of them, in the same manner as from flints, fire is excited.

The dolphin, also, has bones, but not a spine. But other sanguineous animals either have bones not very different from those of the animals hitherto enumerated, as birds; or, though they are without bones, they have something analogous to them, as fishes. For, of fishes, such as are viviparous have a cartilaginous spine, as is the case with the fishes called selache, or of the cartilaginous kind; but such of them as are oviparous have a thorn, which subsists in them in the same manner as the bone of the back in quadrupeds. It is, however, the peculiarity of certain fishes, that within the flesh they have separate thorns. Serpents, also, accord with fishes in this, that in the place of bones they have the spine of the back. With respect, however, to oviparous quadrupeds, the larger among them [as the crocodile] have in the back something which is more similar to bone than a thorn, but the less [as the lizard] have in the back something which more resembles a thorn than bone. But all animals that are sanguineous have a spine, which is either bony or thorny; and the other bones that have been enumerated are inherent in some animals, and in others not, just as they have certain members, or are deprived of them. For such animals as have not legs and arms, are also deprived of the bones of the legs and arms; but those which have these parts, or what are analogous to them, have also bones corresponding to such parts, which differ in the more and the less, or in the analogous. The particulars, therefore, respecting the bones, subsist after this manner in animals.

8 Cartilage, however, is of the same nature with the bones, but differs from them in the more and the less; and as a bone when cut off is not increased, this is also the case with a cartilage. But the cartilages in sanguineous animals that are terrestrial and viviparous are not hollow, nor have they marrow within them, in the same manner as there is in bones. In cartilaginous aquatic animals, however, the cartilage which is

517 under the back, in the place of the spine, is hollow, and full of a certain moisture which is analogous to marrow. But in viviparous and pedestrious animals there are cartilages about the ears and the nostrils, and about certain extremities of the bones.

9 Again, there are other genera of parts, which have neither the same nature with these, nor are remote from them; such, for instance, as nails, the horn of the foot, whether the foot is one entire hoof, or is cloven, the horns of the head; and, besides these, the beaks of birds. For these are flexible and fissile, but no bone is either of these, but is alone frangible. The colours, likewise, of horns, nails, and hoofs, follow the colour of the skin and the hair. For the horns and hoofs of animals that have a black skin are black, but of those that have a white skin are white; and the like takes place in the intermediate colours. The nails, also, subsist after the same manner. But the teeth are of the nature of the bones; and hence in black men, such as Ethiopians, and the like, the teeth and the bones are white; but the nails are black, in the same manner as the whole of the skin.

Of horns, however, the greater part is hollow from the root, which is at the mouth, and inwardly proceeds from the head; but the extremity of the horn is solid and simple. And the horns of stags alone are solid throughout, and multifidous. Other horned animals, likewise, do not change their horns; but the stag alone changes them every year, unless he is castrated. We shall, however, afterwards speak about castrated animals. But the horns rather adhere to the skin, than to the bone; and hence in Phrygia, and elsewhere, there are oxen which move their horns in the same manner as some animals move their ears. Of those animals, however, that have nails, (but all animals that have toes have nails, except the elephant, and this animal has undivided toes, indistinct, and
517 imperfect, and has no nails), some have straight nails, as man; but others have crooked nails, as of pedestrious animals, the lion, and of winged animals, the eagle.

10 With respect to hair, however, and what is analogous to hair, and also with respect to the skin, they subsist in the following manner. Such animals, indeed, have hair as are pedestrious and viviparous; but those viviparous fishes alone, that bring forth a friable egg, have scales; for of long fishes, neither the conger, nor the lamprey, have an egg of this kind; but the eel, in short, produces no egg. The thickness, thinness, and magnitude of hair, however, differ according to the situation of the

parts in which they are, and the quality of the skin. For it happens for the most part, that harder and thicker hairs grow in thicker skin; and that more abundant and longer hair grows in the more hollow, and more moist parts of the body, if these parts are adapted to the production of hair. The like, also, takes place in the scales of terrestrial animals, and of fishes. Such animals, therefore, as have soft hair, through the use of better food acquire harder hair; but those that have hard hair, have through this mean fewer and softer hairs. In the third place, hairs differ according as places are hotter or colder. Thus, for instance, human hair is hard in hot, but soft in cold, regions. And again, straight hair is soft, but curled hair is hard.

11 The nature of the hairs, however, is fissile. But hairs differ from each other, in the more and the less; and some are so gradually indurated, that they no longer resemble hairs, but thorns, such as the hairs of the land hedge-hog. The like, also, takes place in nails; for the genus of nails in some animals is so hard, that it does not in any respect differ from bones.

Man, however, has the thinnest skin of all animals, in proportion to the magnitude of his body. But in all skin there is a mucous viscosity; in some, indeed, less, but in others more. Thus, for instance, in the skin of oxen there is a mucous moisture, from which they make glue; and in some places, glue is made from fishes. The skin, also, when cut off, is of itself deprived of sense; and this is especially the case with the skin of the head, because, between this and the bone, [or cranium] nothing of flesh intervenes. But, whenever the skin is by itself, if it is cut, it does 518 not coalesce; as in the thin parts of the cheek, prepuce, and the eye-lids. In all animals, likewise, the skin is continued through the whole of the body, so that its continuity is not interrupted, except where there is a natural evaporation through the pores, and where the mouth and nails are situated. All sanguineous animals, therefore, have skin, but all have not hairs, but those only of whom mention has been already made.

The hair, however, changes its colour in old age, and in men becomes whiter; but in other animals it becomes whiter indeed, yet not very apparently so, except in the horse. Hair, also, becomes white at the extremity [which is the most distant from the skin]; but most hair that is hoary becomes white immediately; from which it is evident, that hoariness is not, as some say, a dryness of the hair, since nothing grows immediately dry. But in that disease of the skin, which is called the white leprosy, all the hairs become white. It happens, likewise, that some when they are ill become hoary; but, the hair falling off, when they recover their health black hair grows in its stead. Hair, too, becomes more swiftly hoary when covered than when exposed to the air. But in men, the temples first begin to be hoary; and the anterior parts of the head become hoary prior to the posterior parts; and the hairs about the private parts become white in the last place.

In man, however, alone among all animals, some of the hairs are connascent, and others are produced afterwards, as he advances in age. The connascent hairs, indeed, are those in the head, in the eyebrows, and eyelids; but those of posterior origin are first the hairs on the private parts, in the next place those of the arm-pits, and in the third place those of the chin. For the places are equal in number in which the connascent hairs and those of posterior origin grow.

But the hairs of the head especially fail, and fall off, and the first of all the hairs; and this is confined to the hairs in the anterior part of the head: for no one becomes bald in the posterior part of the head. The smoothness, therefore, [or defluxion of the hairs] of the head, is called baldness, but that of the eyebrows depilation; and neither of these happens to any one prior to the age adapted to copulation. Baldness, however, neither happens to a child, nor to a woman, nor to those who are castrated; but if any one is castrated prior to the emission of the hairs of posterior origin, he will not emit them; and if he is castrated after the emission of these hairs, such hairs will fall off, except those on the private parts. Hairs, however, do not grow on the chin of a woman, some few women excepted; and this takes place when the menses stop, and resembles that which happens to the priestesses in Caria, which is considered as a sign of divination. Hairs, too, grow about the private parts and under the arm-pits of women, but they are few. There are, likewise, both men and women, who from their birth are deprived of hairs of posterior origin; and those, also, are impotent who are without hairs on the private parts.

Other hairs, therefore, are increased by a certain law of nature, either more or less, especially those in the head; in the next place, those in the beard, and particularly those that are more thin. In some persons, likewise, the eyebrows are so thick with hair, when they become old, that they ought to be shaved; and this is because the moisture, from which the eyebrows are produced, proceeds from the commissures of the bones, which commissures in old age are loosened, and, therefore, a greater quantity of moisture pervades through them. But the hairs of the eyelids are not increased; and they fall off when venery begins to take place, and are more abundant in those that are more addicted to venery. They also become hoary very slowly.

While we are growing, also, the hairs, if they are plucked out, are re-produced, but when we no longer grow, they are no longer re-produced. All hair, too, has a viscous moisture at the root, and therefore, if the hair, as soon as it is plucked out, touches any thing light by its root, it adheres to, and draws it. But animals, which are covered with hairs of many colours, have also a variety of colour in their skin, and in the skin of their tongue. With respect to the beard, however, it happens that some have the chin and upper lip hairy, but others have these parts smooth, and have the cheeks covered with thick hair; and those that have a bipartite chin and lip, are less accustomed to become hoary. The hair, likewise, is increased in certain diseases, and especially in those of the consumptive kind, in old age, and in dead bodies, and instead of being soft becomes harder. The very same things, also, happen with respect to the nails. In the libidinous, however, the connascent hairs more rapidly fall off, but those of posterior origin are more swiftly produced. Those, likewise, whose veins are more tumid than is proper, are less subject to baldness; and some when afflicted with this disease have abundance of hair. Hair, however, when cut off, is not by this mean increased, but growing from the root becomes longer. The scales, also, of fishes become harder and thicker; and in those that are slender and old, they become more hard. Of quadrupeds, likewise, when they grow old, the hair and the wool become thicker, but less in multitude. The horns, too, and hoofs of pedestrious animals, and the beaks of birds, become larger in old age: and hoofs are increased in the same manner as nails.

12 With respect to the feathers of birds, they suffer no change from 519
age, except those of the crane; and this bird has been found when it becomes old to have blacker feathers. In consequence, however, of the passive qualities produced by the different seasons, as when the weather happens to be very cold, sometimes the feathers of birds that are of one colour, such as black or a dark brown, become white, as is the case with the feathers of the crow, the sparrow, and the swallow; but white birds are never seen to become black. Indeed, many birds change their colours at stated times of the year; so that one, who is not well acquainted with this circumstance, would be unconscious of the change.

Some animals, likewise, change the colour of their hairs, from changing the water which they drink; for by one kind of water they become white, and by another black. The like, also, takes place with respect to their copulations; for there are in many places waters, which when the sheep have drank, if they copulate immediately after, they produce black lambs; and such is the water called Psychrus, a river of Assyria, in the country of Chalcis in Thrace. In Antandria, also, there are two rivers, one of which causes the sheep that drink it to be white, and the other to be black. The river Scamander, likewise, is seen to make the sheep that drink of it yellow; and on this account they say that Homer calls this river Xanthus, instead of Scamander.

Other animals, therefore, have hair alone in the prone part, [*i.e.* in the back], but they have not any in the supine part, [*i.e.* in the breast and belly], and they have also hairs externally, and not internally; but the hare alone has hairs within the mouth, in the cheek, and under the feet. Farther still, the fish called *mysticetus* has no teeth in its mouth, but has hairs similar to those of a swine, in the place of teeth. Hairs, therefore, when they are cut, grow from the root, and not from the place where they are cut; but feathers if they are cut, neither grow from the root, nor from the part where they are cut, but fall off. Those animals, also, that have continued wings, and not composed from many feathers, as bees, and the like, have not their wings re-produced if they are plucked out; nor is the sting of the bee re-produced when it is ejected; but bees when they loose their sting shortly after die.

13 There are, likewise, membranes in all sanguineous animals. But a membrane is similar to a strong and thin skin; and it is also of a different genus from skin, since it is neither fissile, nor capable of being stretched. About each of the bones, however, and each of the viscera, there is a membrane, both in larger and smaller animals; but the
519 membranes are immanifest in smaller animals, because in these they are very thin and small. The largest, however, of the membranes are the two which are about the brain; of which the superior, and harder membrane adheres to the bone of the head, [and is called dura mater], but the other which is softer, and is in contact with the substance of the brain, [is called pia mater]. The next to these, in magnitude, is the membrane which covers the heart. But a membrane when cut does not coalesce and become reunited; and bones when deprived of their membranes mortify.

14 & 15 The epiploon, or caul, is also a membrane. And all sanguineous animals have an epiploon; but in some, indeed, it is fat, and

in others lean. In animals, likewise, that are viviparous, and have teeth on both sides, the epiploon originates, and depends from the middle belly, where a certain suture, as it were, exists of the epiploon with the belly; and in animals which have only teeth on one side, the epiploon in a similar manner depends from the greater belly.

The bladder, likewise, is indeed membranous, but it is a membrane of a different genus, since it can be extended. All animals, however, have not a bladder, but all viviparous animals have it; and of oviparous animals the tortoise alone. The bladder, likewise, when cut, does not coalesce, except in that part which is called the neck of the bladder, unless very rarely; for a thing of this kind has happened. In the dead, indeed, no moisture pervades to the bladder; but in living animals, not only moisture, but also dry substances, descend into the bladder, from which the disease called the stone is produced. And in some persons there is such a concretion of dry matter in the bladder, that it does not appear to differ in any respect from shells. After this manner, therefore, vein, nerve, skin,fibre, and membrane subsist, and also hair, nails, hoofs, horns, teeth, beaks, cartilage, and bones, and parts analogous to these.

16 Flesh, also, and that which has a nature similar to flesh, subsists in all sanguineous animals, between the skin and the bone, and the parts analogous to the bones. For as spine is to bone, so is that which is fleshy to flesh, in animals which have bones and a spine. Flesh, however, can be divided every way, [*i.e.* according to length, breadth, and depth], and not like the nerves and veins, according to length only. When animals, therefore, become thin, the flesh is consumed, and the veins and fibres become apparent; but by good nutriment and in great abundance, the flesh is changed into fat. In those, also, that have much flesh, the veins are smaller, the blood is redder, and the viscera and belly 520
are small; but in those that have large veins, the blood is blacker, the viscera are large, and the belly is great, but the flesh is in a small quantity. But fat is produced about the flesh of those that have a small belly.

17 Fat, however, and suet differ from each other. For suet, indeed, is entirely fragile, and is congealed by cold; but fat is fluid, and not adapted to be congealed. The broth, also, which is made of the fat of the horse, and the swine, does not coagulate: but the broth which is made from the suet of a sheep, or a goat, does coagulate. Fat and suet, likewise, differ from the places in which they are produced. For fat is produced

between the skin and the flesh, but suet is produced in the extremity of the flesh. The epiploon, also, or caul, is fat in animals that are fat; but is suety in animals that have suet. And animals that have teeth on both sides [as the swine] have fat; but those that have not [as sheep] have suet. With respect to the viscera, however, in some animals the liver is fat, as among fishes, in those of the cartilaginous kind; for, from the colliquation of their liver they make oil. The flesh of the selache, however, [or of a cartilaginous kind of fishes], is without fat; but they have fat separately in the belly. The suet, also, of fishes is fat, and does not congeal. With respect to all animals, however, some have fat about the flesh, but others have it separate from the flesh. But such animals as have not fat separate from the flesh are less fat in the belly and epiploon; and this is the case with eels, for they have but little suet about the epiploon. Most animals, however, become fat in the belly, and this is especially the case with animals that are not in motion. Of fat animals, also, such as the swine, the brains are fat; but of animals that are suety, the brains are dry.

With respect to the viscera, likewise, animals are especially fat about the kidneys. But the left kidney is always more fat than the right; and though the kidneys should be very fat, yet there is always some part in the middle of them which is not fat. The kidneys, however, of animals which have suet, are more fat than the kidneys of those that have fat, and this is especially the case with the kidneys of sheep; for when the kidneys are entirely covered with fat this animal dies. Sheep, however, become fat about the kidneys through feeding abundantly, as is the case with the sheep in Sicily, about Leontium; and, on this account, the shepherds in that country lead the sheep to pasture late in the day, in order that they make take less nutriment.

18 & 19. The part, however, about the pupil of the eye is common to all animals; for all such as have the pupil in the eyes, and have not hard eyes, have it composed from suet, and not from fat. Fat animals, also, both male and female, are less prolific; and they become more fat when they are elderly than when they are young, but this is especially the case when the body, having obtained its proper length and breadth, begins to be increased in depth.

With respect to the blood, also, it subsists as follows; for this is most necessary, and most common to all sanguineous animals, and is not adventitious, but is inherent in all animals whose body is not in a corrupted state. All the blood, however, is contained in what are called

the veins, as in a vessel, but in nothing else, except in the heart alone. The blood, also, does not in any animals perceive the touch, as neither does the excrement in the belly, nor the brain, nor the marrow. But wherever the flesh is divided in a living body, the blood flows, unless the flesh is in a corrupt state. The blood, however, if it is sane, has naturally a sweet taste and a red colour; but if it is vitiated, either naturally, or by disease, it is of a blacker colour. The best blood, likewise, is neither very thick nor very thin, if it is not vitiated, either naturally, or from disease. And in the animal, indeed, it is always hot and moist; but when it proceeds forth externally, it is congealed: the blood of the stag, and the deer, and any other animal of a similar nature, excepted. The blood, however, of all other animals, when separated from the animal, is congealed, unless the fibres are taken out of it. But the blood of the bull is congealed the most rapidly of all blood. Of sanguineous animals, also, the viviparous have more blood than the oviparous. Those animals, too, whose body is in a good condition, either naturally, or from health, have neither very much blood, as is the case with those that have been recently drinking, nor blood in a small quantity, as is the case with those that are very fat. For those that are fat have, indeed, pure, but little blood; and the less blood animals possess the fatter they become; for the fat is exanguious. And fat, indeed, does 521
not become putrid; but blood and sanguineous substances rapidly putrefy, and this is especially the case with the parts about the bones. Man, however, has the most attenuated and the purest blood; but the bull and the ass have the thickest and the blackest blood of all animals; and the blood becomes thicker and blacker in the inferior than in the superior parts. In all animals, likewise, the blood palpitates, and has at the same time a pulsation in the veins in all directions; and it is the only humour that is dispersed through the whole body in animals, and this perpetually as long as they live. The blood, however, is first produced in animals in the heart, and before the whole body is organized. The loss, also, of much blood is attended with swooning, but of a very great quantity with death; and when it becomes very liquid, the animal is diseased. For sometimes it becomes lymphatic, and is so serous that some persons emit a sanguineous sweat. It also happens, that the blood of some persons is so liquid, that it will not congeal, or that only a certain part of it will congeal separate from the rest. In those that are asleep, likewise, there is a less quantity of blood in the external parts; so that if an aperture is made in the body of these, less blood will flow from it than if they were awake. From lymph, however, when

concocted, blood is produced, and from blood fat. But when the blood is vitiated, a flux is produced called an hæmorrhois, either from the nostrils, or from the fundament, and which produces inflated veins. Pus, also, is generated from the blood when putrid; and from this a white and thick substance is produced, called porus.

But the blood of females differs from that of males; for it is thicker and blacker in females, if both are similar in health and age. Females, likewise, have less blood in the external parts of the body, but more in the internal parts. A woman, also, has the most blood of all female animals; and what are called the menses are more abundant in women than in other animals. The blood, also, of the menses, when vitiated, is called a flux. Women, however, have less of other diseases which originate from the blood than men. For few women are afflicted with a flux of blood from the hæmorroidal veins, or with an inflation of the veins, or with a flux of blood from the nose; and if any one of these happens to them, the menses are not in so good a condition. The blood, likewise, differs according to age in multitude and form [*i.e.* in quantity and quality]. For, in those that are very young it is lymphatic, and abundant; but in those that are old it is thick, black, and in a small
521 quantity; and in those that are in the acme of life it has a middle mode of subsistence. The blood, also, of those that are old is congealed rapidly, and even in the superficies of the body; but this does not happen to young persons. Lymph, however, is unconcocted blood, either because it is not yet concocted, or because, through a defect of heat, it has degenerated into a serous and aqueous matter.

20 In the next place, it follows that we should speak concerning the marrow; for this, in some sanguineous animals, is one of the liquid parts. All such liquid parts, however, as are naturally inherent in the body are contained in vessels, just as the blood is contained in the veins, and the marrow in the bones; but the other liquid parts are contained in membranes, and skins, and bellies. In young animals, however, the marrow is entirely sanguineous; but when they become older, in those that are fat the marrow is fat, but in those that have suet instead of fat the marrow is suety. All the bones, however, do not contain marrow, but only those that are hollow, and even in some of these there is no marrow. For of the bones of the lion some have marrow, and others contain so small a quantity of it, that, as we have before observed, some assert that they are entirely without marrow. In the bones of swine,

likewise, there is but little marrow; and in some of them there is none at all.

These, therefore, are nearly the humid parts of animals, which are always connascent with them; but milk, and the genital seed, are things of posterior origin. Of these, however, that which is secreted from all animals when it is inherent is milk; but genital seed is not secreted in all animals, but instead of it, in some, as in fishes, there is what is called the soft roe. All such animals, however, as have milk have it in the mammæ; and those animals have mammæ that generate an animal in themselves, and bring it forth externally; such as animals that have hair, as man, and horse; and likewise cetaceous animals, such as the dolphin, the sea-calf, and the whale; for these have mammæ and milk. But such animals as alone generate an animal, or an egg, externally, have neither mammæ nor milk; as, for instance, fishes and birds. All milk, however, has an aqueous lymph, which is called whey; and it has also a certain corpulent substance, which is called cheese.

The milk, therefore, of animals which have not teeth on both sides coagulates; and, on this account, from the milk of tame animals cheese is made; but the milk of those animals that have teeth on both sides does not coagulate, as neither does the fat; and it is thin and sweet. The milk, however, of the camel is the thinnest of all milk; the second in thinness is the milk of a mare, and the third is that of the ass, but that of the cow is more thick. Milk, therefore, is not coagulated by cold, but becomes through it more serous; but it is coagulated and thickened by 522
fire. Milk, however, for the most part, is not produced in any animal prior to copulation; but when conception takes place milk is produced indeed, yet so, that both at first and at last it is useless. Certain food, also, will cause animals that are not pregnant to have milk; and these have been elderly women, from whose mammæ when sucked milk has flowed, and in such abundance as to nourish the infant with it. In Mount Oeta, likewise, the shepherds rub the udders of the she-goats that have not copulated with a nettle so vehemently as to cause pain, by which it comes to pass, that in the first place a sanguineous moisture is induced, after which a purulent humour follows, and at last milk is produced, and in no less abundance than from those that have copulated. For the most part, however, milk is not produced in the males of other animals, nor in those of the human species, though it sometimes happens that males also have milk. Thus, in the island of Lemnos, so much milk was induced from the two mammæ of a he-goat, which were situated near the penis, that little cheeses were made from it, and the same thing

also happened to the male offspring of this goat. Things of this kind, however, are considered as prodigies; since the Lemnian deity [Vulcan] said that it was an indication of the acquisition of wealth. In some men, also, after puberty, a small quantity of milk is induced, and from some by sucking a great quantity has been produced. A fatness, likewise, is inherent in milk, which in coagulation becomes of an oily nature [and is called butter]. In Sicily, however, and other regions, where there are a great quantity of she-goats, they mingle the milk of goats with that of sheep [for the purpose of making cheese]. But milk is especially coagulated, not only that which contains the greatest quantity of cheese, but also that which contains the driest cheese. Some animals, therefore, have more milk than is sufficient to nourish their offspring. The milk, also, of sheep and goats is most useful for the purpose of making cheese and reserving; and next to this, the milk of the cow. But the milk of the mare and the ass are mingled in order to make Phrygian cheese. There is, however, more cheese in the milk of the cow than in the milk of the she-goat. For shepherds say, that from an amphora of goat's milk twenty cheeses wanting one are made, for an obolus a piece; but that thirty were made from an amphora of cow's milk. The milk, however, of those animals which have as much as is sufficient to nourish their offspring is useless for the purpose of making cheese; and of this kind are
522 all those animals that have more than two mammæ; for the milk of none of these is either abundant or adapted to make cheese. But the juice and coagulum of the fig coagulate milk. Wool, therefore, imbued with the juice of the fig is washed with a little milk; for this milk, being mingled with a greater quantity of milk, coagulates the whole.

21 This coagulum, however, is milk; for it is contained in the belly of those that suckle. Coagulum, therefore, is milk, containing a fire in itself, which is produced from the heat of the animal, in consequence of the milk being concocted. But all ruminating animals have a coagulum; and among animals that have teeth on both sides, the hare has it. Coagulum, also, is better in proportion as it is more old; for a thing of this kind, and also the coagulum of a hare, are useful in the cure of a diarrhœa. But the best coagulum is that of the fawn.

More or less milk, however, is emulged from animals that have milk, according to the magnitude of their bodies, and the diversity of their food. Thus, in Phasis, though the cows are small, yet they give a great quantity of milk, because they are nourished with the best food; but each of the large cows of Epirus only gives an amphora and a half of

milk from its two mammæ. The cows, however, of Epirus are so large, that they cannot be milked by the shepherds, except when they stand, or are a little inclined; because when they sit they cannot reach to their mammæ. Except the ass, other large quadrupeds also are produced in Epirus; but the largest among them are the oxen and dogs. Since, also, the oxen are large, they require a greater quantity of pasture; but the region abounds with a pasture of this kind, for it has places adapted to the feeding of cattle in every season of the year. As the oxen, likewise, in Epirus are very large, so also are the sheep, called Pyrrhic, from Pyrrhus the king. With respect to nutriment, however, one kind extinguishes [*i.e.* diminishes] milk; such, for instance, as the Median grass, and especially in ruminating animals; but another kind of nutriment increases the milk, such as the cytisus, and the orobus; except that the cytisus, when it is in blossom is of no use, for it inflames; and the orobus is not useful to pregnant animals, because it renders their delivery more difficult. In short, of animals that are able to eat, as the possession of them is advantageous, so likewise they afford a greater quantity of milk, if they are well fed. Certain flatulent substances, also, when given as food produce much milk; such, for instance, as beans, when largely given to sheep, she-goats, cows, and young goats; for it 523
causes the udders to be distended with the quantity of milk. It is an indication, however, that there will be abundance of milk, when the udder, prior to the time of parturition, is pendent towards the earth. But in all animals milk endures for a long time, if they continue without copulation, and have appropriate food; and this is especially the case with sheep, beyond all other quadrupeds; for they are milked during the space of eight months. In short, ruminating animals have much milk, and which is useful to the making of cheese. The cows, however, about Torona, a few days prior to their delivery, are without milk, but at every other time they have milk. But with respect to the milk of women, the milk which is of a livid colour is better for the nourishment of infants than that which is white; and the milk which is of a dun is more salubrious than that which is of a white colour. That milk, therefore, is most nutritious which contains the greatest quantity of cheese; but that is most salubrious to infants which contains the least.

All sanguineous animals, however, emit seed; but what seed contributes to generation, and how it contributes, will be shown elsewhere. Man, however, in proportion to the magnitude of his body, emits the most of all animals. But the seed of animals that have hair is viscous, but of other animals is not viscous; and the seed of all animals is white. The assertion, therefore, of Herodotus is false, that the Ethiopians emit black seed. Seed, also, in its egress, if it is sane, is white and thick, but after its egress is becomes thin and black. It is not, however, congealed in ice, but becomes very thin and aqueous, both in colour and substance. But it is coagulated and condensed by heat; and if it is emitted after it has been for some time in the womb, it comes forth thicker, and sometimes dry, and with contortion. And prolific seed, indeed, subsides in water, but unprolific seed is diffused in the water. That, however, which Etesias writes concerning the seed of elephants is false.

BOOK IV

1 We have, therefore, already discussed such parts of sanguineous animals as are common, and also such as are peculiar, and have considered the dissimilar and the similar parts, and such as are internal, 523
and such as are external; hence it remains, that we should now speak concerning exanguious animals. There are, however, many genera of these; one, indeed, being the genus of what are called the mollia; and these are such as, being exanguious, have a fleshy substance externally, but a solid substance internally, in the same manner as sanguineous animals, and of this kind is the genus of cuttle-fishes. But another genus of this kind is that of crustaceous animals; and these are such as have a solid substance externally, but a soft and fleshy substance internally. Their hard covering, however, is not brittle, but is capable of being bruised; and of this kind are the genera of locusts and crabs. Another genus is that of testaceous animals; and these are such as have a fleshy substance internally, but externally a solid substance, which is brittle, and may be broken, but cannot be bruised; such as is the genus of snails and oysters. The fourth genus is that of insects, which comprehends many and dissimilar species of animals. Insects, however, are, as the name implies, such animals as have certain incisures either in the supine or prone parts, or in both; and they neither consist of a body partly bony, and partly fleshy, but of a body which is between both; for their body both internally and externally is similarly hard. Of insects, also, some are winged, and others are without wings; such as the *iulus*, and the scolopendra. But the winged are such as bees, beetles, and wasps. And insects of a doubtful genus are ants, and what are called glow-worms; for some of these are winged, and others are without wings.

Of those animals, therefore, which are called mollia, the external parts are these; one indeed, what are called the feet; the second, the head, which immediately adheres to the feet; and the third, the cavity, [or belly and alvus], which contains all the rest of the body, and which some improperly denominate κελυφος, a shell.† And besides these are the fins, which are circularly disposed about the cavity. It happens, however, in all the mollia, that the head is between the feet and the belly. All the mollia, likewise, have eight feet, which, according to length, in the bottom part, are furnished with a twofold order of

† In the original κεφαλην, *a head*; but, I have no doubt, it should be κελυφον, as in the translation; for a shell is hard, but the cavity or belly of the mollia is soft.

acetabula,[†] one genus of the polypus excepted. It is the peculiarity, however, of the cuttle-fish, and the small and large calamary to have two long trunks, which in their bottom part are rough, with a twofold series of acetabula; and by these they take their food, and convey it to their mouth. By these, also, in stormy weather adhering to rocks, they fix
524 themselves as by an anchor. But they swim by the fins, which they have about the cavity. The polypus, therefore, uses its flagella, or arms, as feet and hands; for it uses the two arms, which are above the mouth, for the purpose of taking the food and conveying it to the mouth; but it uses the last of its arms for the purpose of coition. And this last arm is the sharpest, is alone of a whitish colour, and in its extreme part is bifurcated, and annexed to the back. That smooth part, however, of the polypus is called the back, from which the order of the acetabula commences. The polypus, also, before the cavity or alvus, and above the arms, has a hollow pipe [or canal] by which it returns the sea-water it has received in its cavity, when it happens to have this water in its mouth. But it moves this pipe, at one time to the right, and at another to the left; and by this, also it ejects its ink. The polypus, likewise, swims obliquely, extending its feet above what is called its head; and it happens that, swimming in this manner, it sees what is before it; for its eyes are above, but its mouth is behind. Its head, however, as long as it lives is hard, and is as it were inflated. It touches, also, and holds with its arms incurvated from the supine part; and a membrane is extended through all the space which is between the feet. But if it falls on the sand, it is no longer able to hold any thing.

The polypi, however, and the above-mentioned mollia differ [in many respects]. For the polypus has a small cavity, or belly, but its feet are long; but the other mollia have a large belly, and short feet, so that they are unable to walk with them. When, however, the other mollia are compared with each other, the small calamary is longer, but the cuttle-fish is broader. But of the two calamaries, those that are called *teuthoi* are much the larger, for they grow to the magnitude of five cubits. Some cuttle-fish, also, grow to the size of two cubits. The arms, likewise, of the polypus become two cubits, and even more, in length. The genus, however, of the great calamary is but few in number. But the small and large calamary differ in figure from each other, for the point of the large calamary is broader. Again, a fin circularly surrounds

[†] The acetabula are knots and joints, within which the parts of the leg or arm enter without intermission.

the whole cavity of the larger calamary, [and this fin is very continued], but the cavity of the small calamary is surrounded with a less continued fin. The small, however, as well as the large calamary, lives in the sea.

After the feet in all the mollia is the head, which is in the middle of the feet that are called *flagella*. But of the head one part is the mouth, 524
in which there are two teeth. And above these are two large eyes, between which is a small cartilage, which contains in itself a small brain. But in the mouth there is a small fleshy substance, which serves the purpose of a tongue, for no one of these animals has a tongue. After the head is the cavity, or belly, the external parts of which are obvious to the sight. The flesh, however, of it is divided, not indeed in a right line, but in a circle. And all the mollia have a skin about this.

After the mouth, also, they have a long and narrow oesophagus; and adhering to this they have a large avenue of the throat, similar to the craw of a bird. In continuation with this is the belly, which resembles one of the ventricles of ruminating animals; and the figure of it is similar to the spirals of a horn. But after the belly a thin intestine succeeds, which again tends upward to the mouth. The intestine, however, is thicker than the stomach. No one of the mollia, likewise, has any viscera, but they only have what is called *mytis*, in which their ink is contained. This ink, also, is possessed in the greatest quantity by the cuttle-fish. All the mollia, therefore, when they are terrified, and especially the cuttle-fish, emit this ink. The *mytis*, also, is placed under the mouth, and through this the stomach extends; but in the inferior part, by which it is joined to the intestine, it contains the ink, and the avenue of the mytis is closed by the same membrane as the avenue of the intestine. By the same avenue, also, the ink and the excrement are ejected. All the mollia, likewise, have certain capillaments in their body.

The cuttle-fish, therefore, and the large and small calamary, have internally certain solid parts, under the back. And of these, the solid part of the cuttle-fish is called *sepium*, or a cuttle-bone, but the solid part of the large and small calamary is called *gladius*, or a sword. These, however, differ from each other; for the sepium is strong and broad, and has a nature between bone and spine, and a certain fungous and friable corpulency; but the gladius of the calamary is thin, and more cartilaginous. The sepium and gladius, also, differ in figure, in the same manner as the abdomen or cavity differs, into which they are received. The polypi, however, are without this solid part [which is analogous to bone], but they have a cartilaginous substance about the head, which becomes hard when any one of the polypi becomes old. The females,

likewise, among the mollia, differ from the males; for the males have an avenue under the stomach, which extends from the brain to the inferior part of the abdomen, and the part to which it tends is similar to a papilla. The females, however, have, as it were, two papillæ in the
525 superior part of the abdomen. And under the papilla, both of the male and female, certain red corpuscles are seen.

The egg, also, of the polypus is externally anomalous and large, but inwardly it contains a moisture, the whole of which is of a uniform white colour, and smooth; and the quantity of this moisture is so great, that it can fill a vessel larger than the head of the polypus. On the contrary, the cuttle-fish has two cavities or abdomens; and there are many eggs in these, which are similar to white hail-stones. The manner, however, in which each of these parts is situated may be seen from anatomical figures. But all these males differ from the females, and this is especially the case with the cuttle-fish. For the males have the prone part blacker than the supine part; they have, likewise, all the parts rougher than the females, and variegated with rods; and the extremity of their body is more acute.

There are, also, many genera of the polypi; and one genus of these is frequently seen to swim on the surface of the water, and is the largest of all of them. Those, however, that live near shores are much larger than those that live in the sea. Besides these, also, there are other small polypi, of a various colour, and which are not eaten. There are, likewise, two other genera, one of which is called *heledone*, and differs from the rest in the length of its feet, and in only having one series of acetabula or joints in its legs; for all the others have two series of acetabula. But the other genus is what some call *bolitœna*, and others *ozolis*. There are, also, two other genera of polypi, which are in shells, one of which is called by some *nautilus*, and by other *nauticus*. This genus, however, resembles a polypus, and the shell of it is similar to a hollow comb, but does not so naturally adhere to the polypus that it cannot be separated from it. This polypus, likewise, frequently feeds near the land, but afterwards is thrown by the waves on the sand, and falling from its shell is caught, and dies on the land. These polypi, also, are small, but in their form resemble the *bolitœnæ*. But there is another polypus, which is covered with a shell like a snail, so that it never leaves its shell, but sometimes outwardly extends its arms. And thus much concerning the mollia.

2 With respect to crustaceous animals, however, one genus is that of locusts; and another genus, similar to this, is that of lobsters, which differ from locusts in having claws, and certain other differences which are not numerous. A third genus is that of squills, and a fourth is that of crabs. There are, however, many genera of squills and crabs. For of squills there are those that are crooked, and the *crangones*, and a small kind of squills, which never become larger. But the genus of crabs is all various, and cannot be easily enumerated. The greatest, therefore, is that which they call *maia*; but the next to these are the *paguri*, and *heracleotic* crabs; and besides these are those which live in rivers. The rest are less than these, and are more anonymous. In the stores about Phœnicia, however, there are certain crabs which they call troopers, because they run so swiftly that it is not easy to catch them; but when opened, they are found to be empty from the want of food. There is, also, another genus, which is as small indeed as crabs, but in form resembles lobsters. All these, therefore, as we have before observed, have externally that which is solid and testaceous in the place of skin; but internally that which is fleshy. They have the supine part of the body, however, [*i.e.* the belly] distinguished by certain tablets, into which the females deposit their eggs.

With respect to feet, indeed, locusts have five on each side, together with their ultimate claws; and, in a similar manner, crabs have ten feet, together with their claws. But of squills, those that are crooked have five feet on each side, which verge to the head, and are acute. They have, also, five others on each side, at the belly, the extremities of which are broad. They have not, however, tablets in the supine parts; but in the prone parts [*i.e.* in the back] they resemble locusts. But the contrary takes place in the crangones; for they have on each side four feet, near the head, and three other thin feet consequent to these, and the rest of the body, which in the greater part is without feet. The feet, likewise, of all the above-mentioned animals, are bent obliquely [*i.e.* towards the body of the animals]; in the same manner as the feet of insects; and those which have claws bend them inwardly [so that the cavity is towards the body]. The locust, also, has a tail, and five fins on each side; but the squills which are called crooked, have a tail, and four fins on each side. The crangones, likewise, have a tail, and fins on each side of it. But the middle fins of the squills of each genus are thorny; except that the middle fins of the crooked squills are broad, but those of the crangones acute. The crab, however, alone, of animals of this kind, is without a tail; and the body, indeed, of locusts and squillæ is long, but that of crabs is round.

526 The male locust, however, differs from the female [in many respects]. For the first foot of the female is bipartite, but that of the male is simple. With respect to the fins, also, in the supine part of the body, the female locust has large fins, but they diminish as they approach to the neck; but the male has all the fins equally less, and not dissimilar by diminution. Farther still, in the last feet of the male there are, as it were, large, and sharp spurs; but in the last feet of the female these are small and smooth. In a similar manner, also, both the male and female locust have two large and rough horns before their eyes, under which there are other less and smooth horns. But the eyes of all these are hard, and are moved inwardly, outwardly, and obliquely. This is, also, similarly the case, and even in a greater degree, with the eyes of most crabs.

The lobster, however, is wholly of a splendid colour, but variegated with black spots. In the inferior part of its body, also, it has eight feet [orderly] disposed as far as to two large feet. But its two great feet are much larger than the rest; and in their extremities are much broader than the extremities of the feet of the locust. Its two large feet, likewise, are anomalous. For the right foot has its extremity broad, oblong, and thin; but the left foot is thick and round. Each foot, however, is bipartite, and resembles a jaw, with teeth above and below; except that the right foot has, as it were, small, serrated teeth, but the left foot has serrated teeth in its extremity, and within, as it were, maxillary teeth. In the inferior part, also, it has four continued teeth; but in the superior part three, which are not continued. Each foot, likewise, moves the superior part, and by it binds the inferior part. In the inferior part, too, each foot is distorted [*i.e.* inflected]; because each is naturally adapted, to lay hold of, and compress. Above these large feet, also, there are two other hairy feet, a little below the mouth; and a little below these there are many hairy gills, about the mouth, which the lobster continually moves. It, likewise, bends and moves to the mouth these two hairy feet; and the feet which are near the mouth have thin beards adhering to them.

The lobster, also, has two teeth in the same manner as the locust; but above these it has horns, which are shorter and thinner than those of the locust; and, besides these, it has four other horns, which are similar in form to these, but are shorter and thinner. Above these, too, it has
526 small and short eyes, and has not, like the locust, large eyes. Again,

above the eyes there is, as it were, a sharp and rough forehead, larger than that of the locust. And, in short, the face of the lobster is sharper, but the thorax is much broader than the thorax of the locust, and the whole body is more fleshy and soft. But of its eight feet, four of them are bifidated; and four of them are not. And, with respect to what is called the neck, it is divided, indeed, externally into five parts, and the sixth and last part is broad, and is divided into five tablets; but the internal parts, into which the females deposit their eggs, consist of four hairy tablets. And in each of the above-mentioned tablets there is a short and erect spine, which extends outwardly. The whole body of the lobster, likewise, about the thorax is smooth, and is not rough, like that of the locust; yet lobsters, in the external parts of the two large feet, have large thorns. There does not, however, appear to be any difference in the male with respect to the female; for both the male and the female have one claw, just as it may happen, larger than the other; but they never have both of them equal.

All crustaceous animals, however, receive the sea-water in their mouth; but crabs return the water which they receive by a narrower part of the mouth; and locusts transmit the received water to the gills; for locusts have many gills. But it is common to all these animals to have two teeth. For locusts have two primary teeth; and in the mouth they have a fleshy substance instead of a tongue. The belly, also, in these animals immediately follows the mouth, except in locusts, which have a small stomach before the belly; and after this the intestine immediately follows. But this in locusts, in lobsters, and in squills, ends in a right line at the tail, where they emit the excrement, and bring forth their eggs. In crabs, however, [which are without a tail], the intestine ends where there is a certain covering, which is opened when the excrement is emitted, and shut after it is emitted; from which part crabs, also, bring forth their eggs, since female crabs have a receptacle of eggs near the intestine. All crustaceous animals, likewise, have that moisture which is called *mytis* or *mecon*, in a greater or in a less degree.

It is, however, necessary, that we should now survey the several differences of crustaceous animals. Locusts, therefore, as we have before 527
observed, have two large and hollow teeth, in which there is a juice similar to the moisture called *mytis*; and between the teeth there is a small piece of flesh, which is a species of tongue. After the mouth, also, there is a short oesophagus, and a membranous belly, adhering to this; and in the mouth of this belly there are three teeth, two of which are opposite to each other; but the third is below the other two. But the

intestine, beginning and proceeding from the side of the belly, is simple, and of an equal thickness throughout, as far as to the egress of the excrement. Locusts, squills, and crabs, therefore, have these parts. Farther still, locusts have, also, an avenue, which depending from the breast is extended, as far as to the egress of the excrement. But this, indeed, is the matrix of the female locust, and in the male is the receptacle of the genital seed. This avenue, however, is situated near the cavity of the flesh, so that the flesh is between this cavity and the intestine. For the intestine is constituted in the convex part, but the avenue is near the cavity; just as they subsist in quadrupeds. Indeed, this avenue in the male differs in no respect from that of the female; for both are thin and white, and contain in themselves a pallid moisture, and both, also, are suspended from the breast. But after this manner squills have eggs, and certain spirals. The male, however, differs from the female in having in the flesh of the breast two white parts, which in colour and figure resemble the proboscis of the cuttle-fish. But these white parts are contorted in the same manner as the trumpet called *mecon*, and they derive their origin from the acetabula or joints, which are under the last feet. The breast, however, contains, also, red flesh, and is of a sanguineous colour; but it is viscous to the touch, and does not resemble flesh [in general]. Near those two white parts, also, which are contorted like a trumpet, there is another spiral of the thickness of a string, under which there are two sandy substances, adhering to the intestine, and destined to the reception of the genital seed. These parts, therefore, the male possesses; but the female squill lays eggs of a red colour, which are involved in a thin membrane, and are annexed to the belly, and each side of the intestine, as far as to the fleshy parts. These, therefore, are the parts as well internal as external of crustaceous animals.

3 It happens, however, that the internal and external parts of
527 sanguineous animals have [appropriate] appellations, [but the genus of exanguious animals wants many of those parts with which sanguineous animals are endued]. For all sanguineous animals have viscera, but this is not the case with any exanguious animal. A belly, however, mouth, and intestine, are common both to exanguious and sanguineous animals. With respect to the claws and feet of crabs, indeed, we have already shown the manner in which they subsist. For the most part, however, all of them have the right claw larger and stronger than the left. We have, also, before spoken about the eyes of crabs, and shown that most

of them look obliquely. But with respect to the mass of the body, the head, and the other parts of the trunk, are one indistinct whole. The eyes, however, are not in all of them in the same position; but some of them have the eyes obliquely situated under the prone part of the back, and very much distant from each other; and others, as the heracleotic crabs, and those called *maiæ*, have the eyes in the middle, and near to each other. Under the eyes is the mouth, and in it there are two teeth, in the same manner as in locusts, except that the teeth of the crab are not round, but long. The teeth, also, of crabs are covered with two valves, as it were, between which there is a certain fleshy part, which is used by them instead of a tongue, in the same manner as in locusts. They receive water, therefore, in the mouth by removing those valves, and return it by the superior avenues of the mouth, shutting the valves through which it entered. These avenues, however, are immediately under the eyes; and when the water is received, they shut their mouth with both the valves, and thus afterwards eject the water they had received. After the teeth a stomach follows, so very small that the belly seems immediately to succeed after the mouth. But from the middle of the belly an intestine originates, which is simple and slender; and the intestine is terminated under the external part of the covering, as was before observed [in Chap.2]. Within the cavity, also, or alvus of the crab, there is a pale [excrementitious] liquor, and together with this certain small, oblong, white substances, and other red substances, variegated with spots. The male, however, differs from the female crab in magnitude, thickness, and covering. For the female has a covering [at the avenue of the intestine] larger, more distant, and more hairy, as is, also, the case in female locusts. After this manner, therefore, the parts of crustaceous animals subsist.

4 With respect to testaceous animals, however, snails, whether of the land or the sea, and all such as are called shell-fish, and besides these, the 528
genus of sea hedge-hogs, if they have a fleshy substance have it similarly to crustaceous animals, for they have it inwardly; but they have the shell externally, and inwardly they contain nothing hard [and analogous to bones]. They have, however, many differences with respect to themselves, both as to their shells, and the internal flesh. For some of them have not any flesh, as the sea hedge-hog; but others have flesh, indeed, yet have it perfectly enclosed in the shell, so that no part is apparent except the head, as is the case with land-snails, and the shell-

fish, which are called by some *cocalia*, the buccina,[†] the murex,[‡] and others of the turbinated kind. But of other testaceous animals, some are bivalves, and others univalves. I call, however, bivalves, those that are contained in two shells; and univalves, those that are contained in one shell, and of which the fleshy part is uncovered; such, for instance, as the *lepas*. But of the bivalves, some can be opened [and shut], as cockles and mussels; for all such as these adhere to one shell, but do not adhere to the other, and can be opened and shut; but others are bivalves, indeed, and adhere to each shell, yet not so that they cannot be opened and shut, as the fishes called *solenes*. There are, also, some that are wholly contained in the shell, so that no part of the flesh is externally naked, as those that are denominated in Greek *tethya* [and in Latin are called *vertibula, tubera,* and *calli*].

Again, in the shells of testaceous animals there is a difference with respect to each other. For some of them are smooth, as is the case with the *solen*, mussels, and certain shell-fish, called by some *galades*; but others are rough, as oysters, the pinna, certain genera of shell-fish, and the buccina. And of these some, indeed, are striated, as the cockle, and a certain genus of shell-fish; but others are not striated, as the pinnæ, and a certain genus of shell-fish. Shells, likewise, differ in thickness and thinness, as well in the whole as in a part, as for instance, about the lips or extremities. For some have thin lips, as mussels; but others thick, as marshy oysters (λιμνοστρεα).

Farther still, some testaceous animals move themselves, [and change their places] as the cockle; for some say that scallops fly, since they frequently leap from the iron instrument by which they are caught. But other testaceous animals are immoveable, and are so fixed to a place that they can only be removed by force, as the pinna. All the turbinated
528 kind, however, are moved, and creep; and the *lepas* searches after nutriment when detached from the place to which it adheres. It is common, however, to these, and to other testaceous animals, to have a shell internally smooth. The flesh, also, as well of the univalves as the bivalves, so adheres to the shell, that it can only be divulsed from it by violence; but the flesh of the turbinated kind is less adherent. It is peculiar, however, to the latter to have the shell turbinated from the head, and to have a spiral inflexion as far as to the bottom. All

† Shell-fish, resembling a horn.

‡ A shell-fish, from the liquor of which the ancients made purple.

testaceous animals, likewise, are covered with a shell from their birth. And farther still, all the turbinated kind are moved according to the right-hand part, not towards the spiral, but in an opposite direction.

Such, then, are the differences of the external parts of these animals; but the nature of the internal parts, and especially in the turbinated kind, is after a manner similar in all of them. For they differ from each other in magnitude, and in the passive qualities pertaining to excess. The univalves, however, and shut bivalves do not much differ; for they have a small difference with respect to each other, but they have a great difference with respect to those testaceous animals that are immoveable. This, however, will be more evident from what will afterwards be shown. But the nature of all the turbinated kind is similar; yet differs, as we have before observed, in excess. For some of them have larger and more manifest parts, but others have parts which are less and more obscure. They, likewise, differ in hardness and softness, and other such like passive qualities. For all of them have the flesh which is most external in the mouth of the shell firm, some, indeed, more, but others less; but from the middle of the shell are the head and two horns. These, however, in larger testaceous animals, are large; but in those that are less, are very small. All of them, likewise, raise the head out of the shell after the same manner; and when they are terrified, again draw it within the shell. Some of them, however, as the snail, have a mouth, and sharp, small, and thin teeth. They have, also, a proboscis, in the same manner as flies; and this proboscis is in the form of a tongue. This proboscis, however, in the purple and buccina is hard. And as ox-flies, and horse-flies, pierce with their proboscis the skin of quadrupeds, in like manner these testaceous animals perforate the shells of others on which they feed.

After the mouth immediately follows the belly, which resembles the 529
craw of a bird, and is similar to the belly of the sea-snail. But in the inferior part they have two white and firm substances, similar to mammæ; such as are, also, found in the cuttle-fish, except that these are of a firmer consistence. After the belly, however, there is a long twofold stomach, which is extended as far as to the mecon [*i.e.* to the membrane containing the excrements]; which membrane is in the bottom [of the navel]. These things, therefore, are manifest in the purple, and buccina, in the spiral of the shell. But the intestine follows the stomach, the whole of which is simple as far as to its egress. The beginning, however, of the intestine is about the spiral of the *mecon*, than which it is broader. For the *mecon* is as it were, for the most part,

in all testaceous animals, full of an excrementitious matter. But afterwards, the intestine being inflected upward, tends again to the fleshy parts, and terminates near the head; from which place all turbinated testaceous animals, as well terrestrial as marine, eject their excrements. In large sea-snails, however, under the belly, a membrane is annexed to the stomach, from which a long and white avenue extends, and which is similar in colour to those fleshy parts which have the form of mammæ. This avenue, also, has incisures similar to those which are in the egg of the locust, except that the colour of this avenue is white, but the egg of the locust is red. This avenue, however, has no egress, but is contained in the narrow cavity of a thin membrane. From the intestine, also, certain black and rough parts extend downward in a continued series, such as are in the tortoise, except that they are less black. Other snails, however, have these parts white, except that they are less in smaller snails.

The univalves, also, and bivalves, are partly similar to snails, and partly dissimilar. For they have a head, horns, and a mouth, and a certain part which has the form of a tongue; yet in less animals of this kind these parts, on account of their smallness, are immanifest; and in such of them as are dead, or which are not moved, they cannot be perceived. All of them, however, have the *mecon*, yet not in the same place, nor of an equal magnitude, nor similarly manifest; but the *lepas* has it downward in the bottom, and the bivalves in the cavity next to the commissure of the two shells. All these, likewise, have certain capillaments circularly
529 disposed, in the same manner as in cockles. And with respect to what is called an egg, it is contained, when they have it, in one side of the periphery of the circle, in the same place in which snails contain their white substance; for in them this has a similar subsistence. All such parts, however, as we have mentioned, are indeed manifest in large testaceous animals; but in those that are small they are either not at all, or scarcely visible. Hence, these are especially apparent in large cockles; and these are such as have one of the valves broad, and which serves them as a covering. With respect, however, to the egress of the excrement, in others it is obliquely; but what is called the egg [though it is a certain excrement], yet there is no avenue for its egress in any of these animals, but it comes forth from the flesh in the form of a swelling. Hence it is not in the same part in which the intestine is, but the egg, indeed, is in the right-hand, and the intestine in the left-hand, parts. Such, therefore, in other testaceous animals, is the egress of the excrement; but in the savage *lepas*, which some call the marine ear,

[because it resembles the human ear], the avenue through which the excrement is ejected is in the bottom part of the shell. It is, however, evident that the belly in this animal is immediately after the mouth, and, also, those substances in the form of an egg [which we have before mentioned]. But the situation of the parts in all these animals must be surveyed from dissections.

With respect, however, to what is called the small crab, it accords in some things with crustaceous, and in others with testaceous, animals. For it has a nature similar to that of locusts, and is born unconnected with any shell; but in consequence of entering into, and living in a shell, it resembles testaceous animals; so that it seems to partake of the nature both of testaceous and crustaceous animals. And, in short, the form of these animals resembles that of spiders, except that the part under the head and thorax is larger than in the spider. They have, also, two slender red horns, and two small eyes under these, which do not recede inwardly, nor incline, like the eyes of crabs, but are direct. Under the eyes, likewise, is the mouth, which is surrounded, as it were, with numerous capillaments. And consequent to these are two bifurcated feet, by which they convey food to their mouth. There are, also, two other feet on each side, and a third small foot. The inferior part, however, of the thorax, is wholly soft, and within it a certain matter of a pale colour is contained. One avenue, likewise, extends from the mouth to the belly; but the avenue by which the excrement is emitted is not apparent; and the feet and the thorax are hard, indeed, but less so than in crabs. They do not, however, adhere to the shell in the same manner as the purple and the buccina, but they easily liberate themselves from it. Those small crabs, also, that dwell in turbinated shells are longer than the neritæ, *i.e.* than those that dwell in sea-shells (*εν τοις νηριταις*). Though, however, the small crabs called *neritæ* are different from the turbinated kind, and, also, differ from them in length, yet in other respects they are not dissimilar to them. For in the neritæ, of the bifurcated feet which are under the mouth, the right foot is small, but the left is large; and, therefore, in walking they more frequently use the left than the right foot. Among the shell-fish, however, which after the manner of testaceous animals adhere to the shell, there is one which they call the little lame fish. But the neritæ have a smooth, large, and round shell, and a form similar to the buccina, except that they have not the *mecon* black, like the buccina, but red. They, also, firmly adhere to the middle of the shell. In fair weather, therefore, they leave their shell, and search for food; but when the wind blows [violently] the small crabs betake themselves to rocks, and there quietly remain. But the neritæ adhere to rocks after the manner of the *lepas*, and the *aporrhais*, and others of the like kind. All these, however, adhere to rocks, by removing the shell which serves them for a covering. For what both shells are to the bivalves that one shell is to those of the turbinated kind. Within the shell, however, there is a fleshy substance, and in this there is a mouth. The like, also, takes place in the *aporrhais*, in the purple, and all such like animals. Such of them, however, as have the left foot larger than the right, these never dwell in the shells of the turbinated kind, but dwell in the shells of the neritæ. But there are certain snails which contain in their shells animals similar to the small lobsters which are produced in rivers. They differ from them, however, in having the inward part of the shell soft. But the nature of their form must be surveyed from dissections.

5 The sea hedge-hogs, however, have not a fleshy substance internally, like other testaceous animals; for all of them are deprived of this, and instead of it, they have a certain black substance. But there are many genera of sea hedge-hogs; one of which, indeed, is esculent. And these contain within themselves what are called eggs, which are large, and fit to be eaten, and are both in the larger and smaller sea hedge-hogs; for they have these immediately from their birth. But what are called the *spatages* and *brysses* form the two other genera of sea hedge-hogs. These, however, dwell in the deep sea, and are rarely seen. Besides these, also, there are what are called *echinometræ*, or *hedge-hog-mothers*, and these are in magnitude the greatest of all the sea hedge-hogs. And, in addition to these, there is another small genus, which are covered with large and hard prickles; but these are produced at the depth of many fathoms in the sea, and are used by some as a remedy in stranguries. About Torone, however, there are white sea hedge-hogs, which have shells, prickles, and eggs, but they are greater in length than other sea hedge-hogs; though their prickles are neither large nor strong, but soft. From the mouth, however, there are a great number of black parts, which are extended as far as to the external avenue, [or excretory duct], but are unconjoined with each other, and are separated by certain limits. Those sea hedge-hogs, likewise, that are esculent, are moved more frequently, and more rapidly, which is always indicated by their bristles [to which some extraneous matter constantly adheres]. All of them, therefore, have eggs, but some are very small, and not fit to be eaten.

It happens, however, that what are called the head and mouth in the hedge-hog are in the bottom part; but the part through which the excrement is emitted is upward; and the very same thing happens in all testaceous animals of the turbinated kind, and in the *lepas*. For these animals search for their food in the ground, so that their mouth is towards the earth, where their nutriment is; but the excrement is in the upward part, and in the prone parts of the shell [*i.e.* in the parts directed towards the heavens]. The hedge-hog, also, has five teeth, which are inwardly hollow; and in the middle of these there is a fleshy substance, which performs the office of a tongue. After this follows the stomach, and then the belly, which is divided into five parts, and is full of excrement. All the cavities of the belly, likewise, unite in one at the egress of the excrement, *i.e.* at the avenue where the shell of the hedge-hog is perforated. What are called eggs, however, are contained under the belly in another membrane, and these are equal in number in all of them; for they are five, and are, therefore, an odd number. But above the eggs are those black parts, which depend from the root of the teeth, and are bitter, and not fit to be eaten. In many animals, however, a thing of this kind, or that which is analogous to it, happens; for it happens in tortoises, in the hedge-frog, in the common frog, in turbinated shell-fish, and in the mollia. But eggs of this kind differ in colour, and are either entirely unfit to be eaten, or more so than others. The body of the hedge-hog, therefore, is continued from the beginning to the end, but resembles a lamp, which is without a circular covering. The hedge-hog, also, uses his prickles instead of feet for supporting himself, and being moved on these, he changes his place.

6 Those animals, however, which are called *tethya*, [and in Latin are by some denominated vertibula, or calli, and which, by following the etymology of the Greek word, may be called mammillæ] have a certain most peculiar nature; for the whole body of these animals alone is concealed in a shell; but the shell is of a substance between skin and shell, and on this account may be cut like a hard hide. This animal, therefore, adheres by its shell to rocks. But it has two avenues, distant from each other, very small, and not easily seen, through which it emits and receives moisture; for it has no apparent excrement, like other testaceous animals, such as that which is in the hedge-hog, or in that animal which is called *mecon*. When, however, they are opened, in the first place, they are found to have a nervous membrane about the shelly part, and in this the fleshy substance of this animal is contained, but it does not resemble the flesh of other animals. This flesh however, is

similar in all the *tethya*. This flesh, also, adheres in two places to the membrane and the skin obliquely; and where it adheres, it is in each place narrower. In these places, likewise, there are two avenues, through which these animals receive and eject nutriment and moisture, just as if one of these avenues was a mouth, but the other the passage through which the excrement is emitted. And one of these avenues, indeed, is thicker, but the other thinner, but within these is a cavity on each side, between which a certain small continued substance intervenes, and in one of the cavities moisture is produced. They have not, however, any other part, either organic, or destined to sensation, nor is any excrement found in them, as it is in other testaceous animals, as we have before observed. The colour, also, of this animal is partly pallid, and partly red.

The sea-nettle, also, constitutes a peculiar genus of animals. It adheres, however, to rocks in the same manner as some testaceous animals, but
531 is sometimes liberated from them. It, likewise, has no shell, but the whole of its body is fleshy. But it is sensitive, and seizes the hand, to which also it adheres [of any one that touches it], in the same manner as the polypus with its arms, so as to cause the flesh of the hand to swell. It has, also, a mouth in the middle of its body, and lives in a rock just as testaceous animals live in a shell. It, likewise, takes, as if it were a hand, any little fish that may fall in its way; and if it meets with any thing esculent it eats it. A certain genus, also, of the sea-nettle is liberated from rocks, and eats the hedge-hogs and cockles that may fall in its way. These animals, however, do not appear to have any excrement, but in this respect they resemble plants. There are, also, two genera of sea-nettles; one of which, indeed, is less, and more eatable; but the other is large and hard, such as are those which are produced about Chalcis. In winter, therefore, their flesh is firm, and on this account they are then sought after, and are eatable; but in summer they are dissolved. For then they become fluid, are rapidly dispersed by the touch, and in short cannot be plucked. But when they suffer very much from the heat, they betake themselves into the more interior parts of rocks. And thus much concerning the parts, as well exterior as interior, of the mollia, and of crustaceous and testaceous animals.

7 Let us, however, after the same manner, speak concerning insects. But this genus contains many species in itself; though to some, which are allied to each other, no common name is assigned; as, for instance, to the bee, the drone, the wasp, and others of the same kind. This is, also, the case with such as have wings enclosed in a sheath, as the green beetle,

the beetles called pilularii [who make a ball from dung], the cantharides, and the like. Of all insects, therefore, there are three common parts, *viz.* the head, the cavity about the belly, and the third, that which is between these, which corresponds to the breast and back of other animals. This middle part, however, which is between the head and the belly, is in most insects one [and is not distinguished by any incisures]; but in insects that are long, and have many feet, the intermediate parts are nearly as numerous as the incisures. All insects, likewise, when divided have life, except those which are very cold, or which on account of their smallness are soon refrigerated [when divided], since wasps, also, live when they are divided. The head, therefore, and the belly live together with the middle part [when the insect is divided], but without 532
this part the head does not live. But such insects as have a long body, and many feet, live for a long time when divided, and the part which is cut off is moved to either extremity, for it moves both towards the head, and towards the tail, as is evident in what is called the scolopendra.

All insects, likewise, have eyes, but no other organ of sense is apparent; except that some have a tongue, which all testaceous animals have, and this they employ both for tasting their food, and drawing it to them. The tongue, however, in some insects is soft, but in others, as in the purple, is very hard and strong. The ox-fly, also, the horse-fly, and nearly most other insects, have a strong tongue. For in all insects that have not a sting behind, the tongue is as it were a weapon; but such as have a sting have no teeth, a certain few excepted; since flies by inserting this [*i.e.* their tongue] draw blood, and gnats pierce with this. Some insects, however, have stings. But some, indeed, contain the sting in themselves, as bees, and wasps; and others have it externally, as the scorpion, which is the only insect that has a long tail. This insect, also, has claws; and, likewise, the little kind of scorpion, which is engendered in books.

Insects that fly, however, in addition to other parts, have wings. And some of them, indeed, have two wings, as flies; but others have four wings, as bees. No flying insect, however, that has a sting behind has only two wings. Farther still, some flying insects have their wings enclosed in a certain sheath, as the beetle; but others have them uncovered, as the bee. But no flying insect steers its course by its tail, and the wing has not any stem, or scissure. Again, some insects have small horns before their eyes, as the butterfly, and the grasshopper; but in such of them as are formed for leaping, the hind are larger than the fore legs, and the legs by which they leap are concave before, and convex behind, in the same manner as the legs of quadrupeds. All

insects, however, have the prone different from the supine parts, in the same manner as other animals; but the flesh of their body is neither testaceous, nor such as that which is contained under the shell of
532 testaceous animals, but has a middle nature between these. Hence they have neither spine nor bone, nor any part analogous to bone, such as that is which is called sepium in the cuttle-fish; nor are they surrounded by a shell. For their body on account of its hardness preserves itself, and does not require any other support. They have, however, a skin, but this is very thin.

Such, therefore, are the external parts of insects, and after this manner they subsist. But inwardly, immediately after the mouth is the intestine, which in most insects is straight and simple, as far as to its exit; but in some few it has convolutions. No insect, however, has either viscera, or fat, as neither has any other exanguious animal. But some of them have a belly, and from this the remaining intestine is either simple, or convolved, as in the locust. The grasshopper, also, is the only insect, and, indeed, is the only animal, that has not a mouth; but, instead of a mouth, has a long, compact, and undivided part, similar to that tongue-like substance which is in animals that have a sting behind. Through this part the grasshopper is alone nourished with dew; but it has no excrement in the belly. There are, however, many species of these insects, and they differ in magnitude and parvitude. The grasshoppers, also, which are called canorous, are divided under the transverse septum, in which they have a visible membrane, which small grasshoppers have not.

There are, however, certain peculiar animals in the sea which, because they are rare, cannot be referred to any certain genus. For now some experienced fishermen say, that they have seen in the sea animals resembling black and round pieces of wood, of an equal thickness; and also other animals resembling shields, of a red colour, and furnished with numerous wings. They, likewise, report that they have seen some animals similar in form and magnitude to the pudendum of a man, except that, instead of testicles, they have two wings; and that such an animal was once caught by the top of the instrument [used by fishermen], which consists of many hooks. After this manner therefore, the external and internal parts of all animals in each genus, as well peculiar as common, subsist.

8 Let us, however, now speak concerning the senses. For they are not similarly present with all animals, but some have all the senses, and others not all. But the greatest number of the senses, and besides which

there does not appear to be any other sense, is five; *viz.* the sight, the hearing, the smell, the taste, and the touch. Man, therefore, and terrestrial viviparous animals, and besides these, sanguineous viviparous 533
animals, are seen to possess all these senses, except some mutilated genus, such as that of moles; since this is deprived of sight. For the mole has not eyes externally apparent; but if a thick skin is taken off from the head, and which is in the place of the eyes, certain corrupted eyes will be seen within, which have all the parts of perfect eyes. For they have the black part of the eye, and within the black what is called the pupil, and the white circle [which is called the white of the eye]. These, however, are less than in animals which have eyes [externally] apparent. But no one of these parts is externally visible, on account of the thickness of the skin with which the eyes of moles is covered from the first, as if nature was mutilated in their generation. For their eyes originate from the brain; and from that part in which the brain is conjoined with the optic nerve two nervous and strong avenues extend to the seat of the eyes, and are terminated in the upper teeth of the mole, which project beyond the mouth. Other animals, however, have a sensible perception of colours and sounds, and besides these of odours and sapours. All other animals, likewise, have that fifth sense which is called the touch.

In some animals, therefore, the sensoria are most apparent, and especially the sensorium of vision. For they have the place of the eyes and the hearing definitely assigned; since some of them have ears, and others [instead of ears] have visible avenues. The like, also, takes place with respect to the smell; for some, indeed, have nostrils, but others, as the genus of birds, have the avenues of smelling. In a similar manner, also, they have a tongue, as the sensorium of sapours. Aquatic animals, however, which are called fishes, have indeed a tongue, the sensorium of sapours, but they have it obscurely; for it is bony, and not detached [but adhering to the inferior part of the mouth]. In some fishes, also, the palate is fleshy; as, of river fishes, in the carp; so that to those who have more accurately surveyed these fishes, the palate has appeared to be a tongue. That fishes, however, have the sense of taste is evident; for they are very much delighted with peculiar sapours, and they are especially caught by the bait of the fish hamia, and by the fat of fishes, as delighting in a taste of this kind, and in food formed from such like baits. But they have not any apparent sensorium of hearing and 533
smelling. For though, about the place of the nostrils, they have certain avenues which to some may appear to be the organs of smelling, yet

such avenues do not extend as far as to the brain, but in some they are blind, and in others extend as far as to the gills.

But that fishes hear and smell is evident. For they are seen to fly from very loud sounds, such as those which are made by the rowing of the three-ranked galleys; so that, in consequence of their flight, they are easily taken in their caverns. For a sound which out of the water is but little, yet to all those that hear it in the water is found to be violent, loud, and oppressive; which also takes place in fishing for dolphins. For fishermen, when they have collected the dolphins about their boats, by making a noise in the sea cause the affrighted dolphins to leap on shore, where they are taken, their head being oppressed by the sound; though neither have dolphins any apparent sensorium of hearing.

Farther still, fishermen in catching fish are especially careful not to make any noise with their nets or oars. Hence, when they understand that many fishes are collected together in a certain place, they drop their nets at such a distance as they think is sufficient to prevent the fishes from hearing them, being cautious lest any sound should arrive at that place, either from the oars, or the motion of the boat. They, likewise, admonish all the sailors to row with the greatest silence, till the fishes are entirely surrounded with the nets. Sometimes, also, when they wish the fishes to assemble together, they adopt the same method as in fishing for the dolphin. For they make a noise by throwing stones, that the fishes being terrified may assemble in one place, and that thus they may surround them with their nets. And before they enclose them, as we have said, they forbid the sailors to make a noise; but when they have surrounded them with their nets, they order the sailors to make a loud noise; because the fishes, hearing the noise and the tumult, fall into the net through fear.

Again, when fishermen see from a great distance a collected multitude of fishes feeding and swimming on the surface of a tranquil sea, and when they are likewise desirous to perceive their magnitude, and of what kind they are, if they sail near to them without any noise, they latently catch them, while they are yet swimming on the surface; but if prior to
534 this any noise happens to have been made, their visible flight is the consequence.

Farther still, in rivers there are little fishes under rocks, which some call *κοιτοι*, gulls; and these are caught by certain persons through striking the rocks in which they are concealed with stones; for these fishes on hearing the sound come forth, having their head oppressed with the noise. From particulars of this kind, therefore, it is evident that fishes hear. There are, however, some who assert that fishes hear the most acutely of all animals; and this they confirm by the authority of those who swell near the sea, and who assert that they have found this to be true by repeated experiments. But the fishes which excel in acuteness of hearing are the mullet, the *chremps*, the sea-wolf, the stock-fish, the *chromis*, and others of the like kind. Other fishes, however, hear less acutely than these and hence they principally live near the bottom of the sea.

In a similar manner, also, it may be proved that fishes have the sense of smelling. For the greater part of them is only to be caught with a recent bait; and all fishes are not caught with the same baits, but with particular baits, which they know by the smell. Some, indeed, are caught by fetid baits; as, for instance, the stock-fish by dung. Again, many fishes live in caverns; and fishermen when they wish to catch them anoint the mouth of the cave with briny odours, by which the fishes are soon allured [and taken]. After this manner, also, eels are caught. For fishermen, in order to catch them, use a certain vessel containing brine, in the mouth of which they place the collar of a bow-net.

In short, all fishes are allured by the smell of roasted flesh; and hence fishermen roasting the flesh of the cuttle-fish, use the smell of it as a bait; for fishes are thus more easily caught. They also say that fishermen put into their nets roasted polypi, on no other account than for the sake of the smell. Farther still, the fishes called *ruades*, when the lotion or the filth of fishes is discharged, fly from it, in consequence of smelling the fetid odour. They, likewise, say that fishes quickly smell the blood 534
of those of their own species. This, indeed, is evident by their flying to distant places when the blood of fishes is poured forth. And, in short, if putrid baits are put into the nets, fishes will neither be willing to enter into nor approach the nets; but if the bait consists of fresh roasted meat, they will immediately tend to it, though at a distance, and enter into the nets. What has been said, however, is especially evident in dolphins. For these have no apparent sensorium of hearing, but they are caught in consequence of being oppressed with a heaviness in the head from sound, as was before observed. Neither, likewise, have they any apparent sensorium of smelling; and yet they smell acutely.

That these animals, therefore, have all the senses is evident; but the remaining genera of animals have a fourfold division, which comprehends the mollia, crustaceous and testaceous animals, and besides these insects. Of these, however, the mollia, crustaceous animals, and insects, have all the senses; for they have sight, the smell, and the taste. For insects, as well those that have as those that are without wings, perceive at a distance. Thus, for instance, bees, and the *cnipes* [a kind of gnat] have a sensible perception of honey; for they perceive it at a distance, as knowing it by the smell; and many insects perish from the smell of sulphur. Farther still, if the powder of sulphur and wild marjoram is sprinkled over the habitation of ants, they leave their holes; and the greater part of such kind of insects fly from the odour of the burnt horn of a stag; but they especially avoid the fume of storax. The cuttle-fish, likewise, the polypi, and the lobster, are caught by baits; and the polypus so pertinaciously adheres to a bait, as to suffer himself to be cut in pieces rather than be divulsed from it. If, however, any one presents to him the herb flea-bane, it is said, that as soon as he smells it, he will drop the bait.

535 The like, also, takes place with respect to the taste. For sometimes insects pursue a different food, and all of them are not delighted with the same sapours. Thus, for instance, bees to not approach to any thing fetid, but to what is sweet; and the *conops*, or gnat, does not fly to any thing sweet, but to what is acid. To have a sensible perception, however, by the touch is, as we have before observed, common to all animals.

But testaceous animals have, indeed, the smell and the taste; and this is evident from the baits by which they are caught; as, for instance, in the purple fish. For this animal is allured by putrid substances, and approaches to a bait of this kind, as having a sensible perception of it at a distance. That testaceous animals, also, have a sensation of sapours is evident from the same arguments; for each of them is delighted with the sapour of those substances to which it is allured by the smell. And again, such as have a mouth receive either pleasure or pain from the contact of sapours.

With respect, however, to the sight and hearing of testaceous animals, nothing stable nor very apparent can be asserted. But the *solenes* are seen to withdraw themselves, and to subside lower, when they perceive the iron instrument by which they are caught; for a small part of their body only is seen, but the rest is as if it were in a place of retreat. And with respect to cockles, if any one presents his finger to them they open and shut themselves, as having the sense of sight. And those who catch the shell-fish called *neritæ* do not approach them with a bait when the wind blows, nor speak, but are silent, because these animals both smell and hear; for if those that are employed in catching the neritæ speak, it is said they withdraw themselves. Of testaceous animals, however, which have a progressive motion, the hedge-hog, but of those that are without motion the tethya, and the sea acorns (βαλανοι), appear to have the sense of smell in the smallest degree. After this manner, therefore, the sensoria subsist in all animals.

9 Voice, however, differs from sound; and the third of these is speech. No animal, therefore, is vocal by any other part than the pharynx; and hence such animals as are without lungs are not vocal. But speech is the articulation of voice by the tongue. The voice, therefore, and the larynx emit what we call vowels, but the tongue and the lips consonants; from which true speech consists. Hence such animals as have not a tongue, 535
or have it adhering to the inferior part of the mouth, do not speak. But the other parts of the body are capable of producing a sound. Insects, therefore, have neither voice nor speech. They produce a sound, however, by the internal, but not by the external, air. For no insect respires; but some of them utter a humming sound, as bees, and winged insects; and others are said to sing, as grasshoppers. All these, however, produce a sound by the attrition of the internal air, which subsists within a membrane under the division called the transverse septum, as, for instance, the genus of grasshoppers; and this is, likewise, the case with flies, bees, and all the rest, in consequence of raising and depressing that membrane when they fly. For [such] sound is produced by the attrition of the internal air. But crickets produce a sound through the attrition of the legs by which they leap. Neither any one of the mollia, however, nor any crustaceous animal has a voice, or produces any natural sound. Fishes, also, are without voice, since they have neither the trachea nor the pharynx. Those, however, that are said to be vocal emit certain sounds and noises, as the fish called the lyre and the *chromis*; for they emit as it were a certain grunting; and this is, also, the case with that fish, in the river Achelous, called the wild-boar; and with the *chalcis* and cuckoo. For the first of these fishes produces a kind of whistling; but the second emits a sound similar to that of the bird called a cuckoo; whence, also, it derives its name. All these fishes, however, that appear to utter a voice, produce it either by the attrition of the more rigid gills which they possess, or of the air which they contain in the cavity of their belly. For in each of these there is a hot air, the attrition and motion of which produce sounds. Those cartilaginous fishes, also, called the selache appear to produce a stridulous sound. These, however, cannot be properly said to have a voice, but to produce a sound. For coskles when they are seen floating on the surface of the water, which they call flying, make a crashing noise; and this is, also, the case with the swallows of the sea. For these fly high, and do not touch the sea, their wings being broad and long. As, therefore, the sound

produced by the wings of flying birds is not voice, so neither is the sound produced by any thing of the above-mentioned kind. The
536 dolphin, likewise, produces a stridulous sound, and murmurs when he comes into the air, yet not in a manner similar to the before-mentioned fishes. For the sound emitted by the dolphin is voice; since he has lungs and a trachea, but his tongue is not loose, and he has no lips, so that he cannot produce articulate voice.

But of those animals which have a tongue and lungs, such as are oviparous and quadrupeds, emit a voice, which is, however, feeble. And some, indeed, produce a hissing sound, as serpents; but others a slender and feeble voice; and others a small hissing, as tortoises. The frog, however, has a peculiar tongue. For the anterior part, which is detached from the other parts of the tongue, adheres after the manner of the tongue of fishes; but the parts of the tongue near the pharynx, by which the frog emits his peculiar voice, are detached and expanded. The male frogs, also, produce a croaking in the water when they call females to coition. For all animals have peculiar voices significant of the desire of coition, as is evident in swine, goats, and sheep. Frogs, however, produce that [venereal] croaking by making the lip of the lower jaw to be equally under the water, and extending the upper jaw above the water. But their jaws becoming luminous from the extension, their eyes have the appearance of lamps; for they are seen to copulate for the most part by night.

The genus of birds, however, is vocal; and such as have a moderately broad tongue, and, also, such of them as have a thin tongue, have especially the faculty of speech. Some of them, therefore, as well male as female, have the same voice; but others have a different voice. Less, also, have a louder voice, and are more loquacious than larger birds; and every bird at the time of coition is more loquacious than at any other time. Some birds, likewise, as the quail, are vocal when they fight; others, prior to the combat, call on their opponents to engage [as the partridge]; and others are vocal when they have conquered, as the cock. And with some genera of birds, indeed, the males sing as well as the females; as for instance, both the male and female nightingale sing; except that the female nightingale ceases to sing during the time of her incubation, and while she rears her young. But in some genera of birds, as cocks and quails, the power of voice belongs rather to the males, and the females do not sing.

536 Different viviparous quadrupeds, however, have a different voice, yet none of them has the faculty of speech; but this is the peculiarity of man. For such animals as have the faculty of speech are also vocal; but

not all such as are vocal have the faculty of speech. All such, too, as are deaf from their birth are likewise dumb; and though they are vocal, yet are entirely deprived of speech. But children as they cannot at first manage the other parts of the body, so neither can they use the tongue. The tongue, also, at first is imperfect, and it is some time before it is detached; so that children, for the most part, lisp and stutter.

Voice and speech, also, differ according to places. Voice, therefore, is, indeed, especially manifest by acuteness, and gravity; nor is there any specific difference in [inarticulate] voices of the same genera. But the difference in articulate voice, which may be called as it were speech, differs according to places in different animals, and, also, in animals belonging to the same genus. Thus, with respect to partridges, the voice of some of them repeats the syllables, *cac, cac,* but of others *tri, tri*. And some small birds, if they are taken from their parents, so that they cannot be educated by them, do not emit the same voice as their parents, but become accustomed to the manners and singing of other birds. A nightingale, also, has been seen instructing her young in singing; so that speech and voice are not similarly natural, but the former may be acquired by instruction. Men, likewise, have, indeed, the same voice, but not the same speech. The elephant, however, is vocal, not through the nose [or proboscis], but emits a windy voice through the mouth, such as men emit when they breathe, and at the same time speak; but with his proboscis he emits a hoarse sound, which resembles the sound of a trumpet.

10 With respect, however, to the sleep and wakefulness of animals, it is obvious to sense that all such as are pedestrious, and sanguineous sleep, and are wakeful; for all such as have eyelids sleep by closing them. Again, not only men are seen to dream, but, also, horses, dogs, and oxen; and besides these, sheep, and goats, and every genus of viviparous quadrupeds. Dogs, also, evince this by barking when they are asleep. With respect, however, to oviparous animals, it is immanifest whether they dream; but that they sleep is evident. In like manner aquatic animals sleep, such as fishes, the mollia, crustaceous animals, locusts, and 537
others of the same kind. All such animals, therefore, though they are seen to sleep, yet sleep but for a short time. But an indication of this cannot be obtained from their eyes, for they have no eyelids, but from the quiet state of their bodies. For fishes are easily caught with the hand while they are asleep, unless they are roused by certain animals, which are called pulices and pediculi. Now, however, if fishes sleep for a long time during the night, they are devoured by these animalcula, which are

very numerous. They are, likewise, produced in the profundity of the sea, and are so numerous that the bait made from a fish is consumed by them, if it continues for any length of time on the land. And it frequently happens that fishermen, when they draw the bait out of the sea, find these animalcula collected about it, so as to form as it were a sphere.

It may, however, in a greater degree be inferred that they sleep from such circumstances as the following: For fishermen frequently attack fishes so latently as to take them in their hand, or strike their harpoon into them, without their being conscious of it. But during the time of their sleep they are very quiet, and only gently move their tails. Their agitation, also, if any thing is moved while they are quiet evinces that they sleep; for they shake, as if roused from sleep. Farther still, fishes are likewise taken in rocks, in consequence of being asleep in them. Frequently, also, fishermen enclose the tunny in their nets while they are asleep; but it is evident that they are asleep from their being taken in a state of rest, and gently disclosing the white of their eyes. The tunny, however, sleep by night rather than by day, and so profoundly as not to be moved by the throwing of the harpoon. For the most part, also, fishes sleep either on the sand, or on the land, or adhering to some stone in the profundities of the sea, or concealing themselves under a rock or bank; and flat fishes conceal themselves while they sleep in the sand. They are detected, however, by the figure of the sand, and are caught by striking them with a trident. With this instrument, likewise, the sea-wolf, the gold-fish, the mullet, and others of the same kind, are caught in the day, in consequence of being asleep; but it does not appear that any of these are caught with the trident, unless they are asleep. But those cartilaginous fishes called selache sometimes sleep so profoundly as to be caught by the hand. The dolphin, however, the whale, and
537 such fishes as have a pipe [through which they eject the water] sleep, elevating this pipe, through which they respire, above the sea, and gently moving their fins. Some, also, assert that they have heard the dolphin snore. The mollia, likewise, sleep after the same manner as fishes; and similarly with these, crustaceous animals.

That insects, also, sleep, is evident from such indications as the following: For they are sometimes manifestly quiet and without motion, and this is especially evident in bees; for by night they are still, and cease to hum. And this is manifest in those insects with which we are especially conversant [*viz.* in flies]. For they are quiet by night, not only because they do not then acutely see, for all animals that have hard eyes see obscurely, but when near the light of a lamp they are nevertheless seen to be quiet.

Man, however, dreams the most of all animals. And infants, indeed, and very young children do not dream; but, for the most part, children begin to dream about the fourth or fifth year of their age. There are, also, some men and women that have never known a dream. Some of these, however, as they have advanced in years have begun to dream, after which a change has taken place in their body; in some, indeed, the change terminating in death, but in others in disease. And thus much concerning sense, and sleep, and wakefulness.

11 The male and female, however, are inherent in some animals, but not in others; though, according to a certain similitude, they are said to bring forth and to be pregnant. There is, indeed, no distinction of male and female in stable animals [*i.e.* which are fixed to one place], nor, in short, in testaceous animals. But in the mollia, and in crustaceous animals, there is the distinction of male and female, and also in pedestrious animals, in bipeds and quadrupeds, and in all such as from coition produce an animal, or an egg, or a worm. In other animals, therefore, this is either simple or not simple. Thus, for instance, in all quadrupeds there is the distinction of male and female, but in testaceous animals there is not; but as in plants, some are prolific, and others are 538
sterile, so likewise in these. In insects, however, and fishes, there are some which are so deprived of this distinction of sex that they cannot generate an offspring, and therefore are neither male nor female, and such are eels. But those who assert that eels are sometimes seen to have capillaments and worms adhering to them [internally], speak inconsiderately, not having previously surveyed where these worms subsist. For no animal of this kind contains within itself an animal, unless it has first conceived an egg; but no eel has ever been found to have an egg. Such animals, also, as are viviparous, contain the fœtus in the womb, and not in the belly; for if it were contained in the belly, it would be concocted in the same manner as nutriment. But the difference which they speak of between the male and female eel, *viz.* that the male has a larger and longer head, but the female a more depressed head, is not to be ascribed to the difference of sex, but of genus.

There are, however, certain fishes which are called *epitrageæ*, [from the sterility of a goat,] and such among river fishes are the carp, and the *baris*; but such fishes have neither an egg, nor any prolific seed, and such of them as are solid, and fat, have a small intestine, and are excellent food. Farther still, as in the genus of testaceous animals and plants, so in the genus of fishes, there are some which bring forth of themselves and generate, though they do not copulate; and of this kind are plaice, the roach, and cockles, and all such as these are seen to have eggs.

In pedestrious and sanguineous animals, therefore, which are not oviparous, the males are for the most part larger, and live longer than the females, the mule excepted; for among mules, the females live longer, and are larger than the males. But in oviparous and viviparous animals, as in fishes and insects, the females are larger than the males, as is evident in serpents, in the spiders called phalangia, in lizards, and frogs. The like, also, takes place in fishes, as in the small selache, in the greater part of fishes that live in shoals, and in all that are called saxatiles [or
538 living among rocks]. That among fishes, however, the females live longer than the males is evident from the females that are caught being older than the males.

The males, however, in all animals, have the superior and anterior parts better, stronger, and more vigorous; but the females, those parts which may be called posterior and inferior. And this is the case, not only in the human species, but also in other pedestrious and viviparous animals. The female, likewise, is less nervous and more infirm than the male, and the hair in such females as have hair, or that which is analogous to hair, is thinner. The flesh, also, of females is more humid, their knees are more liable to crack, and their legs are thinner; but their feet are more elegant, *viz.* in such animals as have these parts. With respect to voice, too, the voice of all females, except that of the cow, is more slender and acute: for the cow has a graver voice than the bull. But with respect to the parts which nature has given to animals for their defence, such as teeth, fangs, horns, spurs, and other such like parts, in some genera, indeed, the male has these, but the female has not. Thus for instance, the female stag has no horns; and among such birds as have spurs, the females of some of them are entirely without spurs. In a similar manner, also, the females among swine have no fangs. In some genera of animals, however, these parts exist both in the male and female, but they are more excellent in the former than in the latter. Thus, the horns of bulls are stronger than those of cows.

BOOK V

1 In the preceding books, therefore, we have spoken concerning the internal and external parts of animals; and farther still, concerning their senses, voice, and sleep, and the difference between the males and the females. Hence it remains, that we should now direct our attention to their generation, and in the first place discuss such particulars as are first. 539 The generations, however, of animals are many, and possess much variety; being, indeed, in one respect dissimilar, but in another similar, to each other. Since, also, the genera of animals was first divided by us, we must now endeavour that the speculation shall be formed after the same manner; except that, then, in speculating about the parts, we made the beginning from man, but now we shall speak about man in the last place, because the generation of man requires the most ample discussion. But we shall first begin from testaceous animals; afterwards speak of such as are crustaceous; and then discuss the rest after this manner, in an orderly series. And these are the mollia and insects. After these, we shall direct our attention to the genus of fishes, both the viviparous and the oviparous; and then consider the genus of birds. And, in the next place, we shall speak concerning the generation of pedestrious animals, as well viviparous as oviparous. But some quadrupeds are viviparous, and man is the only biped that is viviparous.

A common circumstance, therefore, happens in the generation of animals, as well as in the generation of plants. For of plants some are generated from the seed of other plants, but others are generated spontaneously from analogous causes. And of these some receive their nutriment from the earth, but others are ingenerated in other plants, as we have shown in the speculation concerning plants. Thus, also, with respect to animals, some indeed are generated from animals, according to an alliance of form; but others have a spontaneous origin, and not from kindred natures. And of these some are generated from putrid earth and plants, as happens to be the case with many insects; but others are generated in animals themselves, and from the excrements in the parts of animals.

Animals, however, which are generated from kindred animals, if there is the distinction in them of male and female, are generated from coition. But in the genus of fishes, some are generated neither male nor female, and though they accord in genus yet differ in species from other fishes, and some are entirely peculiar. There are, also, some fishes, among which there are females without males, and from which, in the same manner as in birds, windy eggs are produced. All such eggs of birds, however, are unprolific, but the nature of these birds is able to

539 accomplish the generation of an egg, unless some other mode of communication with the males happens to them; which particulars, however, will be more accurately discussed hereafter. But it happens to certain fishes, when they generate spontaneous eggs, that animals are produced from them, except that some produce these animals from themselves, but others not without the co-operation of the male. The manner, however, in which this is effected will be evident in what follows; for nearly the same thing takes place in birds. But from such animals as are generated spontaneously either in the earth, or in plants, or in the parts of these, if they have the distinction of male and female, from the copulation of these something is produced, from which, however, nothing else can be generated. Thus, for instance, from the copulation of lice what are called *lendes* are generated; from the copulation of flies certain worms are produced; and from the copulation of fleas, worms in the shape of eggs. From these, however, neither any other animal, nor any thing else, but things of this kind alone are generated. In the first place, therefore, we must speak of coition, and of the animals which generate from coition; and afterwards discuss other particulars, which follow in an orderly series, and the accidents both peculiar and common which pertain to their generation.

2 Those animals, therefore, copulate in which there is a distinction of male and female. The copulations, however, are neither similar in all animals, nor have they a similar mode of subsistence. For in all sanguineous animals that are viviparous and pedestrious, the males have organs adapted to such a generative act, yet all of them do not similarly approach to the female; but those that discharge the urine backward, as lions, hares, and lynxes, copulate backward;† and with respect to hares, the female frequently gets on the back of the male. The mode of copulation, likewise, is the same in the greater part of other animals. For the usual mode of copulation with most quadrupeds is for the male to get on the female; and this mode is adopted by the whole tribe of birds. There are, however, certain differences with respect to birds. For with some, while the female sits on the ground, the male gets upon her, and this is the case with the bustard, and the cock; but with others, as with cranes, the female does not sit on the ground, during the act of coition; for among these the male leaping is conjoined with the female,

† *i.e.* The buttocks of the one being in an opposite direction to the buttocks of the other.

and the copulation is rapidly accomplished, in the same manner as with sparrows.

Of quadrupeds, however, bears lying on the ground copulate in the 540
same manner as other animals who rest on their feet during the act, the supine parts of the male being in contact with the prone parts of the female; but land hedge-hogs copulate erect, having the supine parts in contact with each other. Among large viviparous animals, however, the female stags do not suffer the males to have connection with them, unless rarely, nor the cow the bull, on account of the stiffness of the penis, but as they withdraw themselves from the male they receive the genital seed; for this has been seen to happen in tame stags. But wolves copulate after the same manner as dogs; and cats do not copulate backward, but the male is erect, and the female places herself under him. Females, however, are naturally libidinous, incite the males to copulation, and cry out during the act of coition. But camels copulate the female sitting, and the male covers her, not having his buttocks in an opposite direction, but in the same manner as other quadrupeds. They continue, likewise, in the act for a whole day; but when they copulate withdraw into solitude, and it is not then safe for any one but the herdsman to approach them. The penis also of the camel is so nervous that the string of a bow may be made from it. Elephants, likewise, copulate in solitary places, and especially near rivers, and where they are accustomed to dwell. But the female elephant copulates lowering her body, and removing her legs; and the male gets upon her back. The sea-calf, also, copulates after the same manner as animals that discharge the urine backward, and, as well as dogs, are a long time in the act of coition. The males, likewise, have a large penis.

3 Among pedestrious animals, also, oviparous quadrupeds copulate after the same manner. For, just as with viviparous animals, the males of some of these get on the back of the female, and this is the case with the land and sea tortoise. But the avenues have something with which they are conjoined, and to which they approximate in coition, as is evident in the turtle-dove, the frog, and every such like genus of animals.

4 But animals without feet, and which have a long body, as the 540
serpent and the lamprey, copulate by enfolding the supine parts with the supine [*i.e.* the belly with the belly]. And serpents are so closely enfolded in the embraces of each other, that the whole body appears to

be one serpent with two heads. This is, also, the case with the genus of lizards; for they copulate with a similar circumplication.

5 All fishes, however, except the broad selache, copulate by uniting the supine parts [*i.e.* the belly] with the supine. But the broad selache, and which, also, have a tail, such as the thornback, the pastinaca, and others of this kind, not only copulate by uniting the supine parts, but by the supine parts of the male being placed on the back of the female, unless the tail is an impediment on account of its thickness. The skate, however, and such as have a thick tail, copulate by the attrition alone of the supine with the supine parts. There are, also, some who report, that they have seen certain selache copulate behind, in the same manner as dogs. But in every class of the selache the female is larger than the male; and nearly, also, in other fishes, the females are larger than the males. The fishes, however, above-enumerated are comprehended among the selache, and also the sea-ox, the lamia, the eagle, the torpedo or cramp fish, the sea-frog, and all the species of sea-dogs. All the selache, therefore, have been clearly seen by many to copulate after this manner. For the connection of all viviparous is of a longer duration than the connection of oviparous animals. Dolphins, likewise, and all animals of the cetaceous kind, copulate after the same manner; for the male copulates by uniting his supine parts with those of the female, and is neither a short nor a very long time in the act. But the males of some of the selache differ from the females, because in the males two appendices hang from the orifice through which the excrements are discharged, and of which the females are deprived; as may be seen in the sea-dog.

Neither fishes, therefore, nor any other animal without feet, have testicles. But serpents and male fishes have two avenues, which become full of genital seed about the season of coition, and all of them emit a milky moisture. These two avenues, however, unite in one, in the same
541 manner as in birds. For birds have testicles internally, and every thing, also, which is possessed by oviparous animals that have feet. This, therefore, is enclosed, and extends to the place and receptacle of the female. But the same avenue of the seed and of the moist excrement exists externally in viviparous and pedestrious animals; and internally there is another avenue, as we before observed in explaining the difference of the parts of animals. In those animals, however, that have no bladder the same avenue of the dry excrement exists externally, and internally there is another avenue, both which are near to each other.

These, also, subsist similarly both in the males and females of this genus of animals. For, except the tortoise, these animals are without a bladder; and the female tortoise have one avenue, though she has a bladder. But with respect to tortoises, they rank among oviparous animals.

The copulation, however, of oviparous fishes is less manifest, and on this account many are of opinion that the females become prolific by swallowing the genital seed of the males; since they are frequently seen to do this. For about the season of copulation, the females, following the males, do this, and strike their belly with their mouths; but the males in consequence of this emit their seed, and in greater abundance. At the time of delivery, however, the males follow the females, and devour the eggs which they lay; and from those that are left fishes are generated. But about Phœnicia fishes are caught through the assistance of each other. For the fishermen take the male and present them to the female mullets, and the females to the males. In consequence, therefore, of this being frequently seen, an opinion has been formed that the females become pregnant from swallowing the seed of the males. This very same thing is, also, done by quadrupeds. For about the season of copulation, both males and females discharge a seminal moisture, and smell the private parts of each other. But with respect to partridges, if the females stand opposite to the males when the wind blows, they become pregnant; and they frequently become pregnant even from the voice of the male, if they happen to be enflamed with the desire of coition. The same effect is also produced from the male flying above the female, in consequence of her inbibing the odour which then proceeds from the male. Both the male and female, likewise, during the act of coition have their mouths open, and their tongues hanging out. The true copulation, however, of oviparous fishes is seldom seen, because they do not continue long in the act; since their copulation has been seen to be effected after this manner.

6 The mollia, also, the polypus, the cuttle-fish, and the calamary, after 541
the same manner copulate with each other; for they are connected with each other by the mouth, at the same time adapting their arms to each other. The polypus, therefore, when it has fixed what is called its head against the earth, and has unfolded its arms, another polypus adapts itself to the extension of the arms, and they make the acetabula to be in continuity with each other. Some, also, say that the male has a kind of penis in one of its arms, in which there are two of the largest acetabula; that a thing of this kind is as it were nervous, and is extended as far as

to the middle arm; and that the whole of it enters into the nose of the female. But the cuttle-fish and the calamary swim at the same time that they are connected with each other; and co-adapting their mouths and arms, they swim in a direction opposite to each other. The nose, likewise, as it is called, of the one adheres to the nose of the other. And with respect to the swimming, that of the one is in a backward direction, but that of the other is towards the mouth. But they bring forth in that part of the body which is called the fistula, by which part, also, it is said by some that they copulate.

7 Crustaceous animals, however, such as locusts, lobsters, squills, and others of the same kind, copulate after the manner of quadrupeds that discharge the urine backward, the female causing her tail to be supine, and the male placing his tail upon it. They copulate, also, near the land, at the beginning of the spring; for the copulation of all such animals has been detected. Sometimes, also, they copulate when figs begin to be ripe. After the same manner, likewise, lobsters and squills copulate. But crabs copulate by their anterior parts, adjusting to each other their envelopments, which consist of many folds. In the first place, therefore, the less gets on the larger crab from behind, and afterwards the larger crab turns himself obliquely. Hence the female differs in nothing else from the male, except that the envelopment of the female is more ample, detached, and hairy, in which she brings forth her eggs, and discharges the excrement. Their copulation, however, is not attended with the insertion of one member into another.

542 8 But insects approach to each other behind, and afterwards the less gets on the larger insect; and the less is the male. The female, however, from beneath, extends an avenue to the male, who is above her; but not the male to the female, as is the case with other animals. And this part, indeed, in some insects is seen to be larger than the proportion of the whole body requires, and is, likewise, very small; but in others is less than the proportion of the body demands. This, however, will be evident, if any one dissects flies when in the act of copulation. Insects, also, are with difficulty liberated from the embraces of each other; for the copulation of animals of this kind continues for a long time. This is evident in insects with which we are familiar, as in flies and beetles. For all these copulate after the same manner, *viz.* flies, beetles, *sphondyles*, the spiders called phalangia, and every other insect of this kind which copulates. But the phalangia that weave copulate as follows:

when the female has drawn one of her extended threads, the male again draws it in an opposite direction; and when they have frequently done this, they come together, and are united in their posterior parts; for in consequence of the roundness of their belly, this mode of copulation is adapted to them. After this manner, therefore, the copulation of animals is effected.

The time, also, of the copulation of animals, and the age at which they copulate, are definite. For the intention of nature is, that this connection should take place with most animals about the same time, *viz.* during the change from winter to summer. And this is the season of the spring in which most animals, the winged, the pedestrious, and those that swim, are impelled to copulation. Some animals, however, copulate and bring forth in autumn and winter; as, for instance, certain genera of aquatic and winged animals. But man especially, and many animals that live with man, copulate at every season of the year, in consequence of having at all times sufficient warmth and abundant nutriment. The time, also, of the gestation of these animals is short, and this is the case with the swine, the dog, and such winged animals as bring forth frequently. Many animals, likewise directing their attention to the nurture of their offspring, copulate in that season of the year which is best adapted to this purpose. But in the human species, the male is more impelled to coition in the winter, and the female in the summer.

With respect, however, to the genus of birds, the greater part, as we have already observed, except the halcyon, copulate and bring forth at the commencement of the summer; but the halcyon brings forth about the winter solstice. Hence when the winter solstice is tranquil, the days are called halcyon. Halcyon days, however, are fourteen, seven before, and seven after, the solstice; as, also, Simonides sings, when he says, "As when Jupiter in the winter month produces fourteen mild days, which mortals call the period in which the winds are still, the sacred nurse of the variegated halcyon." But these tranquil days take place when the winter solstice happens to be southern, and the [setting] of the pleiades northern.† It is, also, said that the halcyon in the first seven days builds her nest; but in the remaining seven brings forth and nurtures her young. About these places, therefore, [*i.e.* in Greece] it does not always happen that the halcyon days are at the time of the winter solstice; but

† *i.e.* They happen in November.

about the Sicilian sea, it nearly always happens. The halcyon, likewise, lays about five eggs.

9 But the plungeon and sea-gull bring forth their eggs in the rocks about the sea; and the multitude of them is two or three. The sea-gull, however, lays its eggs in summer, but the plungeon at the beginning of the spring, immediately after the winter solstice, and their incubation is the same as that of other birds. Neither of these birds, however, hides itself. But the halcyon is seen the most rarely of all birds; for it is scarcely ever seen except about the setting of the pleiades, and the winter solstice. When, likewise, it enters a port, and has flown about as much space as a ship occupies, it immediately disappears; as, also, Stesichorus the poet sings in his verses.

The nightingale, also, lays her eggs at the beginning of the summer; but she lays five or six eggs. And she hides herself from the autumn to the spring.

But with respect to insects, those which do not hide themselves copulate in the winter, and are produced when the weather is tranquil, and the wind southern; such, for instance, as flies and ants.

Most wild animals, however, which have no superfetation bring forth once in the year, as for instance hares. In a similar manner most fishes
543 bring forth but once in the year, as the *chutoi*, which are so called because they are caught in nets in great abundance; such, for instance, as the tunny, the pilcher, the mullet, the *chalcis*, the *colias*, the *chromis*, and the like, except the sea-wolf, who is the only one among these fishes that beings forth thrice in one year. The posterior progeny, however, of this fish is more imperfect. The *trichias*, also, and the fishes that live among rocks, bring forth twice in a year; the barbel alone brings forth thrice in a year. But this may be inferred from its progeny; for in some places it is seen to have offspring thrice. The marine scorpion, however, brings forth twice; and the sargus, likewise, brings forth twice, *viz.* in spring and autumn; but the stock-fish brings forth only once, and that in autumn. The tunny, also, brings forth once in a year, but because she emits into light one part of her offspring maturely, and another part late, she appears to bring forth twice. But her first delivery is in the month of December, in the winter solstice, and her second in the spring. The male tunny, however, differs from the female in this, that the female has under its belly a fin which they call *aphareus*, but the male has not.

10 Of the selache, however, the *lima* or *file* alone brings forth twice; for she brings forth at the beginning of autumn, and about the setting of the pleiades [*i.e.* about the autumnal equinox]; but she brings forth more successfully in the autumn. Each time, also, she is delivered of seven or eight offspring. Some of the sea-dogs, however, the star-fish for instance, bring forth twice in a month; but this happens because the eggs of this fish are not all of them perfect at once. Some fishes, also, bring forth at every season of the year; as, for instance, the lamprey. But this fish brings forth many eggs, which rapidly increase from a small to a large size, in the same manner as the eggs of that kind of lobster called *hippurus*; for these from being very small become rapidly very large; except that the lamprey brings forth at all times of the year, but the hippurus in the spring. The *murus*, however, differs from the *muræna* or lamprey. For the lamprey, indeed, is variegated, and is more imbecil; but the murus is of one colour, is strong, and its colour is similar to that of rosin. It has, also, teeth externally and internally. They, likewise, say that the murus is the male, and the muræna the female; there being this distinction of sexes in them, in the same manner as in other fishes. Each of these fishes, however, comes on dry land, and is frequently caught.

It happens, therefore, that nearly all fishes rapidly increase, and particularly the little fish called the small crow. But this fish brings forth near the land, in such places as are shady and abound with plants. 543
The fish, likewise, called *orphus*, from being small becomes rapidly large. The pilcher and the tunny bring forth in the Euxinus Pontus, or black sea, and no where else; but the mullet, the gold-fish, and the sea-wolf, principally bring forth at the mouths of rivers. The *orcines*, however, the *scorpiones*, and many other genera, bring forth in the deep sea.

11 Most fishes, likewise, bring forth in the three months of March, April, and May; but a few, such as the stock-fish, the *sargin*, and others of this kind, bring forth in the autumn, a little before the autumnal equinox; and this is, also, the case with the torpedo and the file. Some fishes, however, as was before observed, bring forth both in winter and summer; as, for instance, in winter, indeed, the sea-wolf, the mullet, and the needle-fish; but in summer, the tunny, in the month of June, about the solstice. The offspring of this fish, however, resembles a sack, in which many small eggs are contained. The *ruades*, also [or fishes that go in bands] bring forth in summer. Of mullets, likewise, those that are called *labeones, sargones, mucones,* and *capitones*, begin to be pregnant in the month of December, and their pregnancy continues for thirty days.

Some mullets, however, are not generated from copulation, but originate from mud and sand. Generally, therefore, most fishes are pregnant in the spring, though some, as we have said, are pregnant in summer, autumn, and winter; yet this does not similarly happen to all of them, nor simply, nor according to each genus, nor to most, nor in those other seasons are they pregnant with such a numerous progeny as in the spring. In short, it is necessary not to be ignorant that as places produce a great difference in plants and quadrupeds, not only as to the good condition of the body in other respects, but, also, as to their greater or less fecundity; so, likewise, with respect to fishes, places produce a great difference in them, not only as to their magnitude and proper nourishment, but, likewise, as to their progeny and copulation, so that the same fishes in one place generate more and in another less frequently.

544 12 The mollia, also, bring forth in the spring, and rank among the first of marine animals that spawn; but the cuttle-fish brings forth in every season, and continues laying her eggs for fifteen days. As soon, however, as the female has laid her eggs, the male following pours his seminal liquor on the eggs, which renders them firm. But the male cuttle-fish is more variegated than the female, and his back is blacker. The polypus, however, copulates in winter, but brings forth in the spring, and is concealed for about two months. The egg, also, which it lays is as it were a tuft of hair, and resembles the fruit of the white poplar. This animal, however, is very prolific; for from the egg of the polypus an infinite multitude of polypi are produced. But the male differs from the female, both in having a longer head, and because it contains that white substance in its arm which fishermen call the pudendum. The female, also, incubates the eggs which she brings forth; and hence the polypi at that time are in an ill condition of body, because they do not then take any nutriment.

The purples, likewise, are generated about the spring, and the buccina at the end of the winter; and, in short, testaceous animals appear in the spring and autumn to have what are called eggs, except those sea hedge-hogs that are fit to eat; for these though they have eggs especially in spring and autumn, yet they are at no time without eggs, and they particularly abound with them during the full moon, and in warm days. Those hedge-hogs, however, must be excepted, that live in the Pyrrhæan straits; for these are better in winter. And they are, indeed, small, yet full of eggs. In a similar manner, likewise, all snails are seen to be pregnant in the same season of the year [the spring].

13 Of birds, however, those that are wild, as we have before observed, for the most part copulate and bring forth once. But the swallow and the blackbird bring forth twice. The first progeny, therefore, of the blackbird are destroyed by the winter; for she brings forth the first of all birds; but she nurtures her second progeny, and causes them to arrive at perfection. With respect, however, to domestic birds, or such as can be domesticated, these bring forth frequently. Thus, for instance, pigeons and hens bring forth during the whole of the summer. For cocks and hens copulate perpetually, and the latter bring forth at all seasons of the year, except during the winter solstice. Of pigeons, however, there are many genera. For one kind is of a livid colour, and this is the stock-dove, or wood-pigeon. The wood-pigeon, therefore, is less than the pigeon, and is, also, less capable of being domesticated. But the wood-pigeon is black, small, and has red and rough feet; and hence no one rears a wood-pigeon. The ring-dove, therefore, is the largest of pigeons; but the next to this is the vinous pigeon; and this is a little larger than the pigeon properly so called; but the turtle-dove is the smallest of birds of this kind. Pigeons, however, properly so called, bring forth and rear their young at all seasons of the year, if they have a warm place and proper food; but if not, they are pregnant in summer only. The offspring of the pigeon, also, that are produced in spring are better than those that are produced in autumn; but those are the worst of all that are produced in summer, and in very hot weather.

14 Animals, also, differ with respect to copulation, according to age. In the first place, therefore, most animals do not at one and the same time begin to emit their seed, and possess a generative power, but the latter of these is posterior to the former. For the young of all animals are at first unprolific; and if they happen to be prolific, their offspring are smaller and more infirm. This, however, is especially evident in men, quadrupeds, and birds. For the two former, when they are young, produce smaller children, and young birds lay smaller eggs. In the next place, animals of the same species are, for the most part, adapted to copulation at the same age, unless some premature circumstance prevents them, either arising from something prodigious and monstrous, or from their nature having sustained some injury. In the human species, therefore, the aptitude to copulation is indicated by the mutation of the

voice, and the pudenda, not only in magnitude, but also in form, and in a similar manner in the mutation of the mammæ, and especially by the hairs of the private parts. But man begins to have seed about the fourteenth year of his age, and possesses a [perfect] generative power about the age of twenty-one. Other animals, however, have no hairs in their private parts. For some, in short, have no hair, but others have no hair in the supine parts, or have less than in the prone parts. The change of the voice, also, is in some animals manifest; but in others, other parts of the body indicate the commencement of the seed and prolific power.

With respect to voice, also, that of the female in most animals is more
545 acute, and of the younger than that of the older. For among stags, the males have a graver voice than the females; and the males, indeed, are vocal at the season of copulation, but the females when they are terrified. The voice of the female, however, is short, but that of the male is long. The voice, likewise, of dogs is more grave when they become old.

The voices, also, of horses differ. For as soon as they are born, indeed, the females emit a slender and small voice, but the males a small voice, yet louder and graver than that of the female; and in progress of time they have a still louder voice. But when they become two years old, and begin to copulate, the male, indeed, emits a loud and grave voice, but the female a louder and clearer voice than she emitted before, and this is the case, for the most part, till the twentieth year of their age. After this period, however, the voice both of males and females is weaker.

For the most part, however, as we have said, the voice of males differs from that of females, as being more grave, in such animals as have an extension of voice. Yet this is not the case in all animals; but in some, as in oxen, the contrary takes place. For in these the voice of the female is graver than that of the male, and the voice of calves than that of perfect oxen. Hence, bulls when they are castrated change their voice into that of the female, since through this their voice becomes more acute.

The times, however, of copulation, according to age, subsist in animals as follows: Female sheep and goats, therefore, copulate and are pregnant at the age of one year; but this is more especially the case with goats. And the males, also, copulate at the same age. But the offspring of sheep and goats differ in these and other respects; since the males are better when they are produced in the second year, or when they are produced by a male and female of an advanced age.

Both the male and female swine, however, begin to copulate in the eighth month of their age; but the female brings forth at the age of one year; for thus the time of her pregnancy happens to take place. The male, also, generates in the eighth month of his age, but the offspring which he produces before he is one year old is bad. As we have observed, however, the ages of the copulation of animals are not every where similar. For in some places, both male and female swine copulate when they are four months old; so as to produce and nurture offspring in the sixth month. But in other places, boars begin to copulate in the tenth month of their age, and for three years retain a perfect generative power.

Dogs, however, both male and female copulate, for the most part, when they are one year old; but it sometimes happens that they copulate in the eighth month of their age; though this rather takes place with the females than with the males. The females, also, are pregnant during sixty-one, or two, or three days at farthest; but they are not pregnant for a less time than sixty days. And if they happen to be delivered prior to this period, the progeny never arrive at perfection. After they are delivered, likewise, they copulate again in the sixth month, and not before.

With respect to horses, also, both the male and female begin to copulate when they are two years old, and at this period are able to generate; yet the offspring which at that time they produce is small and feeble. For the most part, however, they copulate when they are three years old, and every succeeding continues to be more excellent than the preceding offspring, till they are twenty years old. But the male copulates till he is thirty-three years old; and the female till she is forty; so that horses copulate nearly during the whole of their life. For the male, for the most part, lives thirty-five years, and the female more than forty years. A horse, also, has been known to live seventy-five years.

With respect to the ass, however, both the male and female copulate in the thirtieth month of their age; yet they do not for the most part generate at this period, but when they are three years old, or three years and six months. There has, however, been an instance of an ass which has conceived when but a year old, and has reared her progeny. This has, also, been the case with a cow, the progeny of which has been reared, and arrived at its proper magnitude.

The principles, therefore, of generation in these animals subsist after this manner. But with respect to the human species, the male generates till he is seventy years old; and the woman till she is fifty. This, however, is rarely the case; for few have offspring at these ages; but, for the most part, men cease to generate in the sixty-fifth year, and women in the forty-fifth year of their age.

But sheep bring forth till the eighth year of their age; and if good care is taken of them, they bring forth till they are eleven years old; for it
546 happens that they copulate nearly through the whole of their life. Goats, however, when they are fat are less prolific; on which account vines, also, when they are barren, are said to be goatish. But when goats become lean, they are able by copulation to generate. Rams, however, first copulate with older sheep, and afterwards with those that are young; but, as we have before observed, young sheep produce fewer offspring than those that are more advanced in age.

The wild boar, also, is good for the purposes of copulation till the third year of his age. But the progeny of those that are older than this are of an inferior kind. For the wild boar after this period receives no increase, and becomes destitute of strength. It is usual, also, with this animal to copulate when he is full, and when he has not recently copulated; for, if this is not the case, his connection with the female continues but a short time, and the offspring which he begets are small. The sow, however, at the first birth brings forth a very few offspring; but the second birth is more abundant; and when she becomes old she brings forth in a similar manner, but copulates more slowly. When, however, they are fifteen years of age they no longer generate, but become wild. If they are well fed, likewise, they are more rapidly impelled to copulation, both when they are young and when they become old. But when the sow is pregnant, if she is very fat, she has less milk after her delivery. And with respect to her progeny, they are then the best when she is in the acme of her age, and when they are begot at the beginning of winter; but those that are begot in summer are the worst, for they are small, thin, and moist. But the male swine, if he is well fed, is able to copulate at all seasons, and both day and night; but if he is not well fed, he principally possesses the power of copulating in the spring. And when old, as we have before observed, he is always less able to copulate. Frequently, however, such of them as are unable to copulate rapidly, through age or imbecility, approaching to each other, copulate reclining, through the infirmity of the female. The sow, however, then especially becomes pregnant, when, being furiously eager for copulation, she hangs her ears; for if this is not the case, she does not become pregnant, but the seed is ejected by the womb.

Dogs, however, do not copulate through the whole of life, but only as far as to a certain acme of their age; but, for the most part, their copulations and pregnancies continue till they are twelve years of age. Some dogs, however, have been known to possess the power of generating, and some bitches of being pregnant, even at the ages of eighteen and twenty; though these, as well as other animals, are deprived by old age of the ability of generating and bringing forth.

But the camel discharges its urine backward, and copulates in the manner before explained. The time, however, of its copulation, in Arabia, is in the month of September. This animal, also, is pregnant during the space of twelve months, and only brings forth one at a birth; for it is uniparous. Both the male and female, likewise, begin to copulate when they are three years old; and the female again copulates in one year after her delivery.

The female elephant, however, begins to copulate at the earliest period, indeed, in the tenth year of her age, and at the latest period, when she is fifteen; but the male copulates, when he is five or six years of age. The time, also, of their copulation is the spring. The male, likewise, again copulates after an interval of three years, but is never a second time connected with a female that he has made pregnant. The time, however, of the gestation of the female is two years, and she brings forth only one offspring; for she is uniparous. The embryo, also, of this animal is about the size of a calf of the age of two or three months. And such are the particulars concerning the copulation of animals.

15 We must now, however, discuss the generation both of animals that are produced by copulation, and of those that are produced without it, and in the first place the generation of testaceous animals; for the whole of this genus is, as I may say, produced without copulation. The purples, therefore, in the spring, collecting together in one place, make what is called their wax. But this is, as it were, the honey-comb of bees, except that it is not such an elegant production, but resembles a multitude of the shells of white peas collected in one heap. No one of these shells, however, has any avenue; nor are the purples generated from these, but both the purples and other testaceous animals are generated from putrefaction and mud. This wax, therefore, or honey-comb is, as it were, a certain purgation both of the purples and buccina; for the buccina, also, make wax. Hence those testaceous animals that make wax are generated after the same manner as other testaceous animals, and, indeed, they are more easily generated this way, when

homogeneous substances pre-exist [in the putrid matter]. When the purples, therefore, make their wax they emit a certain mucous viscosity, from the concretion of which those substances resembling the shells of white peas are formed. But these shells, being analyzed by putrefaction, the putrid mucus contained in the shells is diffused into the earth, and in the place into which it flows small purples are formed, which are found when they are caught adhering to larger purples, and some of
547 them which thus adhere are not yet perfectly formed. If, however, the larger purples are caught before they have produced that mucus from which the small purples originate, they do not bring forth in any casual place, but in mats and hurdles, in the same manner as they bring forth in the sea, and from the narrowness of the hurdle in which they are delivered, their fœtus resembles a grape.

There are, however, many genera of purples; and some of them indeed, are large, as those about Sigeum and Lectum; but others are small, as those in the Euripus, and about Caria. And those, indeed, which are found in gulfs are large and rough, and most of them have a black flower, but some of them have a red and small flower. Some of them, likewise, are large, and weigh twelve ounces and a half. The small purples, however, are found in shores, and about coasts, and have a red flower. Again, in the northern parts, they are for the most part black, but in the southern parts red. They are, likewise, caught in the spring, when they make their wax; but during the dog-days they are not caught, for then they are not fed, because they retire and conceal themselves. But that part of them which is called the flower, [and with which wool is dyed] is between the neck and the mecon [or papaver]. That, likewise, which unites these two parts is compact, and the colour of it resembles a white membrane, which is taken off. When the flower, also, is pressed it tinges and dyes the hand, and extends through it like a vein. This, however, appears to be the flower; but the remainder of that which unites the neck and the mecon resembles alum. The flower, likewise, of the purples is then in the worst condition when they make their wax. But [in order to collect the flower of the purple] they bruise the small purples together with their shells; for it is not easy to collect the flower in any other way. They divest, however, the larger purples of their shells, and then take away the flower, on which account the neck and the mecon are, also, separated; for the flower is between these, above that part which is called the belly. The flower, therefore, being taken away, the neck and mecon are necessarily divided. But those that collect the flower of the purples endeavour to bruise them while they are alive, for if they should die before they are bruised, they cast out the flower together with their life; on which account they are preserved in bow-nets, till they are collected and quiet. The ancients, therefore, together with a bait did not employ nets in catching the purple, and, therefore, it frequently happened that the purples escaped; but fishermen of the present time fasten the bait to a bow-net in order that if the purples should drop off, they may not escape. Then, however, they especially drop off when they are saturated; but if they are empty, they are with difficulty divulsed from the bait. These, therefore, are the peculiarities which happen to the purple.

The buccina, also, are generated after the same manner as the purples, and at the same season of the year. Both the buccina and purples, likewise, and other turbinated animals, are from their birth covered with a shell which adheres to them after the same manner and in the same position; but they are fed by stretching out what is called their tongue, under the shell. And with respect to the magnitude of the tongue in the purples, the tongue is longer than a finger, by which they are fed, and by which they pierce shell-fish, and their shell. Both the purple, however, and the buccinum live long; for the purple lives about six years, and every year the increase of them is evident by the intervals of the spiral in the shell, [*i.e.* by the increase of the number of the spirals].

Mussels, also, make wax. But with respect to what are called marshy oysters (λιμοστρεα), wherever there is mud there the origin of them consists. And periwinkles, the *chames*, the *solenes*, and cockles, receive their formation in sandy places. But the erect pinnæ [which are testaceous immoveable animals] are formed in the bottom of the sea in sand and mud; and they contain within their shell, as a guardian, a small crab, or squill, of which when they are deprived they soon perish. In short, all testaceous animals are generated spontaneously in the matter which is at the bottom of the sea, and are different according to the difference of this matter; oysters, indeed, being generated in mud, but periwinkles, and the above-mentioned genera, in a sandy matter. The *tethya*, likewise, and the *balani* or glandes, and those which appear on the surface of the water, such as the *lepas*, and the neritæ, are generated in the hollows of rocks. All such testaceous animals, therefore, have a rapid increase, and especially purples and cockles; for these become perfect in one year. In the shells, however, of certain testaceous animals very small white crabs are formed; most of them, indeed, in broader mussels [which are called *foliati*, because they have a broad shell like a leaf], and in the next place those crabs which are called the guardians of the pinnæ are produced in the shells of the pinnæ. They are, likewise, produced in cockles, and in marshy oysters; but they do not receive any apparent increase. Fishermen, however, say that these guardians of the pinnæ are generated together with the pinnæ. But cockles, in the same manner as purples, disappear for a time in the sand.

After this manner, therefore, testaceous animals are formed. But some
548 of them are formed in shallow water, others in shores, and others in muddy places. Some, likewise, are produced in hard and rough, but others in sandy, places. And some, indeed, change the places of their abode, but others do not. And of those which do not change their habitation, the pinnæ are rooted to a place; but the solenes and periwinkles remain fixed, without being rooted; yet when they are divulsed are no longer able to live. The testaceous animal, however, which is called the star-fish, is naturally so hot as instantly to burn and consume every thing which it touches. But they say that the species of this animal which is found in the Pyrrhæan straits is of the most destructive nature; and that its form resembles that of a painted star. Those marine animals, also, that are called *pulmones* or *lungs*, are spontaneously produced. But the shell-fish used by painters is very thick, and the flower is external to the shell, [and not like that of the purple within it]. Shell-fish of this kind, also, are principally found in the places about Caria. The small cancer, however, derives its origin from the earth and mud; and afterwards enters into empty shells. When it has increased, likewise, it again enters into another larger shell; as, for instance, into the shell of the neritus, or trumpet, or some other of this kind; but it frequently enters into the shells of the small buccina. When, however, it has entered any shell, it carries it about, and in this is again nourished. When, also, it is again increased, it betakes itself to another larger shell.

16 Animals, also, that have no shell, such as the sea-nettle, and the sponge, are generated after the same manner as testaceous animals, in the caverns of rocks. There are, however, two genera of sea-nettles; for some of them adhere to the cavities of rocks, so that they cannot be liberated from them; but others adhere to smooth and broad places, so that they can be liberated from them, and change their habitation. In the second place, sponges have certain cavities. And in these cavities certain animalcula are generated, which have the same analogy to the sponges as the guardians of the pinnæ [above-mentioned] to the pinnæ, from which guardians deriving their appellation they are called

pinnophylaces. But these animalcula dwell in the cavities of the sponges, just as small spiders dwell in their cavities. They, likewise, cause the sponge to open when small fishes are about to enter into it, and to shut when they have entered, these little fishes being their prey. There are, however, three genera of sponges; one of which is thin, the second dense, and the third, which is called Achillœan, is most thin, dense, and 548
strong. Sponges, also, of the last kind are put under the helmets and greaves of soldiers, and cause a blow to produce a less sound; but they are very rare. Of sponges, however, those that are very hard and rough are called *goats*. All sponges, likewise, are either produced near rocks or shores; but they are nourished in mud, of which this is an indication, that when they are taken they are found to be full of mud. For other animals that adhere to any substance derive their nutriment from that to which they adhere. The dense, however, are more imbecil than the thin sponges, because the adherence of the former is not so tenacious as that of the latter.

The sponge, also, as they say, is sensitive; of which this is an indication, that it contracts itself at the approach of the person who plucks it, so that the divulsion of it is difficult. It, likewise, does the very same thing when there is much wind, and the weather is tempestuous, in order that it may not be removed from its situation. There are, however, some who are dubious with respect to this circumstance, as for instance, the inhabitants of Torona. For they say that the sponge nourishes certain animals within itself, such as worms and the like; which, when the sponges are divulsed, are devoured by the fishes that live in rocks, together with the remaining roots of the sponges, but that if the sponges are not radically extirpated, they grow again from what remain, and are again filled.

The largest sponges, therefore, are those that are thin, and of which there are a great number about Lycia; but the dense sponges are most soft; for the Achillœan sponges are more compact than these. In short, the sponges which are found in places where the water is deep, and of a mild temperature, are most soft, for wind and cold harden them, in the same manner as other productions of the earth, and prevent their increase; on which account, the sponges in the Hellespont are dense and rough, and, in short, those which are beyond and within the promontory of Malea differ in softness and hardness. Sponges, however, will not bear immoderate heat, for they become putrid through it in the same manner as plants. Hence the sponges which are found near shores where the sea is deep are the best; for they acquire a proper

temperament, on account of the depth of the water. But while they are unwashed and alive they are black. They adhere, however, to rocks, neither by one part only, nor by all their parts; for between the parts which adhere there are certain void spaces, and a membrane, as it were, is extended about the roots; but they adhere by many of their parts. But
549 the other void spaces in the upper part are closed. Four or five of them are, however, apparent; on which account some assert that these are the avenues by which they receive their food.

There is, also, another genus of sponges which they call *aplysia*, because it is incapable of being washed. These sponges have, indeed, large avenues, but all the rest of their body is dense. And when they are dissected they are found to be more dense and viscous than other sponges, and the whole of their substance resembles the lungs. It is, however, universally acknowledged that this genus of sponges is sensitive and lives for a long time. They are, likewise, easily distinguished in the sea from other sponges, because others are found to be white when the mud subsides, but these are always black. And thus much concerning sponges, and the generation of testaceous animals.

17 Of crustaceous animals, however, locusts are pregnant after copulation, and carry their eggs during the three months of May, June, and July; after which they deposit them in the folds under their belly, and in these the eggs increase in the same manner as worms increase. The same thing, also, takes place in the mollia, and in such fishes as are oviparous; for the eggs of all these are increased. The egg, therefore, of locusts are friable, and is divided into eight parts. For to each of the covers which exist at the side of the locust a certain cartilaginous substance is united, to which the eggs adhere, and the whole resembles a grape; for each of these cartilaginous substances is divided into many branches. And these, indeed, when divided are apparent; but when they are beheld undivided, they seem to form one mass. The largest eggs, however, of the locusts are not those which are near to the avenue of their egress, but those which are situated in the middle; and the least are those which are situated in the last place. The size of the small eggs, also, is that of a grain of millet; and the eggs do not immediately adhere to the avenue, but there is something between them and the avenue. For the eggs on each side, *i.e.* as well from the superior part of the breast
549 as from the inferior part of the tail, are divided by a certain interval; since the covers are thus naturally disposed. They are not, therefore, able to enclose the eggs laterally, and hence the intervals in the

extremities of the eggs are so disposed as to cover all the eggs. The locust, also, when it brings forth its eggs places them in the cartilaginous folds, at the same time expanding its tail, and compressing and bending itself. At the time of its delivery, likewise, the cartilaginous folds are increased, and become the recipients of the eggs; for the locust then deposits its eggs in these folds, in the same manner as the cuttle-fish deposits its eggs near the branches of vines, and other substances which it meets with in the sea. After this manner, therefore, the locust brings forth. The locust, however, after it has brought forth its eggs, concocts them in its cartilaginous folds for the space of twenty days, and afterwards ejects them into a fit place, so as that they are seen to be conglomerated in a heap; and from these eggs locusts are produced in about fifteen days, and they are frequently less than the length of a finger. Locusts, therefore, bring forth their eggs before, but eject them after, the rising of Arcturus.

But the pregnancy of crooked squills continues about four months. And locusts, indeed, are produced in rough and rocky, but lobsters in smooth, places. Neither of them, however, are produced in muddy places. Hence lobsters are produced in the Hellespont, and near the island of Thasus; but locusts about Sigœum and Athos. But fishermen, from the shores and adjacent coasts, conjecture whether the places in which they intend to fish are rough or muddy. In the spring and winter, however, locusts and lobsters are found near the land; but in summer they live in the deep sea. For at one time they are in search of heat, and at another of cold. Those crustaceous animals, also, which are called bears bring forth in a manner similar to locusts. Hence, in winter, and before they begin to lay their eggs in spring, they are the best as food; but when they have brought forth they are the worst. Locusts and crabs, also, put off their shell in the spring soon after they are born, and during every succeeding year, in the same manner as serpents put off their skin, which is called their old age. But all locusts are long-lived.

18 The mollia, however, from copulation have a white egg. But this in process of time becomes friable, in the same manner as the eggs of crustaceous animals. And the polypus, indeed, brings forth either in caverns, or shells, or other hollow places, and its progeny resembles the tufts of the wild vine, or the fruit of the white poplar, as was observed before. The eggs, also, which the polypus lays depend from the walls of
the cave in which they dwell in such abundance, that when collected, 550
they fill a vessel much larger than the head of the polypus, in which

these eggs are contained. Polypi, therefore, are produced from the bursting of these eggs, in about fifty days after they are brought forth, and they creep on the ground very numerous, like the spiders called phalangia; in which, though the nature of the several members cannot be distinguished, yet the whole form is sufficiently apparent. This great multitude, however, of polypi perishes, on account of their smallness and imbecility; and some of them are so small, that no parts can be distinguished, though from the touch it is evident that they move.

The cuttle-fish, also, brings forth, and its eggs resemble large and black myrtle-berries; for she sprinkles them with her ink. These eggs, likewise, adhere to each other in the form of a grape, and the whole is so connected by one certain substance, that the eggs cannot easily be separated. For the male pours on the eggs a certain mucous moisture, by the viscosity of which they adhere to each other, and increase. And these eggs, indeed, as soon as they are emitted are white, but when the cuttle-fish sprinkles them with her ink, they become larger and black. But when a little cuttle-fish is generated, it is wholly formed within the egg from the white of it, and the membrane of the egg bursting, it comes forth. When, however, the female has sprinkled the egg with her ink, the interior white part of the egg becomes like a drop of hail; and from this the little cuttle-fish is produced, the head being first formed, in the same manner as birds, and the belly being connected with the head. The mode, however, in which the navel of the cuttle-fish is connected with that part of the egg from which the little cuttle-fish originates has not yet been discovered, except that as the little cuttle-fish increases the white of the egg always becomes less; and at length, in the same manner as the yellow in the eggs of birds, this white entirely disappears. But, as in other animals, so likewise in these, the eyes at first are seen to be large. As for instance, let A be the egg, BC the eyes, and D the little cuttle-fish.† The cuttle-fish, also, is pregnant in the spring; but she brings forth in fifteen days [after conception]. And when she is delivered of her eggs, in the space of another fifteen days they become like the small stones of a grape; on the bursting of which little cuttle-fish come forth. If, however, any one cuts the membrane of the eggs before they spontaneously burst, the little cuttle-fish, being not yet completely formed, emit dung, and change their colour through fear, so that from being white they become red.

† See the figure in Sealiger's edition of this work.

Crustaceous animals, therefore, differ from the mollia in this, that the 550 former hatch their eggs conjoined to themselves; but the polypus, the cuttle-fish, and other animals of the like kind, hatch their eggs separated from themselves, wherever they may be, and this is especially the case with the cuttle-fish, whose belly during her incubation is frequently seen near the ground. But the female polypus at one time hatches her eggs by incubation, and at another time, placing herself at the entrance of her hole, she extends her arm over the eggs, and thus hatches them. The cuttle-fish, however, brings forth near the earth, among weeds, branches of trees, reeds, heaps of stones, or any other accumulation of this kind. Hence fishermen deposit, in opportune places, bundles of twigs, [in order that they may allure the cuttle-fish to bring forth in them]; for the cuttle-fish in these bundles lays that long series of eggs, adhering to each other, which resembles a lock of hair. The cuttle-fish, also, when she lays her eggs, after some respite, sprinkles them with her ink, because the act of bringing them forth is painful. But the calamaries bring forth in the deep sea; and their eggs are in a connected series, like those of the cuttle-fish.

Neither the calamary, however, nor the cuttle-fish live long; for except some few of them, they do not live the space of two years. This is, also, the case with the polypi. But from one egg one small cuttle-fish is produced; and the like, also, takes place with the calamaries. The male calamary, however, differs from the female; for if the belly of the female is dissected, it will be seen that she contains in it two red substances, which resemble small mammæ, and which the male has not. The cuttle-fish, also, has this difference; and besides this, the male is more variegated than the female, as was before observed.

19 With respect to insects, also, we have before shown that the males are less than the females; that in copulation the male gets on the back of the female, and that they are a long time before they are separated. Most, however, of those insects that generate from coition bring forth very soon after copulation. All insects, likewise, bring forth worms, except a certain genus of butterflies, and these bring forth a hard substance resembling a grain of bastard saffron, but which internally is liquid. An animal, however, is not generated from a certain part of worms, in the same manner as it is from a certain part of eggs; but the whole is increased, and becomes a distinctly formed animal. Some insects, also, are generated from kindred animals. Thus, for instance, the spiders called phalangia, and other spiders, are generated from phalangia

and spiders, and the *attelabi* (a species of locusts), crickets, and grasshoppers. But other insects are not generated from animals, but are
551 produced spontaneously. And some, indeed, are generated from the dew which falls on leaves; and these insects are naturally generated in the spring, but they are likewise frequently produced in winter, when for a long time the weather has been mild, and the wind southern. Others, again, are generated in putrid mud and dung, but others in wood; some, indeed, in the wood of plants, and some in dry wood. Some, likewise, are generated in the hairs, but others in the flesh, and others in the excrements, of animals. And of these, some are generated when the excrements are ejected, but others, as what are called *lumbrici*, or belly-worms, while the excrements are yet in the animal. There are, however, three genera of these; those which are denominated broad, and the round, and the third kind, what are called *ascarides*. From the round, therefore, and the ascarides, nothing else is generated; but the broad kind, which alone adhere to the intestines, produce something similar to the seed of a cucumber; by the emission of which, physicians conjecture that this broad worm is latent in the intestines.

What are called butterflies, however, are generated from caterpillars; but caterpillars are generated from green leaves, and especially from *raphanos*, which some call cabbage. And at first, indeed, something less than a grain of millet is produced; afterwards small worms originate from this; and these increasing, in the space of three days are formed into small caterpillars. Such caterpillars, also, when increased, cease from motion, change their form, and are called *chrysalides*, or *aureliæ*. In this state, likewise, they are enclosed in a hard shell; but move if they are touched. The chrysalides are, also, enclosed in cavities, which resemble the webs of spiders, but they neither have a mouth, nor is any other part apparent. In a short time, too, the shell bursts, and winged animals fly out of it, which we call butterflies. At first, therefore, while they are caterpillars, they are nourished and emit excrement; but when they become chrysalides, they neither taste any thing, nor emit any excrement. After the same manner, also, that butterflies are generated from caterpillars other animals are generated from worms, whether such worms originate from the copulation of animals, or are produced without copulation. For the worms which are afterwards changed into
551 bees, drones, and wasps, when they are recently produced are nourished, and are seen to have dung; but when from worms they pass into another form, they are called nymphs, indeed, but in that state neither receive any nutriment, nor have any dung; and being enclosed, they remain

without motion, till they are formed into another animal. For then the substance in which they are enclosed bursting, they come forth [changed into bees, drones, or wasps].

The *penies*, also, and the *hyperes*, are produced from certain caterpillars which have an undulating motion in their progression. For as the sea during its fluctuation alternately raises and depresses the water, so these caterpillars, as they move along, raise their body in a bow-like manner, and afterwards, as they proceed farther, depress it. But every genus of animals which is produced from caterpillars has the colour of the caterpillars from which it originates.

From a certain large worm, also, which has, as it were, horns, and differs from other worms, in the first place, indeed, the worm being changed, a caterpillar is produced, afterwards a *bombylius*, and from this the *necydalus*;† but the worm undergoes all these metamorphoses in the space of six months. From this animal, also, certain women unfolding silk draw it out into a thread, and afterwards weave it. But Pamphyle, the daughter of Latois, is said to be the first that wove with the threads of the silk-worm, in the island Co.

From the worms, likewise, which are in dry wood beetles are generated, after the same manner. In the first place, indeed, the worms being immoveable, a certain membrane is produced; afterwards, the shell in which they are enveloped bursting, beetles come forth. But from bee-hives the *prasocourides*‡ are generated, and these also have wings. And from the broad little animals which swim on the surface of rivers horse-flies (οιστροι) are produced; and hence there is a great quantity of these flies about rivers, where such animals are generated. But from black and hairy, though not large caterpillars, in the first place glow-worms, which do not fly, are produced; and these again change, and winged animals are generated from them, which are called locks of hair.

The *empis*, however, or *gnat*, is generated from the *ascarides*; but the ascarides are worms which are produced in the mud of wells, and wherever there is a conflux of water in which there is an earthy sediment. At first, therefore, the mud itself becoming putrid receives a white colour, afterwards it is black, and last of all it acquires the colour 552
of blood. When, however, this is the case, substances resembling sea-weeds, very small and red, are produced. These, after a time, begin to

† A worm, which in the space of six weeks is changed into a silk-worm.

‡ The *prasocouris* is an insect which feeds on lettuces in gardens, and particularly on leeks.

move, adhering to the mud from which they originate; but afterwards, being freed and separated from the mud, they move on the water, and are called ascarides. After a few days, also, they stand erect on the water, and become immoveable and hard. In the next place, the shell in which they are enclosed bursting, a gnat emerges, and sits on the water, till the sun or the wind moves him; for then he flies away. *The principle, however, of the generation of all other worms, and of animals which burst forth from worms, appears to be derived from the sun or the wind.* But the ascarides are generated more rapidly and in greater abundance in water, the sediment of which is various and promiscuous, such as is the sediment with which the land of Megara is manured; for a substance of this kind becomes rapidly putrid. The ascarides, also, are more numerous in autumn, for then there is less moisture, [by an abundance of which they are destroyed].

The ticks, however, [or lice of dogs and sheep] are generated from putrid grass; but green beetles are produced from the worms which are found in the excrements of the ox and the ass. And the beetles called pilularii, who make certain balls from dung which they roll, in winter hide themselves in these balls, and bring forth worms, from which beetles are generated. In leguminous substances, also, worms are generated, from which winged animals are produced, similar to those above-mentioned. But flies are generated from the worms found in the excrement which is separated from straw;† on which account, also, those who are employed in this work endeavour to separate the different parts of the dung, and then say they have rendered it fit for use. The principle, however, of these worms is small. For at first, indeed, a little red spot is seen, and from being immoveable, it begins to be moved, as it were, adhering to the dung; afterwards it becomes an immoveable little worm; then being moved, it again becomes immoveable. But from this state it becomes a perfect fly, and it moved when excited by the wind or sun.

Ox-flies, however, are generated from [putrid] wood; but the *orsodacnæ mordellæ*, or blister-flies, from transformed worms; and these worms are produced in the stalks of cabbage. The cantharides are generated from
552 caterpillars in the fig-tree, the pear-tree, and the pine-tree; for in all these worms are produced. The cantharides, also, are generated from the

† *i.e.* They are generated from the dung employed by husbandmen in dunging land, after they have separated it from the straw, not yet become putrid, with which it is mixed on dunghills.

worms in the sweet-briar. They, likewise, are impelled towards fetid substances, because they originate from a matter of this kind. But the *conops*, or gnat, is produced from worms which are generated in the dregs of vinegar; for animals are generated in substances which appear to be most free from putrefaction. Thus, for instance, worms are produced in old snow. But old snow is of a redder colour; and hence the worms that are formed in it are red and hairy. The worms, however, that are generated in the snow in Media are large and white; but all of them move with difficulty.

But in Cyprus, where the stone chalcitis [a stone of a brazen colour] is calcined, after the calcination has continued for many days, winged animals are generated in the fire, in a small degree greater than large flies, which walk through the fire, and leap. These animals, however, and the worms which are produced in snow, die; the former when separated from the fire, and the latter when separated from the snow. But that it is possible for the bodies of certain animals to remain in fire without being burnt is rendered evident by the salamander. For this animal, as they say, while it walks through the fire extinguishes it.

About the river Hypanis, however, which is near to the Cimmerian Bosphorus, at the time of the summer solstice, a kind of shells are found in the river, larger than grape-stones, from the bursting of which a winged quadruped comes forth. This animal lives, and flies, till the afternoon; but when the sun declines it wastes away, and dies when he sets, its life not being extended beyond one day; on which account, also, it is capped ephemeron. Most animals, however, that originate from caterpillars and worms are at first enclosed in certain substances which resemble the webs of spiders. Such, therefore, is the generation of these animals.

20 But the wasps which are called ichneumons, and which are less than other wasps, after they have killed the phalangia [these spiders being their prey], carry them to a wall, or something else of this kind which has a hole, and closing the hole with mud, they deposit their eggs in it, from which ichneumon wasps are generated. But some small and anonymous animals, whose wings are enclosed in a crustaceous substance as in a sheath, make small caverns of mud, either in sepulchres, or walls, 553
and in these bring forth worms [from which they are, afterwards, propagated]. The time, however, of their generation, from beginning to end, is nearly, for the most part, measured by three or four weeks. Worms, therefore, and animals in the form of worms, are, for the most

part, generated in three weeks, but oviparous insects in four weeks. The egg of these insects, also, increases, and is perfected in the first week after copulation; but in the three remaining weeks they sit on their eggs, and liberate them from the shell, as may be seen in the spider, and other insects of this kind. But the transformations happen to most in three or four days; so that the same days which are critical in diseases are also critical in the mutation of such animals.

Such, therefore, is the mode of the generation of insects; but they perish when their members are contracted and wrinkled, in the same manner as larger animals perish through old age. The wings, also, of such insects as are winged become contracted [through dryness] about autumn. But ox-flies perish through a dropsy in their eyes.

21 With respect to the generation of bees, however, there are various opinions. For some say that bees neither bring forth nor copulate, but bring their offspring, some, indeed, from the flower of the honey-suckle, but others from the flower of the *calamus*; and others assert that they bring it from the olive-tree, of which they say this is an indication, that when there is an abundance of olive-trees, then, also, the swarms of bees are most numerous. Others, again, assert, that the drones bring their offspring from some one of the above-mentioned flowers, but that the honey-bees derive their origin from the leaders [*i.e.* from the kings[†]]. Of the leaders, however, there are two genera, one of which is red, but the other is black, and more variegated. The magnitude, also, of the leaders is double that of the labouring bees, and in these, the inferior part, after the transverse septum, is to the superior part in the ratio of three to two. By some, likewise, they are called mothers, because they are thought to produce the honey-bees; of which they say this is an indication, that the drones are produced without the leaders, but the progeny of the honey-bees are not. But others assert that bees are
553 generated from copulation, and that the drones are the males, and the honey-bees the females. The generation, however, of the ordinary bees is in the cavities of the honey-comb, but the leaders are generated under the honey-comb, and remain suspended, separately, six or seven in number. The honey-bees, also, have a sting, but the drones have not; and the kings or leaders have, indeed, a sting, but they do not use it; on which account some are of opinion that they are without a sting.

† For, with respect to bees, what the moderns call the queen is called by Aristotle the king-bee, as may be seen in his Treatise *On the Generation of Animals*.

22 There are, however, many genera of bees; and the best kind, indeed, is small, round, and variegated; the second is long, and resembles a wasp; but the third is called a thief, and this kind is black, and has a broad belly. In the fourth class, also, are the drones, which in magnitude, indeed, are the largest of all, but are without a sting, and are spurious; on which account certain persons entwine something at the bottom of the hive, so as to leave room for the honey-bees to enter into it, but prevent the drones, because they are larger. As we have before observed, however, there are two genera of the leaders. But in each hive there are many leaders, and not one only; and the hive perishes if it does not contain a sufficient number of leaders; not because it would then be without rulers, but, as it is said, because the leaders contribute to the generation of the honey-bees. The hive, also, perishes if the leaders are very numerous, for they divulse it. When, therefore, the spring is late, the season is dry, and blights are prevalent, then bees are produced in less abundance. In dry seasons, however, they produce more honey; but in rainy seasons they are more prolific; and on this account it happens that at one and the same time there is a fertility of olives and of bees.

But in the first place bees form their hive, and afterwards place their offspring in them; as some say, emitting their progeny from their mouth; and this is the opinion of those who assert that they bring their offspring from a different place than the hive. Afterwards they make honey, for the purposes of nutriment, both in summer and autumn; but the honey is better which is made in autumn. The hive, also, is constructed from flowers; *but they derive their wax from the tears of trees; and their honey is that which falls from the air, and especially during the rising of the stars, and when the rainbow is extended over the earth.* And, in short, honey is not produced before the rising of the pleiades. The wax, therefore, of bees is from flowers; and that they do not make the honey, but carry away that which falls from the air, is indicated by this circumstance, that the keepers of bees find their hives full of honey in 554
the space of one or two days. And farther still, flowers, indeed, are produced in autumn; but honey is no longer produced, if, during that season, it is taken from the hives. The honey, therefore, which is now made being taken away, and no nutriment, or but little, remaining for the bees, they would again make it, if it were produced from flowers. Honey, also, [is at first liquid, but afterwards] becomes thick, and acquires a consistence; for at first it is like water, and for some days is in a fluid state. Hence, if during these days it is taken from the hive, it is not thick, but in twenty days especially acquires a consistence. It

may, also, be immediately known by the taste whether the honey is perfectly concocted; for if it is, it is thicker and sweeter. Bees, however, collect honey from all plants that have flowers in a cup, and, also, from other things which are sweet; nor do they injure any fruit, but receiving the juices of flowers by an organ resembling a tongue, they carry them to their hives. The time, also, of taking the honey from the hives is when the wild fig-tree begins to exhibit its fruits; but bees produce the best offspring at the time in which they collect their honey. They, also, carry the wax and *erithace* [which is the thicker part of the wax] with their legs, but they put the honey from their mouth into the cells. And they sit upon their offspring when they are delivered of it, in the same manner as birds.

This offspring, also, is at first a little worm, which lies at the side of the hive; but afterwards it raises itself, and is nourished, and so adheres to the hive as to seem to be detained by it. The seed, however, both of the honey-bees and drones, from which little worms are produced, is white, and these worms, when increased, become honey-bees and drones; but the seed of the kings is of a reddish colour, and its consistence resembles that of thick honey. Its bulk, also, is immediately similar to that which is produced from it. A worm, however, is not first produced from the seed of the kings, but a honey-bee is immediately seen. When bees, also, bring forth, the honey is disposed in the opposite part of the hive. Sometimes, likewise, the fœtus of bees is not perfected, and then mutilated bees are produced, *viz.* such as are without legs and wings; but when the fœtus is perfect, then the membrane bursting in which they
554 are enclosed, [perfect] bees come forth, which immediately fly. As long, also, as they are in the form of a little worm bees emit excrement, but not afterwards; unless, as I have before observed, they have not come out of the membrane in which they are enclosed. But if any one amputates the head of the fœtus before it has wings, the bees themselves devour it; and if any one plucks off the wings of a drone bee, and sends it into the hive, the bees will eat off the wings of the other drones. The extent, also, of the life of bees is six years, though there are some that live seven years. But when a hive of bees has continued nine or ten years, those that keep bees consider themselves as fortunate.

In Pontus, however, there are very white bees, which make honey twice in every month. But the bees in Themiseyra, near the river Thermodon, make their honey-comb as well in cavities of the earth as in hives. These hives, however, do not contain much, but very little wax, but the honey in them is thick, and the honey-comb is smooth and equable. Yet these bees do not make honey at all times of the year, but in winter only; for there is much ivy in this place, which flourishes in the winter; and from the flowers of the ivy they collect their honey. White and very thick honey, also, is brought to Amisum from mountainous places; and this honey is made by bees, without hives, in the cavities of trees. Honey, likewise, of this kind is made in Pontus. There are, also, bees that construct triple hives [*i.e.* distinguished by a triple order of cells], in the cavities of the earth; and these hives, indeed, contain honey, but have no worms [from which bees are produced]. Neither, however, are all hives such as these, nor do all bees make such hives.

23 But hornets and wasps make hives for their offspring, when they are without a leader, and when they wander and cannot find one. The hornets, indeed, construct their hives in some elevated place, but wasps in a hole. But when hornets and wasps have a leader, they build their hives under the earth. All the hives, therefore, of these, in the same manner as those of bees, are hexagonal; but they are not composed from wax, but from a matter which is of the nature of bark and sand. The hive, also, of hornets is much more elegant than that of wasps. But they deposit their fœtus or seed, which is of the size of a drop of liquor, in the same manner as bees, in one side of their cells, to the wall of which it adheres. The fœtus, however, is not of the same size in all the cells, but in some it is so large as to be able to fly; in some, also, there are nymphæ, [*i.e.* fœtus enclosed in membranes]; and in others there are worms. Among the fœtus, also, of hornets and wasps, the worms alone emit excrement, in the same manner as among bees; and as a long as the fœtus are nymphs, so long they remain immoveable and enveloped in membranes. Opposite, likewise, to the cell of the fœtus, as much honey as is equal in size to a drop of liquor is produced in the hives of hornets. The offspring, however, of hornets and wasps are not generated in the spring, but in the autumn, and they especially receive an apparent increase during the full moon. The fœtus, also, and the worms of these insects do not lie at the bottom of the cell, but adhere to the side of it.

24 Some silk-worms, also, in a stone, or something of this kind, construct nests of mud or clay, which terminate in an acute angle, and which they cover with plaster, so as to give them the appearance of salt. And these nests are so thick and hard, that they can scarcely be perforated with a lance. In these, likewise, they bring forth white

worms, involved in a black membrane. But, separate from the membrane, wax is produced in the clay. And this wax is much paler than the wax of bees.

25 Ants, also, copulate and bring forth; but the little worms which are their offspring do not adhere to any thing when they are born. These worms, however, as they increase, from being at first small and round, become long and distinctly formed; but the generation of these is in the spring.

26 Land scorpions, likewise, bring forth many little worms in the form of eggs, which they incubate. But the little scorpions, when they are perfect, expel and destroy their parents; which is, also, the case with spiders. And frequently about eleven in number are produced.

27 With respect to spiders, all of them, indeed, copulate in the manner already explained [in Chap. 8]. At first, however, they generate small worms; for the whole of these worms, and not a part, is changed into spiders [though animals which are produced from an egg are not produced from the whole egg, but from a part of it]. But these worms are round at first; and when the spiders bring them forth, they incubate them, and in the space of three days the worms become distinctly
555 formed. All spiders, likewise, bring forth in a web; but some in a thin and small, and others in a thick, web. And some, in short, bring forth in a round cavity; but others contain their offspring in their web, to a certain extent. Spiders, however, are not all generated at once; and as soon as they are born they leap, and begin to form a web. But if the worms of spiders are bruised, a thick and white moisture is found in them, such as is in spiders when they are just born. The spiders, however, which live in meadows bring forth in a web, the half of which is within, and the other half without the spiders themselves, and in this they incubate and produce animals. But the spiders called phalangia weave a thick web, in which they bring forth and incubate their offspring. And such spiders, indeed, as are less hideous bring forth a less numerous progeny; but the offspring of the phalangia are very numerous; and when they are adult surround their mother, kill and cast her out of the web. Frequently, also, they kill the male, if they can catch him; for the male, sometimes, incubates together with the female. From one of these spiders, likewise, a hundred small phalangia are

produced. But spiders, from being small, become perfect in about the space of four weeks.

28 Crickets, also, copulate after the same manner as other insects, the less getting upon the larger, for the male is less than the female. But the female brings forth on the ground, fixing in it the tube which she has at her tail, and which the males have not. A multitude of crickets, likewise, assembling together, bring forth at once, in the same place; so that there is, as it were, a hive of them. Afterwards, when they have brought forth, worms in the form of eggs are produced in the place of their delivery, which are enveloped in a certain thin earth, as in a membrane; and from this, when it bursts, they come forth, and fly. The fœtus, however, of the cricket is so soft that it is crushed even if it is touched. But they do not deposit their offspring on the surface of the earth, but a little below it. And when the offspring have arrived at maturity, small and black crickets come forth from that terrene substance in which they are enclosed; and afterwards, the skin bursting, they immediately become larger. They, likewise, bring forth at the end of the summer; and when they have brought forth, die. For, at the time of their delivery, worms are produced about their neck [which strangle them]; and the males die about the same time. But they come forth from the earth in the spring. Crickets, however, are not generated in mountainous places, nor in barren ground, but in plains, and crevices of the earth; for they bring forth their offspring in crevices. Their eggs, also, remain during the winter in the earth; but, together with the summer, crickets are produced from the fœtus of the former year.

29 In a similar manner, also, the *attelabi* [grass-worms, or a kind of locusts] bring forth, and having brought forth, die. But their eggs are destroyed by the autumnal rains, when they are abundant; though, if it happens to be a dry autumn, a greater number of *attelabi* are generated, because they are not similarly destroyed. For the destruction of these insects appears to be without order, and their generation to be casual.

30 Of grasshoppers, however, there are two genera; some, indeed, being small, which are the first that are seen, and the last that perish; but others are large, which also sing, and these are produced last, and are the first that perish. The grasshoppers, however, that sing, whether they are small or large, have similarly an incision at the transverse septum; but those that do not sing have not this incision. Some, also, call the large

and singing grasshoppers *achetæ*, but those that are small *cicadulæ*. Some of these small grasshoppers, also, that have an incision at the transverse septum sing a little. Grasshoppers, however, are not produced where there are no trees, on which account they are not generated in the plains of Cyrene, but there are many of them in the environs of the city; and they are especially numerous where there are olives, because these trees are not the occasion of much shadow. For grasshoppers are not produced in cold places, and hence neither are they produced in shady thickets.

The large and small grasshoppers, also, similarly copulate with each other, in a supine position. And the male inserts a part of himself into the female, in the same manner as other insects; but the private parts of the female have a fissure. Grasshoppers, likewise, bring forth in uncultivated land, excavating it with the sharp point which they have behind, in the same manner as the species of locusts called *attelabi*; for the attelabi, also, bring forth in uncultivated ground; and, on this
556 account, they abound in Cyrene. Grasshoppers, likewise, bring forth in the reeds with which they support vines, for this purpose perforating the reeds; and, also, in the stalks of the squill. These fœtus, however, fall to the ground. But they are numerous in rainy seasons. When, also, the worm of the grasshopper increases in the earth, it is called *tettigometra*, or the matrix of the grasshopper, and then the grasshoppers are most pleasant to the taste, before they burst through the bark or matrix in which they are enclosed. But about the time of the summer solstice they come forth by night, the matrix then immediately bursting, and grasshoppers are produced from the tettigometra. They immediately, however, become black, more hard and large, and begin to sing. But in both genera of grasshoppers [*i.e.* both in the larger and less kind], the males sing, and not the females. And at first, indeed, the males are sweeter food, but after copulation, the females; for they have white eggs. When grasshoppers, however, at the approach of any one, fly away, they emit a liquid like water, which husbandmen say is their urine; and they also assert, that, as they are fed with dew, they emit excrement analogous to dew. If any one, however, bending and unbending his finger, gradually approaches to grasshoppers, they will remain where they are sooner than if he immediately extends his finger. They will, also, get upon his finger; for, in consequence of seeing obscurely, they conceive the finger to be a leaf agitated by the wind.

31 All such insects, however, as are not carnivorous, but feed on the juice of living flesh, such as fleas, lice, and bugs, generate from coition what are called *lendes*, from which nothing else is again produced. But of these insects, fleas, indeed, are generated from the least putridity; for they are formed where they is dry dung; bugs are generated from the vapour which arises from the skin of animals; and lice are generated from flesh. For when lice are nearly generated, certain pustules without pus are formed, and if these are opened with a needle lice come forth. Some men, also, are afflicted with disease when there is an abundance of 557
humidity in the body, and some have after this manner perished, as they say was the case with the poet Aleman, and Pherecydes of Syria. And in certain diseases great quantities of lice are produced. There is, likewise, a kind of lice which are called fierce. These are harder than the common sort, and are with difficulty separated from the body.

Lice, therefore, are produced in the heads of children; but in a less quantity in the heads of men. Women, also, are more subject to lice than men; but those that have lice in the head are less afflicted with a pain in the head. Lice, likewise, are generated in many other animals. For birds have them, and what are called pheasants are destroyed by them, unless they roll themselves in the dust; and other birds whose wings have a quill have lice. Other animals, also, that are hairy have lice, except the ass, who has neither lice, nor ticks; but oxen have both. Sheep, however, and goats have ticks, but not lice; and swine have large and hard lice; but in dogs, what are called *cynoræstæ*, or *dog-flails*, are generated. All lice, however, are generated from those animals that have them. But animals that wash themselves, and produce lice, then generate a greater quantity of them when they change the water in which they wash. In the sea, also, lice are generated in fishes; not, however, from the fishes themselves, but from the mud. These lice too, in their form resemble wood-lice, except that they have a broad tail. There is one form, however, of marine lice, and they are generated every where, but especially in caverns. But all these animals have many feet, are exanguious, and insects. About the fins, however, of some of the tunny a horse-fly is produced, which resembles a scorpion, and is as large as a spider. But in the sea which lies between Cyrene and Egypt, there is a certain fish about the dolphin which they call a louse; and this fish is the fattest of all lice, in consequence of its being furnished with abundance of nutriment from the hunting of the dolphin.

32 Other small animals, likewise, are generated, as we have before 557
observed, some, indeed, in wool, and in such substances as are formed

from wool; as, for instance, moth-worms, which are principally produced in dusty woollen substances, but especially if a spider is enclosed in them; for if there is any moisture in the wool, the spider absorbs it, and renders the wool dry [which in this state is adapted to the generation of moth-worms]. This worm, also, is produced in garments. In old wax, also, in the same manner as in old wood, an animal is generated, which appears to be the smallest of all animals, is white, and is called *acari*. In books, likewise, other animals are generated; some of which, indeed, are similar to those in garments, but others to very little scorpions, without a tail. And, in short, in all dry substances when they become moist, and in all moist substances when they become dry, little animals are generated [on account of the change of the temperament, from which corruption follows].

There is, likewise, a small worm, which is called *xylopthorum*, or the *wood-destroyer*, which is no less singular than any of the before-mentioned insects. For a variegated head comes forth from the shell in which it is enclosed, and its feet are at one extremity of its body [*i.e.* near the head], like those of other worms. But the rest of its body is enclosed in a garment of the nature of a spider's web, which is surrounded with sprigs, so that it seems to adhere to this garment when it walks. These sprigs, however, are connascent with the garment, in the same manner as the shell in which snails are enclosed naturally adheres to the snails; nor do they fall off, but, as being connascent, can only be separated by divulsion; and if any one takes away the garment of this insect it becomes useless and dies, in the same manner as a snail when divested of its shell. In the course of time, also, this worm becomes a chrysalis, in the same manner as caterpillars, and lives without any [visible] motion. But the winged animal which is generated from it has not yet been seen.

The wild fig-tree, also, produces an insect which is called *psen*. This insect is at first a small worm; but, afterwards, the skin in which it is enclosed bursting, the *psen* leaving it flies away, enters into the fruit of the wild fig-tree, and opening it, is the cause that the fruit does not fall off. Hence, husbandmen append branches laden with wild figs near to those that are fit to eat, and plant the wild in the neighbourhood of domestic figs.

558 33 With respect, however, to sanguineous and oviparous quadrupeds, their generations, indeed, are in the spring; yet they do not all copulate at the same season; but some in the spring, others in the summer, and others in autumn, according as each finds the following season is adapted

to the generation of its offspring. The tortoise, therefore, brings forth eggs enclosed in a hard shell, and distinguished by a twofold colour, in the same manner as the eggs of birds; but, having brought them forth, she digs a hole in the earth, in which she buries them, and having made the surface of the ground smooth, she then incubates her eggs, and in the following year discloses them. But the water-tortoise, coming out of the water, excavates in the earth a ditch in the form of a tub, in which she brings forth, and leaves her eggs. After thirty days, however, have elapsed, she again digs up the earth, rapidly discloses her eggs, and immediately brings her young into the water. The marine tortoises, also, bring forth in the earth eggs similar to those of tame birds; and, burying them in the ground, incubate them at night. But they bring forth a great multitude of eggs; for they bring forth a hundred.

Lizards, also, and both land and river crocodiles, bring forth in the earth. But the eggs of the lizard are spontaneously disclosed in the earth. For the lizard does not live the space of a year; since the period of its existence is said to be six months. The river crocodile, however, brings forth many eggs, at most about sixty, and of a white colour; she also incubates them for sixty days; for the life of this animal is long. But from the very small eggs of the crocodile a very large animal is produced; for the egg is not larger than that of a goose, and the little crocodile which is born from it is in proportion to the size of this egg. It increases, however, to the magnitude of seventeen cubits. Some, likewise, say that it increases as long as it lives.

34 Of serpents, however, the viper produces an animal external to herself, having first produced an egg within herself. But the egg, in the same manner as that of fishes, is of one colour, and the shell of it is soft. The young viper, also, which is not enclosed in a hard shell, as neither are the young of fishes, is contained in the superior part of the womb. But the viper brings forth small vipers, which are enveloped in membranes, which burst on the third day [and the young vipers come forth]. Sometimes, however, it happens that the little vipers eat through the mother, and come forth. The viper, likewise, brings forth one at a time in one day; but she brings forth more than twenty little vipers. Other serpents, however, produce their eggs externally, and these eggs 558
are connected with each other, like the necklaces of women. But when they bring forth, they deposit their eggs in the earth, and there incubate them. These eggs, also, they disclose in the following year.

BOOK VI

1 After this manner, therefore, the generations of serpents and insects, and besides these of oviparous quadrupeds, subsist. With respect to birds, however, all of them are oviparous, but the season of their copulation and bringing forth their young is not the same in all. For some, as I may say, copulate and bring forth at all times, such as the hen and the pigeon. The hen, indeed, during the whole year, except two months in the winter solstice [*i.e.* in December and January]. But some hens, that are of a generous breed, are so prolific, that they lay sixty eggs before they begin to incubate them; though hens of a generous breed are less prolific than those of a common breed. The hens, however, which are called Adrianic, are, indeed, of a small size, but they bring forth every day; but they are choleric, and frequently destroy their young. They have, likewise, all various colours. Domestic hens, also, bring forth twice in the same day. And some of them on account of their very great fecundity soon die. Hens, therefore, bring forth, as I have said, continually; but the ring-dove, the turtle-dove, and the vinous pigeon [or stock-dove], bring forth twice every year, and not ten times, like domestic pigeons.

Most birds, however, bring forth in the spring, and some of them, indeed, are multiparous. But they are multiparous in a twofold respect; some, indeed, because they bring forth frequently, as pigeons; but others because they bring forth many at a time, as hens. All birds, however, with crooked claws are pauciparous, except the kestrel. And this bird is the most prolific of all those that have crooked claws; for she has been seen to lay four eggs, and she has brought forth more than four.

Other birds, indeed, bring forth in nests; but those which do not fly
559 much, as partridges and quails, do not bring forth in nests, but in the earth, collecting together twigs in which they deposit their eggs. This is, likewise, the case with the lark, and the *tetrix* [which, according to some, is a species of lark]. These birds, therefore, build their nests in the open air; but the bird which the Bœotians call *merops* is the only bird that builds its nest in caverns of the earth. Thrushes, however, form their nests from mud, in the same manner as swallows, in lofty trees, and they frame them in such successive connection with each other, as to resemble from their continuity a chain of nests. But the lapwing is the only bird of its kind which does not build a nest; but entering into the hollow trunks of trees, it there brings forth without any preparation. And the cuckoo builds its nest both in edifices and in rocks. But the *tetrix*, which the Athenians call *ourax*, neither builds its nest in the earth, nor in trees, but in low plants.

2 The egg, however, of all birds has a hard shell, if it is produced from seed, and is not corrupted; for some hens bring forth soft eggs. The eggs, also, of birds have a twofold colour; being, indeed, externally white, but internally yellow. But the eggs of birds which dwell in rivers and marshes differ from those which dwell on dry land; for in the eggs of aquatic birds there is a much greater proportion of the yellow than of the white. And the colours, also, of the eggs differ according to the genera of the birds. For of some, as of the pigeon and the partridge, the eggs are white; of others they are yellow, as of the birds that dwell in marshes; and of others, as of the turkey-hen and the pheasant, they are marked with points. But the eggs of the kestrel are red, like vermilion. There is, also, a difference in the parts of the egg; for in one part it is acute, but in the other more broad; and on its egress from the bird the broad part takes the lead. Small and acute eggs, likewise, are those from which females are produced; but males are produced from such as are round and circular towards the acute part.

Though eggs, therefore, are naturally disclosed by the incubation of birds, yet they are sometimes disclosed spontaneously, as in Egypt; where from eggs buried in the earth in dung chickens are hatched. And in Syracuse, they say that a certain toper, having put eggs under a mat in the earth, continued drinking without intermission till the eggs were hatched. Eggs, also, placed in warm vessels have been so concocted that chickens have been spontaneously disclosed from them.

The genital seed, therefore, of all birds is white, in the same manner as the seed of other animals. But the female having by copulation received the seed in the superior part of the womb, which is near the transverse septum, conceives an egg, which at first appears to be small and white, but afterwards is red and sanguineous; and when in the course of time it is increased, the whole becomes red and yellow. When, also, it becomes more thick it is distinguished by two colours; and the yellow is inward, but the white externally surrounds it. But when it becomes perfect, it is liberated from the womb [to which it was fixed], and comes forth into the light. At that time, likewise, the shell from being soft becomes hard; so that the egg in proceeding from the womb has not a firm consistence, but when it has proceeded from it, it immediately becomes firm and hard, unless it is diseased.

An egg, also, has been seen to come forth entirely yellow, so as to resemble the young bird which is afterwards produced from it. Such eggs, likewise, are found in a hen, when she is dissected, under the transverse septum; for in that place, to which the eggs usually adhere, eggs are seen having the magnitude of a perfect egg, but entirely of a yellow colour; which is considered as a prodigy. Those, however, who assert that *windy* [or addle] eggs are the relics of preceding eggs which were produced from copulation, do not assert what is true; for it has been sufficiently ascertained, that the young of hens and geese that have not yet copulated have brought forth windy eggs. But windy eggs are not so large as, are less sweet, and moister than, those that are prolific, but they are more numerous. These windy eggs, however, if they are put under the incubating bird, are not thickened in their moist part, but both the yellow and the white remain as they were before. Windy eggs, also, are produced from many birds, *viz.* from the hen, the partridge, the pigeon, the peacock, the goose, and the goose-fox.

The young of birds, however, are more rapidly hatched by incubation in summer than in winter; for hens in summer disclose their young in eighteen days; but in winter they are sometimes five-and-twenty days before they are disclosed. Birds, also, differ from birds in this, that some are more adapted to incubation than others. But if it happens to thunder during the incubation of a bird the eggs are corrupted. The eggs, however, which are called by some *cynosura* [or the urine of a dog] and *urina*, are produced in summer more than in winter. But windy eggs are denominated by some *zephyria* [or the eggs of zephyr], because birds in the spring are seen to receive the western gales. A like effect, however, is produced by titillating the vulva of birds with the hand. Windy eggs, also, may become prolific, and those which have been now produced from coition are changed into another genus, if before the yellow is separated from the white the female which has windy eggs copulates with the male, or is pregnant from the seed of another bird; for then the windy eggs become prolific, and the matter which was formed by the prior male subsists according to the nature of the male with which the female afterwards copulated. But if the female copulates after the yellow is separated from the white, then the copulation contributes nothing to the fecundity of the windy egg, or to that which was conceived by a prior copulation, passing into the genus of the other male with which the female is afterwards connected. If, also, coition is intermitted when the eggs have begun to be formed, they do not at all increase; but if copulation again takes place, they rapidly arrive at their proper magnitude.

The yellow and the white of an egg, however, have a nature contrary to each other, not only in colour, but, also, in power. For the yellow acquires a firm consistence by the cold, but the white does not, but is rather liquefied by cold. By fire, however, the white acquires a firm consistence, but the yellow does not, but continues to be soft, unless it is burnt; and it acquires a firmer consistence, and becomes more hard when boiled, than when exposed to a naked fire. The yellow and the white, also, are separated from each other by a membrane.

But the globules which are seen near the beginning of the yellow do not contribute any thing to the generation of the young, as some are of opinion they do. These globules, also, are two; one, indeed, in the inferior, but the other in the superior, part of the egg. It happens, likewise, with respect to the yellow or yolk and the white of an egg, that if any one takes several of these out of the shell, and boils them in a saucepan over a gentle fire, all the yolks will come together into the middle, and the whites will surround them in a circle.

The young, however, of hens first bring forth immediately on the commencement of spring; and they bring forth a more numerous progeny than older hens, though the offspring of the younger are of a less size than those of the older. In short, if birds do not incubate, they die or become diseased. Among birds, also, the females after copulation are seized with terror, shake their feathers, and frequently scatter small pieces of straw; and sometimes they do the very same thing when they bring forth. But the female pigeon at that time [*i.e.* after copulation] raises her tail; and geese plunge themselves in the water. Most birds, however, are rapidly pregnant, both with windy and prolific eggs, as is particularly evident in the partridge, when she is stimulated to copulation. For if she snuffs the male she becomes pregnant, and is immediately useless for fowling; since the partridge appears to possess the sense of smelling in an exquisite degree.

The generation of the egg, however, after coition, and again the generation of the infant bird from the egg when concocted, are not in all birds accomplished in an equal space of time, but the times differ according to the magnitude of the generating birds. The egg of the hen, indeed, acquires a consistence after copulation, and is for the most part perfected in ten days; and the egg of the pigeon is completed in a somewhat less time. The female pigeon, also, has the power of retaining the egg in the womb, even at the time of parturition. For if she experiences any disturbance, either with respect to her nest, or from the loss of a feather, or is afflicted from any other circumstance, she retains her egg, and defers her delivery to a more convenient time. Pigeons, likewise, have this peculiarity with respect to copulation, that the male will not get on the female unless they have first mutually kissed each

other. The senior pigeon, indeed, at first copulates, having previously kissed, but afterwards he is connected with the female without kissing; but younger pigeons never copulate without a kiss preceding. This, therefore, is their peculiarity; and still farther, when the male is not present, the female pigeons get on each other, in the same manner as the males get on the females; and though the one emits nothing into the other, yet they conceive and bring forth more eggs, from which, indeed,
561 no young one is produced, but all such eggs are windy.

3 The generation, however, from an egg, as has been before observed, happens, indeed, to all birds after the same manner, but the times of its arriving at perfection differ. In hens, therefore, after three days and three nights have elapsed, indications of the future bird present themselves; but in larger hens these indications appear more slowly, and in those that are less more swiftly. In this time, therefore, the yellow ascends to the acute part of the egg, where the principle of the egg subsists, and the egg is disclosed; and then, in the white of the egg, the heart, like a sanguineous point, presents itself to the view. But this point, as if animated, leaps and is moved; and from it, when it is increased, two veiny avenues, sanguineous and convolved, tend to each of the surrounding tunics. About the same time, also, a certain membrane appears, distinguished by sanguineous fibres, and which arising from the two above-mentioned avenues of the veins [surrounds the white of the egg]; and shortly after, the body of the young bird may be distinguished, which at first is very small and white. But the head is very manifest, and the eyes especially are inflated; and do not till after a long time decrease, and so contract themselves as to have a due proportion with the other parts of the body. At first, also, in the inferior part of the body no member is seen distinct from the rest. But of the two avenues which extend from the heart, the one tends to the membrane which surrounds the white, but the other to the yellow, performing as it were the office of a navel. The principle, therefore, of the infant bird is from the white, but its nutriment is derived from the yellow, through the navel.

On the tenth day, however, the whole infant bird, and all its parts, become manifest. But it still has the head larger than the rest of the body, and the eyes larger than the head, not as yet possessing sight. The eyes, also, about this time being prominent, are larger than beans, and black; but if the skin of them is taken off, a white and cold liquor is seen, which is very splendid when exposed to the light, but has nothing of solidity.

Such, therefore, is the manner in which the eyes and head are disposed. 561
At the same time, too, the viscera, the belly, and the nature of the intestines, become apparent; and the veins which are seen to extend to the heart now become situated near the navel. From the navel, also, two veins are extended, one to the membrane which contains the yellow; and the yellow at this time is liquid, and more so than is natural to it; but the other vein extends to the membrane which contains the whole membrane in which the infant bird is enclosed, and, also, the membrane of the yellow, and the liquid which is between these. For the infant bird gradually increasing, one part of the yellow becomes situated in the superior and the other in the inferior place; but in the middle the white liquid is situated. The white, however, as at first, is situated under the inferior part of the yellow. But on the tenth day the whole of the white subsides under the yellow, and then the white is in a small quantity, is viscous and thick, and is of a yellowish colour. For each of the parts within the egg is arranged after the following manner: In the first and last part of the egg contiguous to the shell there is a twofold membrane, the one native, and conjoined to the shell, the other adventitious, and produced when the infant bird begins to be formed. This second membrane, also, is under the first, and in it the white is contained, and, also, the infant bird, together with the membrane in which it is enclosed, and which separates it, lest it should immediately touch the white, or the liquid. Under the infant bird, also, is the yellow, to which one of the before-mentioned veins extends, but the other vein tends to the surrounding white. Another membrane, however, contains all these [*viz.* the infant bird, the yellow, and the veins], together with a moisture which is of the nature of lymph. Afterwards, as has been before observed, another membrane surrounds the embryo itself, and separates it from the liquid. And under this is the yellow, involved in another membrane, to which the navel, originating from the heart and the great vein, extends; so that the embryo is not in either of the liquids [*i.e.* is not in either the white or the yellow].

On the twentieth day, however, the infant bird begins to chirp, and if any one separates the shell and touches it, it will move itself in a small degree. Now, also, when twenty days have elapsed after incubation, and the bird is disclosed, it is covered with soft feathers. But it has its head above the right leg, near the flank; and the wings are above its head. At this time, also, the chorion, or membrane, which is after the last

562 membrane of the shell, and to which one of the navels extends, becomes apparent; and in this the infant bird is then wholly situated. In a similar manner, also, the other membrane becomes apparent which surrounds the yellow, and to which the other navel is extended. Each navel, however, originates from the heart and the great vein. At this time, likewise, [*i.e.* about the twentieth day] the navel, which extends to the membrane that surrounds the white, is loosened from the animal, and falls; but the other navel, which extends to the yellow, adheres to the infant bird by a thin intestine, and through it much of the yellow passes to the young bird, and a yellow sediment is emitted in the belly. About this time, likewise, the young bird discharges its excrement towards the external chorion. And both in the belly of the bird and external to it there is a certain white excrement. But, at length, the yellow always becoming less is wholly consumed by the bird; so that if it is cut open ten days after it is disclosed, a little of the yellow will be found still remaining in the intestine. The bird, however, is now liberated from the navel, and nothing intervenes; but the whole of the moisture is consumed. During the before-mentioned time, also, the young bird sleeps, but is sometimes excited from sleep, looks up and chirps; and the heart is inflated, together with the navel, as if respiring.

After this manner, therefore, the generation of birds from an egg subsists. Some birds, also, bring forth unprolific eggs, and these produced from copulation. But no bird is generated from these by incubation. This, however, has been observed particularly to happen with pigeons.

Double eggs, however, have two yolks, which sometimes that they may not be confounded are separated by a thin interstice of the white; and sometimes the two yolks are in contact with each other without this interstice. There are, also, some hens that bring forth all their eggs double, so that in these, also, the above-mentioned circumstance happens respecting the yolk. For a certain hen having brought forth eighteen double eggs, disclosed two chickens from each, those eggs excepted
562 which were unprolific. Two chickens, also, were disclosed from each of the double eggs, but one of the chickens was larger than the other. But the last chicken that was disclosed was a monster.

4 All birds, however, of the genera of pigeons, such, for instance, as the ring-dove and the turtle-dove, bring forth two eggs; but the most which the turtle and ring-dove bring forth are three eggs. And the pigeon, indeed, brings forth, as we have before observed, at all seasons,

but the ring and turtle-dove bring forth in the spring, and not oftener than twice. When the former eggs, likewise, which they have produced are destroyed, they lay the second eggs; for many birds destroy their eggs. The ring and turtle-dove, therefore, as I have said, sometimes bring forth three eggs, yet they never disclose more than two young ones from them; and sometimes they only disclose one; but the remaining egg is always unprolific. Most birds, however, that are but one year old generate nothing. But all birds, when they once begin to bring forth, always contain eggs in the womb; though in some birds the eggs are so small that they can scarcely be seen. The pigeon, also, for the most part, produces a male from one egg, and a female from the other; and, for the most part, she produces the male prior to the female. Besides this, likewise, after she has brought forth in one day, she omits many days, and again brings forth another egg. The male, however, incubates in a part of the day; but the female at night; and the young bird which was produced from the first of the eggs is disclosed, and flies within twenty days. But pigeons perforate the egg the day before they disclose it; and both the male and female warm [with their vivific heat] the young for a certain time, in the same manner as they warm their eggs. With respect to rearing the young, also, the female conducts herself with more severity than the male, in the same manner as other animals after delivery. They, likewise, bring forth ten or eleven times in a year; and in Egypt twelve times. But both the male and female pigeon copulate within the year in which they are born; for they copulate in six months after their birth. Some, however, say that the ring-dove and the turtle-dove copulate and generate when they are three months old, of which they consider the multitude of them as an indication. Their gestation, likewise, continues for fourteen days, and their incubation for the same number of days; but in another fourteen days the young are covered with feathers, and fly so much that they cannot easily be caught. It is, also, said, that the ring-dove lives forty years, and partridges more than 563
sixteen years. But the pigeon having hatched her young, brings forth again in the space of thirty days.

5 The vulture, however, builds her nest in inaccessible rocks; and on this account the nest and the young of the vulture are but rarely seen. Hence, Herodotus, the father of Brysson the sophist, says, that vultures come from another land, unknown to us; of which, he says, this is an indication, that no one has seen the nest of the vulture, and that many of them suddenly appear following armies. Though it is difficult,

however, to see their nest, yet it has been seen. Vultures, also, bring forth two eggs. All carnivorous birds, therefore, accord in this, that they are not seen to bring forth more than once in a year; except the swallow, which, though a carnivorous bird, builds her nest and brings forth twice in a year. And if any one pierces the eyes of the offspring of swallows while they are yet young, they will afterwards be restored to a sane condition, and again possess the power of seeing.

6 The eagle, however, brings forth three eggs, but discloses two of these, as it is said in the verses which are ascribed to Musæus:

"Three eggs she lays, two hatches, and one bird she rears."

For the most part, therefore, this happens to be the case; yet three young have been known to be disclosed from the eggs of the eagle. But when the two young ones are grown up, the eagle ejects one of them, in consequence of being wearied with their nurture. At this time, also, it is said she remains without food, because she does not then seize the whelps of wild beasts. Her nails, also, are then distorted for a few days, and her wings become white; so that at that time eagles become cruel to their offspring. But the osprey receives and nurtures the ejected young eagle.

The eagle, also, incubates her eggs for thirty days; and as much time as this is, also, spent by other large birds in incubation, as by the goose and the bustard; but by birds of a middle size, as the kite and the hawk, twenty days are employed in incubation. The kite, however, for the most part, brings forth two eggs, and sometimes she discloses from them three infant birds. But the kite called *Aetolius* sometimes brings forth four eggs, and discloses four young. The raven, however, not only brings forth two eggs, as some say, but many. She, also, incubates them twenty days, and expels the young from her nest. This very same thing is, likewise, done by other birds. For frequently those birds that bring forth many young expel one of them. All the genera of eagles, however, are not similarly disposed with respect to their offspring, but the eagle called *pygargus*[†] is cruel, and the black eagle is benevolently disposed, towards her young. For, in short, all birds that have crooked claws expel their young from the nest as soon as they are able to fly, and compel them by blows to depart. Nearly, likewise, most other birds, as we have before observed, do the same thing; and except the crow,

† So called from having a white tail.

having nourished, take no farther care of their young. The crow, however, pays attention to them for a certain time; for she nourishes them even when they are able to fly.

7 With respect to the cuckoo, some are of opinion that this bird is a transformation from the sparrow-hawk, because about the time in which the cuckoo is seen the sparrow-hawk, which resembles the cuckoo, disappears. About the time, however, in which the cuckoo begins to sing, not only the sparrow-hawk is not to be seen, but this is also the case with every other species of hawk; except that they are only visible for a few days. But the cuckoo is seen for a short time in summer, and disappears in winter. The sparrow-hawk, however, has crooked claws, but the cuckoo has not. Farther still, neither does the head of the cuckoo resemble that of the sparrow-hawk, but both its head and its claws are more similar to those of the pigeon, and it alone resembles the sparrow-hawk in its colour, except that the colour of the latter is variegated like lines, but that of the cuckoo like points. The cuckoo, also, in its magnitude and flying resembles the smallest species of the sparrow-hawk, which, for the most part, is not seen at the time in which the cuckoo appears; though both of them have been seen at the same time. The cuckoo, also, has been seen to be devoured by the sparrow-hawk; though no birds of the same kind are known to do this. It is, likewise, said, that no one has ever seen the young of the cuckoo. The cuckoo, however, brings forth, indeed, but does not build a nest; and sometimes she brings forth in the nest of lesser birds, the eggs of which birds she eats. She more particularly, however, brings forth in the nest of the ring-dove, and eats the eggs of this bird. For the most part, 564
likewise, she brings forth one egg, but very seldom two. She, also, brings forth in the nest of the hen linnet, who incubates the egg, and nurtures the young of the cuckoo. But the cuckoo at the time in which she brings forth is most fat, and her flesh is sweet as food. The young, likewise, of sparrow-hawks are very fat, and their flesh is very sweet. But there is a certain genus of sparrow-hawks who build their nest on steep rocks.

8 A great number of birds, however, incubate, as we have observed concerning pigeons, the males alternately succeeding the females in this employment during the time in which the female is engaged in the search of food. Among geese, however, the females alone incubate, and when they have begun to do this, continue sitting without intermission.

The nests, also, of all birds that live in marshes are constructed in swampy and grassy places, on which account, being undisturbed while they incubate, they are able to impart a certain nutriment to their young, and are not entirely without food. Of crows, also, the females alone incubate, and continue sitting on their eggs without intermission. But in the mean time the males nourish the females, bring them food, and feed them with it. With respect to pigeons, however, the female beginning to incubate in the evening continues in this employment the whole night, till the hour of breakfast; and the male incubates the remaining time. But partridges make two receptacles of eggs; and in one of these the female, but in the other the male, incubates. Each, also, discloses its young, and as soon as they are disclosed copulate with them.

9 The peacock, however, lives five-and-twenty years, and especially begins to generate at the age of three years; at which time, also, it receives a variety of plumage; and discloses its eggs in thirty or a few more days. But it brings forth once only in a year, and lays twelve, or a somewhat less number of eggs. It, likewise, omits two or three days in bringing forth, and does not lay its eggs in an uninterrupted succession; but its first progeny do not consist of more than eight eggs. Peacocks, also, bring forth windy or unprolific eggs. But they copulate
564 in the spring; and bring forth immediately after copulation. They, also, lose their feathers at the same time that the leaves fall from trees, and they again begin to be replumed when the trees are in blossom. Those persons, likewise, that rear peacocks put their eggs under hens to be incubated, because if the female peacock incubates her own eggs, the male flying on the female breaks them. On this account, also, among certain wild birds, the females [when the time of bringing forth is near], flying from the males, lay and incubate their eggs [secretly]. Two eggs, however, alone of the peacock are put under a hen, because the hen can only disclose two eggs of the peacock by incubation. Care, likewise, is taken that there is no intermission in the incubation of the hen, and that for this purpose she is supplied with food.

Birds, however, at the time of copulation have their testicles manifestly larger; those, indeed, that are more salacious have them more conspicuously larger, as cocks and partridges; but those that do not continually copulate have them less manifestly increased.

10 After this manner, therefore, the conception and generation of birds subsist. But with respect to fishes, that all of them are not

oviparous has been asserted before, for the cartilaginous kind called selache are viviparous. They first, however, conceive an egg in themselves, and nourish it in themselves, the marine frog excepted. Fishes, also, have different wombs, as we have before observed. For those that are oviparous have wombs situated in the inferior part, and divided into two cavities; but the selache have wombs more similar to the wombs of birds. They differ, however, from those of birds, because birds contain their eggs in the superior part of the womb, *i.e.* under the transverse septum; but certain of the selache contain their eggs in the middle part of the womb, near the spine of the back; in which place the eggs, when increased, are transferred to another part of the womb.

The egg, likewise, of all fishes is not of a twofold colour, but of one colour only. But it is more white than yellow, both at first and when it contains a young fish. The generation of fishes, however, from an egg differs from that of birds, because fishes have not a second navel [or umbilical chord] extending to the membrane which is under the shell, but they have in the same manner as birds the avenue which extends to the yellow of the egg. But in other respects the generation of fishes from an egg is similar to the generation of birds. For both birds and fishes are produced in the supreme part of the egg, and the veins in both similarly extend first from the heart; and the head, the eyes, and superior parts are, at first, in both, similarly the largest. As the fœtus, however, 565
increases, the egg always becomes less, and at length disappears and enters into the interior parts of the little fish, in the same manner as the whole of the yellow is absorbed by the infant bird. The navel, also, adheres a little below the entrance of the belly. And the navel in fishes while they are young is long, but as they increase it becomes less, and diminishes till at length, as was before observed in birds, it wholly hides itself within the fœtus. Both the embryo, also, and the egg are contained in a common membrane, and under this there is another membrane which particularly contains the embryo; but between the membranes there is a liquid. The nutriment, also, of the little fishes exists in the belly, in the same manner as the nutriment of infant birds, and is partly white, and partly† yellow.

With respect to the figure of the womb, however, the mode of its subsistence must be surveyed from dissections. But there is a difference in the wombs with respect to themselves, as in those of the sea-dogs with respect to each other, and, also, with respect to the wombs of the

† *i.e.* The nutriment of fishes is white, but of birds yellow.

broad fishes. For in some, as in the little sea-dogs, as we have before observed, the eggs adhere to the middle of the belly, near the spine; but when increased they are liberated from the place to which they adhere; and as the womb is divided into two cavities, the eggs when liberated descend into each inferior part of the womb, and adhere to the transverse septum. The womb, however, both of these and other sea-dogs, a little below the transverse septum, has, as it were, white mammæ, which are not apparent, except when these animals are pregnant. The small sea-dogs, therefore, and thornbacks have in their womb certain testaceous substances which contain an egg-formed liquid; but the figure of this shell resembles the tongue of a flute; and within the shell there are capillary tubes. In those little sea-dogs, therefore, which some call *colts*, when the shell bursts and falls the young ones come forth; but thornbacks bring forth their young enclosed in the shell, and when the shell bursts the young fish emerges. The thorny sea-dog, however, contains its eggs in the superior part of the womb, near the transverse septum, and above the [before-mentioned] mammæ of the womb; but when the egg descends, and is liberated, then a young one is produced
565 from it. The generation of the sea-fox is, also, effected after the same manner.

The sea-dogs, however, that are called smooth have their eggs situated in the middle of the womb, in a manner similar to the small sea-dog; and afterwards the eggs descend into each of the cavities into which the womb is divided. An animal, also, is then generated, which through the navel adheres to the matrix, so that the egg being consumed the embryo appears to subsist in a manner similar to that in which quadrupeds are contained in the womb of their mother. But the navel [or umbilical chord] being long, adheres [by one of its extremities] to the inferior part of the matrix, as if depending from an acetabulum, but by the other extremity adheres to the middle of the embryo, or to that part in which the liver is contained. On dissecting the embryo, likewise, nutriment of the nature of an egg will be found, though the egg no longer remains; but a chorion and peculiar membranes envelop each of the embryos, in the same manner as in quadrupeds. The embryos, likewise, have their head at first in the superior part of the womb; but when they are larger and perfect, they have their head in the inferior part of the womb. Males, however, are generated in the left part of the womb, and females in the right, and sometimes both males and females are generated in the same part. When embryos, also, are dissected, the viscera, such as the liver, *etc.* in the same manner as those of quadrupeds, are found to be large and sanguineous. All the selache, however, have a multitude of eggs in the superior part of the womb, near the transverse septum, which eggs in some are larger, and in others less; but in the inferior part of the womb they contain the embryos. Hence fishes of this kind are thought to bring forth and copulate frequently in a month, because they do not bring forth all that they have conceived at once, but frequently, and for a long time; for while they conceive eggs in the superior part of the womb, the fœtus is matured and arrives at perfection in the inferior part of it.

Other sea-dogs, therefore, and, also, the file, and the torpedo or cramp-fish, emit their offspring from themselves, and again receive them into themselves. A large cramp-fish, also, has been seen which contained in itself eighty embryos. But the thorny dog is the only one of the sea-dogs which does not receive its offspring within itself, in consequence of the thorn [with which the young are armed]. Of the broad fishes, also, neither the pastinaca nor the scate receive their young within themselves, on account of the roughness of their tail. Nor does the marine frog receive its young, on account of their thorns, and the magnitude of their head. For, as we have before observed, the marine frog is the only fish of the cartilaginous kind that is not viviparous.

11 Such, therefore, is the difference of these fishes with respect to each other, and such is the generation of them from eggs. The males, 566
however, at the time of copulation have their ducts so full of genital seed, that when compressed white seed flows externally from them. But these ducts or avenues are bifidated, and originate from the transverse septum and the great vein. At the time of copulation, therefore, the ducts of the males are manifestly enlarged, though at other times they can scarcely be seen. For in some fishes they sometimes become entirely immanifest, as we before observed concerning the testicles of birds. The spermatic ducts, however, of the males differ from those pertaining to the womb in females in other respects, and in this, also, that the ducts of the males adhere to the kidneys, but those of the female are easily moved and involved in a thin membrane. The mode, however, in which the ducts of the males subsist must be surveyed from anatomical descriptions.

Superfetation, also, pertains to the selache, and the longest period of their conception is six months. But that sea-dog which is called the star-fish brings forth most frequently; for it brings forth twice in a month. It, also, begins to copulate in the month of September. Other sea-dogs, however, bring forth twice in a year, the little sea-dog excepted; and this fish brings forth once in a year. All of them, likewise, bring forth in the spring; but the file-fish brings forth the second time in autumn, towards the setting of the winter pleiades; and the first time, in the spring. Its second progeny, also, is more abundant than the first. But the cramp-fish brings forth in autumn. The selache, likewise, bring forth on the land, ascending for this purpose from the sea, and the profundities of the sea, both on account of warmth and the greater security of their offspring. With respect to other fishes, therefore, none are seen to copulate with any other than those of their own kind; but the file-fish and the scate alone appear to do this; for there is a certain fish which is called *rinobatus*, or the *file-scate*. For it has the head and anterior parts of the scate, but the posterior parts of the file, as being formed from both these fishes.

12 The sea-dogs, therefore, and those of the same species, such as the dog-fox, the dog properly so called, and those broad fishes, the cramp-fish, the scate, the smooth scate, and the pastinaca, producing an egg
566 within themselves, are after this manner viviparous. The dolphin, also, the whale, and such other cetaceous animals as have not gills, but a pipe through which by respiring they draw the air, are viviparous; and this is, also, the case with the *pristes*, and the sea-ox. For no one of these is seen to have eggs, but each has immediately a fœtus, from which when evolved an animal is generated, in the same manner as man and viviparous quadrupeds. The dolphin, however, for the most part, only brings forth one fœtus, and sometimes two; but the whale, for the most part, and frequently, brings forth either two or one. And the *phocæna* brings forth similarly with the dolphin; for it resembles a small dolphin, and is produced in the Pontic sea. The *phocæna*, however, differs from the dolphin; for its magnitude is less, its back is broader, and it is of an azure colour. Many, also, say that the *phocæna* is a certain species of the dolphin. All fishes, however, that have a pipe respire; for they have lungs, and receive the air through the pipe by respiring. And the dolphin has been seen to sleep and snore, with his pipe raised above the water. The dolphin, likewise, and the phocæna have milk, with which they nourish their young; and they, also, receive their offspring when little within themselves. But the offspring of the dolphin increase rapidly, for in ten years they acquire a perfect magnitude. The female dolphin, also, is pregnant during the space of ten months; but she brings forth in the summer, and not at any other season. At the rising of the

dog-star, however, it happens that dolphins are concealed for thirty days. They, also, follow their young for a long time; for the dolphin is an animal attached to its offspring. It, likewise, lives many years; for it is certain that some dolphins have lived twenty-five, and others thirty years. For fishermen, having cut off the tails of certain dolphins, have afterwards returned them to the sea, and by this circumstance have been able to ascertain their age.

The phoce, however, or sea-calf, is an amphibious animal; for it does not, indeed, receive water [like fishes], but respires, and sleeps, and brings forth on the land, but near shores, as ranking among the number of pedestrious animals. It lives, however, the greatest part of its life in the sea, from which, also, it derives its nutriment; on which account, it must be considered in conjunction with aquatic animals. The phoce, therefore, immediately conceives an animal in itself, brings it forth enveloped in a chorion, and has milk in the same manner as a sheep. 567
But it brings forth one or two, and at most three, offspring. It, likewise, has two mammæ, and is suckled by its young after the manner of quadrupeds. It brings forth, also, like a woman, at every season of the year, and especially at the time in which goats first bring forth. But about the twelfth day after delivery it brings its offspring to the sea, and this it does frequently every succeeding day, that it may gradually accustom them to that element. The young phocæ, however, do not walk, because they cannot support themselves on their feet, but they creep along in a prone position. The phocæ, also, can collect and contract itself, for it is fleshy and soft, and has cartilaginous bones. It is, likewise, difficult to kill the phoce violently, unless it is struck on the temple; for the body of it is fleshy. But it emits a voice resembling that of an ox. The vulva of the female, also, is similar to that of the scate; and all such animals have a vulva resembling that of a woman.

13 After this manner, therefore, aquatic and viviparous animals generate, and bring forth their offspring. But oviparous fishes, as we have before observed, have their womb bifidated, and situated in the inferior part. All fishes, however, that have scales are oviparous; such, for instance, as the wolf, the mullet, the cephalus or pollard, and the *etelis*; and all those that are called white, together with those that are smooth [because they are without scales], except the eel. Their egg, also, is sandy. It appears, however, to be so, because the whole womb of these fishes is full of eggs, so that in small fishes there appears to be only

two eggs; for the womb in them is immanifest, on account of its smallness and thinness.

Concerning the copulation, therefore, of all fishes, we have before spoken. In most fishes, however, there is a distinction of male and female; but it is dubious whether there is this distinction in the roach and chamæ, for all that are caught of these are found to have eggs. Fishes, therefore, that copulate generate eggs from their copulation. But they, also, have eggs without copulation; and this is evident in some river fishes. For the *phoxini*, which are very small, have eggs, as I may say, as soon as they are born.

All fishes, however, scatter their eggs [through the sea], and, as it is
567 said, the males devour the greater part of the eggs, and some perish in the water; and those alone are preserved which are brought forth and deposited in opportune places. For if all the eggs were preserved, fishes of every kind would be too numerous. Many, also, of those which are preserved are not prolific, but those only which the male sprinkles with his genital seed. For this purpose, however, the male follows the female; and when the female has brought forth, the male sprinkles the eggs with his seed. From all such eggs, likewise, as he sprinkles little fishes are produced; but they are produced from other eggs, just as it may happen. The very same thing, likewise, happens in the mollia. For the male cuttle-fish, when the female brings forth, sprinkles the eggs with his seed; which, also, may reasonably be supposed to happen in the other mollia, but at present is only seen to take place in the cuttle-fish. Fishes, also, bring forth their eggs near the land; gudgeons, indeed, in the cavities of stones, except the egg which is brought forth is broad and sandy. Other fishes, likewise, bring forth in a similar manner. For places near the land are warmer, and afford more abundant nutriment; and the offspring in such places are not so liable to be devoured by the larger fishes. Hence the greater part of fishes in the Pontic sea bring forth near the river Thermodon; for this place is tranquil and warm, and abounds with sweet water. But other oviparous fishes bring forth once in a year, except the small *phycis*; and this fish brings forth twice in a year. The male phycis, however, differs from the female in this, that it is blacker, and has larger scales.

Other fishes, therefore, bring forth their eggs through the avenue or mouth of the womb; but the fish which some call the needle, at the time of bringing forth bursts, and thus the eggs come forth. For this fish has under its belly a certain fissure resembling the fissure of the blind serpent. When the needle-fish, however, has brought forth, it lives, and its belly which was burst again coalesces.

The generation, however, from an egg happens similarly, both in those fishes that produce an egg inwardly and in those that produce it externally. For the fœtus is generated in the summit of the egg, and is enclosed in a membrane, and the eyes, which are large and spherical, are the first parts that are manifest; by which it is evident, that fishes are not, as some say, generated in a manner similar to those animals which are produced from worms. For, on the contrary, in the latter the inferior parts are at first larger, but the head and the eyes are increased 568
afterwards. When the egg, likewise, bursts the little fish appears, just as the nut of an almond is seen when the shell is broke. And at first, indeed, the little fish receives no nutriment, but is increased from the moisture produced from the egg. Afterwards, however, it is nourished with river water, till it is sufficiently increased. But when the Pontic sea purifies itself, a certain substance is carried into the Hellespont, which they call *phucos* [or sea-weed], and which is of a yellow colour. Some, however, say that this is the flower of the *phucos*, from which a paint is made. It is, also, produced at the beginning of the summer. Shell-fish, and the little fishes which live in these places, are, likewise, nourished with it. And some persons that live near the sea say, that the purple-fish derives from this *phucos* its flower.

14 The fishes, however, of lakes and rivers, for the most part, as I may say, conceive in the fifth month of their age; and all of them bring forth within a year. But the females of these, in the same manner as marine fishes, do not emit all their eggs at once, nor do the males discharge the whole of their seed, but the females have always a greater or less number of eggs, and the males a greater or less quantity of genital seed. The females, also, bring forth in an appropriate season; the carp, indeed, five or six times in a year, and especially at the rising of certain stars; but the *chalcis* brings forth thrice; and others once in a year. All of them, however, bring forth on the sides of rivers and lakes, and in places abounding with reeds; as, for instance, the phoxini and perches. But the glanis and the perch bring forth eggs, adhering to each other like a necklace, in the same manner as frogs. So connected, indeed, and entwined are their eggs, that fishermen roll the eggs of the perch round the reed to which their line is fixed.

The larger glanii, therefore, bring forth in deep lakes, and some of them in lakes six feet in depth; but the smaller glanii bring forth in lakes

less deep, and especially near the roots of the willow, or some other tree, and, also, among reeds and moss. Sometimes, also, the large glanii copulate with the small, and applying those avenues to each other which
568 some call navels, the female emits eggs, and the male his genital seed. Such eggs, also, as are mingled with the genital seed immediately become whiter, and, as I may say, in the same day appear to be larger. But in a short time after, the eyes of the little fish are manifest; for these in all fishes, in the same manner as in other animals, are immediately most manifest, and appear to be the largest of all the parts. Those eggs, however, which have not been sprinkled with the seed of the male are useless and unprolific. But when the little fishes are increased from prolific eggs, they are purified as it were from a shell; and this is a membrane which envelops both the egg and the little fish. When, however, the genital seed is mingled with the egg, the egg becomes very viscous and glutinous, and, therefore, adheres to the roots of herbs or other substances in the places where it is brought forth. Where the greatest number of eggs, also, are laid, the male glanis guards them, in the absence of the female that brought them forth. The increase, however, of the glanii from eggs is most slow, on which account the male sits as a guardian of the eggs for thirty or forty days, in order that the progeny may not be devoured by fishes that may happen to meet with them. But the generation of the carp is next to that of the glanis in slowness; and, in a similar manner, its eggs, after they are brought forth, are guarded by the male. The eggs, also, of some smaller fishes so rapidly increase, that in the space of three days they become little fishes. But the eggs which have been sprinkled with the seed of the male receive an increase, both in the day in which they were sprinkled, and, likewise, afterwards. The eggs of the glanis, therefore, increase to the size of the grains of the bitter vetch; but those of the carp, and such like fishes, to the size of a grain of millet.

These fishes, therefore, bring forth and generate after this manner. But the *chalcis* brings forth its eggs thick, and collected together. And the fish which they call *psylon* brings forth its eggs near shores, in places exposed to the wind. This fish, also, collects its eggs together in one place. But the carp, the *balin*, and all other fishes, as I may say, are impelled into shallow water for the purpose of bringing forth; and frequently thirteen or fourteen males follow one female. But when the female after having brought forth her egg departs, the males sprinkle it
569 with their genital seed. Most of their eggs, however, perish; for, because the female as she passes through the water lays her eggs, it is necessary

that the eggs being dissipated by the stream should perish, except those which adhere to some substance they may meet with. For no fish besides the glanis is a guardian of the eggs of the female except the carp, and this fish only then guards them when it finds them collected into one place. All male fishes, however, have genital seed, except the eel; and this fish has neither genital seed, nor an egg. Mullets, therefore, ascend from the sea into lakes and rivers; but eels, on the contrary descend from lakes and rivers into the sea.

15 Most fishes, therefore, as has been said, are generated from eggs. Some, however, are generated from mud and sand; and there are some of this description even among those which are produced from copulation and eggs, which they say sometimes happens to be the case both in other muddy places and in the pools about Cnidos, which become dry through the heat occasioned by the dog-star, so that even all the mud remains without moisture. These places, however, begin to be filled with water with the first rain that falls, and as soon as the water commences, little fishes are generated in it. But these little fishes are a certain species of mullets, which are not produced from copulation, and are of the size of small *mœnidia*; but in these there is neither any egg nor genital seed. In the rivers in Asia, also, where they flow into the sea, other small fishes are generated after the same manner as the before-mentioned, of the largest of which they make sauce.

Some, however, assert that all mullets are generated spontaneously from putrid matter. [But this is false]; for the females of them are seen to have eggs, and the males genital seed. There is, however, a certain genus of them which is spontaneously generated from mud and sand.

That some fishes, therefore, are generated spontaneously, neither from eggs nor from copulation, is evident from what has been said. All such fishes, however, as are neither oviparous nor viviparous are generated either from mud or sand, or any putrid matter which floats on the surface of the water. Such, for instance, is the fish *aphye*, which is called foam, and is generated from sandy earth; and this fish is without increase, and is unprolific. After a long time, also, it perishes, and 569
another is again produced in its stead. Hence, except during a short time, the *aphyæ* are generated in every other season of the year; for they continue to be formed from the autumnal arcturus to the spring. But there are indications that they are sometimes produced from the earth [which is at the bottom of the sea]; for they are not caught by fishermen if it is cold, but only when it is warm, in consequence of ascending from

the earth at the bottom of the sea, allured by the tepid heat. If the earth, also, at the bottom of the sea is agitated, it frequently happens that the aphyæ are caught, and are then better and more numerous. But other aphyæ are inferior to these in consequence of receiving a rapid increase.

The aphyæ, likewise, are produced in shady and warm places, at those times in which the earth becomes warm through the serenity of the weather. Hence they are produced in Athens, at Salamis, and near that part of the Piræus which bears the name of Themistocles, and in Marathon; for in these places foam is generated. But they are seen in such like places, and during such like serenity of weather. Sometimes, likewise, they are generated when there happens to be an abundance of rain, in the foam which is produced by the rain water; and hence they are denominated *foam*. Sometimes, also, they are borne along on the surface of the sea, when the weather is serene, and then they are involved in the foam from which they are produced, in the same manner as small worms in the dung from which they originate. Hence the aphyæ are borne along to many places together with the sea; and they are generated and caught in greater abundance when the year happens to be moist and warm.

Other aphyæ, however, are the progeny of fishes. For the aphye which is called *cobites* is produced from small putrid gudgeons, which descend into the earth. But the aphyæ which are called *arades* are produced from the *phalerice*. From these, also, the *trichides* are generated; and from the *trichides*, the *trichiæ*. From one aphye, likewise, such as that which is found in the port of the Athenians, those little fishes are generated which are called *encrasicholi*. But there is another aphye, which is the progeny of the mœnides and mullets. The aphye, however, which is called foam, and which is unprolific, is moist and continues but for a short time, as was before observed; for at length
570 nothing is left but its head and eyes. But fishermen that they may carry it away with them sprinkle it with salt; for by this means it continues for a longer time.

16 Eels, however, are neither generated from copulation, nor do they bring forth eggs, nor has any eel ever been caught which had either any genital seed or an egg. In the dissection of eels, also, no avenues have ever been found destined to the reception or emission of seed, or to performing the office of a womb; but among sanguineous animals, the whole genus of eels is neither produced from copulation nor from eggs.

It is, however, evident that this is the case; for in some muddy lakes, all the water being exhausted, and the mud extracted, when the rain falls, eels are again generated. But in dry seasons, they are not generated even in perennial lakes; for they live and are nourished by rain-water. That they are neither generated, therefore, from copulation nor from eggs is evident. To some, however, eels appear to have the power of generating, because small worms are ingenerated in certain eels; for they fancy that eels are generated from these worms. This, however, is not true; but they are generated from what are called the entrails of the earth; which are spontaneously formed in mud, and in moist earth. For some eels have been seen proceeding from these entrails of the earth, and other have been found in them when dissected. In the sea, also, and in rivers substances of this kind [*i.e.* such as the entrails of the earth] are generated, especially where there is putrefaction. They are generated, indeed, in the sea, in places where the *phucos* is found; but in rivers and lakes they are generated near the banks; for in these places the bland heat causes putrefaction. Such, therefore, is the generation of eels.

17 All fishes, however, do not bring forth at the same time, nor in a similar manner, nor are all of them pregnant for an equal time. Prior to copulation, therefore, shoals of male and female fishes assemble, but at the time of copulation they so join themselves together, that for the most part one male is united to one female. Some of these, also, are pregnant for not more than thirty days; but others for a less time. All of them, however, continue pregnant during times which may be divided into hebdomads. But those fishes are pregnant the longest time which some call *marini*. The *sargus*, indeed, copulates in the month of 570
December; but is pregnant during thirty days; and among the genera of mullets, that which some call the lip, and, also, the muxon, are pregnant at the same season and during an equal time with the sargus. All fishes, likewise, are in pain during the time of their pregnancy; hence at this period especially they fall on the land, for they are furiously hurried along towards the shore; and, in short, at this time they are in continual motion till they bring forth; and this is especially the case with the mullet. But when they have brought forth they are tranquil. Many fishes, however, cease to bring forth when little worms are generated in their belly; for small and animated substances are produced in them, which expel the fœtus [before it is perfect, and cause an abortion].

Those fishes, indeed, that are denominated *ruades* bring forth in the spring; but most bring forth during the vernal equinox. Others, however, do not bring forth in the same season of the year, but some in the summer, and others during the autumnal equinox. But among fishes of this kind, the first is the *arista*, or ear-of-corn, which deposits its offspring near the earth; and the cephalus is the last. This is evident from the progeny of the arista being first seen, and that of the cephalus being seen the last. The mullet, also, is among the first that bring forth; but the stock-fish, in most places, brings forth at the beginning of summer, and in some places in autumn. The *alopias*, also, which they call *anthias*, brings forth in summer. But after these are the gold-fish, the wolf, the *mormurus*, and, in short, those that are called *runners*. The last, however, of those that live in a shoal are the barbel, and the coracinus;[†] and these bring forth in autumn. But the barbel deposits its eggs in mud; on which account it is late in bringing forth, for the mud is cold for a long time. The coracinus, however, is later than the barbel, and brings forth among sea-weeds, because it lives in stony places. It is, also, pregnant for a long time. But the *mœnides* bring forth after the winter solstice. Most other fishes, however, that dwell in the deep sea bring forth in summer; of which this is an indication, that at this time they are not caught.

The *mœnis*, likewise, is the most prolific of fishes; and of the selaches the marine frog; but they are rare because they easily perish. For the marine frog brings forth all its eggs at once, and deposits them near the
571 earth. In short, the selache are less prolific [than oviparous fishes], because they are viviparous. But these are especially preserved on account of their magnitude.

The fish, also, called the needle is among the number of those that are late produced; and many of them, prior to their bringing forth, are burst by the eggs they contain. The womb, however, of this fish is burst, not so much on account of the number, as on account of the magnitude of the eggs; and the progeny of the needle are scattered round their parent, after the manner of the spiders called phalangia. For as soon as they are brought forth, they adhere to the mother; and if they are touched, they run away. But the fish called the ear-of-corn brings forth by rubbing its belly on the sand.

The belly of the tunny, likewise, bursts through its fat; of which fishermen conceive this to be an indication, that when the female tunnys fail in one year, the tunnys, also, fail in the following year. The tunny,

† A blackish sea-fish, having a head shining like gold.

also, appears to be older by a year than the *pelamys.*[†] But the tunny and the scombrus, or mackerel, copulate in the month of February, and bring forth about the beginning of June. The eggs, also, which they bring forth are, as it were, in a sack. But the increase of the tunny is rapid. For when these fishes bring forth in the Pontic sea, what some call *scordylæ*, but the Byzantines *auxidæ*, are generated from the egg; because they are increased in a few days; and the *scordylæ*, indeed, come forth from the Pontic sea, together with the tunny, in autumn, but return to the sea in spring, being now *pelamides*. Nearly, indeed, all other fishes receive a rapid increase; but all in the Pontic sea increase more rapidly. For every day many hamiæ are manifestly increased. In short, it is necessary to think that the same fishes have not in different places the same time of copulating, being pregnant, bringing forth, and receiving increase; since those fishes that are called coracini, in some places, bring forth about the time of getting in the wheat. In what has been said, however, we have directed our attention to what most usually happens.

The conger, also, generates and contains in itself a fœtus; but this is not in all places similarly manifest; nor is the fœtus very apparent, on account of the fatness of the conger. It contains, however, in its womb eggs connected with each other in an oblong series, in the same manner as the eggs of serpents are connected. And though the eggs of the conger cannot be clearly seen on account of the fat in which they are immersed, yet if fire is applied to the fat, the eggs may be sufficiently distinguished. For the fat is consumed and evaporates; but the eggs leap, and, being separated, crepitate. The eggs, also, of the conger may be distinguished by the touch. For if any one handles the conger and rubs it with his fingers, he will feel that the suet is smooth, and the eggs 571
rough. Some congers, therefore, have suet alone, but no eggs; and others, on the contrary, have no suet, but an egg of such a kind as I have just now mentioned. And thus I have nearly considered all other animals that bring forth an egg, as well the aquatic as the winged and terrestrial; and have, likewise, spoken concerning their copulation, pregnancy, and generation, and other particulars similar to these.

18 Let us now direct our attention to what pertains to such pedestrious animals as are viviparous, and, also, to man. We have already, therefore, considered what pertains to the copulation of all

† A fish produced in the Hellespont, of the genus of the tunny.

animals, both peculiarly and in common. But it is common to all animals to be vehemently stimulated by the desire of, and pleasure arising from, coition. Females, therefore, when they have brought forth are at first more ferocious [through love of their offspring]; but the males are ferocious at the time of copulation. For horses at this period bite other horses, and throw and pursue their riders; and wild boars are most ferocious, though at this time they are most imbecil from the effect of copulation. They, also, then fight with each other in a wonderful manner, arming themselves for this purpose, and hardening their skin by rubbing it against trees, and frequently defiling themselves with mud, and then suffering it to dry. They, likewise, fight with each other so fiercely, separating themselves from the herd to which they belong, that both the combatants frequently die.

After the same manner, also, bulls, rams, and goats are affected. For those among these that at other times feed gregariously in concord fight at the time of copulation, becoming discordant with each other. But the male camel at the time of coition is ferocious, whether a man or another camel approaches him; for he is always hostile to the horse. The like, also, happens to savage animals. For bears, and wolves, and lions, at this time become more ferocious to those that approach them; but they are less hostile to each other, because no animal of this kind is gregarious. The she bear, likewise, becomes more savage from the presence of her young, in the same manner as bitches from the presence of their puppies. Elephants, also, at the time of copulation become fierce and unmanageable. Hence they say, that the Indians do not permit their
572 elephants to copulate. For becoming furious at this time, they throw down the habitations of the Indians, which are badly constructed, and commit other ravages. They say, however, that an abundance of nutriment renders them more mild, and that other elephants are introduced to them who are ordered to strike and tame them.

Those quadrupeds, however, that frequently copulate, and not at one season of the year, such for instance as those animals that live with man, as swine and dogs, are less ferocious at the time of copulation, in consequence of the abundant connection of the male and females with each other.

Female quadrupeds, likewise, are impelled with an ardent desire of copulation, and especially a mare, and in the next place a cow. Mares, therefore, have the *hippomania*, or *an insane desire of coition*; hence, also, libidinous women are metaphorically said to have the hippomonia. Mares, also, at this time are said to become prolific by the wind; and hence in Crete they do not separate the stallions from the mares. When mares, however, are impelled with this desire of copulation, they run away from the other horses. This passion, also, is that which in some animals is called *pruriency* [το καρπιζειν]. But they run neither towards the east nor the west, but towards the north or south. When, also, they are under the influence of this passion, they do not suffer any one to approach them, unless they are either fatigued by labour, or have arrived near the sea; but then they emit a certain humidity which is called hippomanes, in the same manner as that substance which a colt recently born carries in its forehead. And this resembles what is called *capria* [or the virus which flows from the genital parts of a sow]. This, likewise, is especially sought after by all those who are employed in sorcery. At the time of copulation, also, they incline to each other more than before; they, likewise, move their tail frequently, and neigh in a manner more unusual than at another time. A certain moisture, also, flows from their genital parts resembling seed, but much thinner than the seed of the male; and some call this *hippomanes*, though this *hippomanes* differs from that which is produced in colts. But they say it is difficult to collect this moisture, because it flows gradually. At that time, likewise, mares make water frequently, and sport with each other. And such are the particulars respecting horses.

Cows, also, are inflamed with an ardent desire of bulls. But the passion with which they are possessed is so vehement, that cowherds can neither manage them, nor lay hold of them. Both mares, however, and 572
cows plainly show when they are stimulated to copulation; for at this time the genital parts of cows swell, in the same manner as those of mares. Farther still, cows get on the backs of bulls, and always follow and are present with them. But the younger females both among mares and cows are the first that are impelled to copulation. And when the weather is serene, and their bodies are in a good condition, they are more impelled than at other times.

Mares, therefore, when their hair is cut have less of this impulse to copulation, and become more sad. But male horses know the females with which they associate by their odour, though they began to feed together but a few days prior to copulation. And if other mares should be mingled with the troop, the horses bite and expel them, and each horse feeds with the mare of his choice, separate from the other mares. They, likewise, give to each mare thirty stallions, or a few more. And if another horse approaches, the first horse, running in a circle about his own mares, fights with him; and if any mare is allured by this other horse, he bites and restrains her. But the bull at the season of copulation feeds with the cows, and fights with the other bulls. At other times, however, they feed with other bulls separate from the cows, which is called a contempt of the herd. For frequently the bulls in Epirus are not seen for three months. And, in short, all savage animals, or the greatest part of them, do not feed together with the females prior to the season of copulation; but when they become adult, the males are separated, and feed apart from the females.

Swine, also, when they are impelled to copulation, which is called *subation*, attack even men themselves. But a passion of this kind in bitches is called *catulition*. In the females, likewise, when they are stimulated with lust, the genital parts swell and become wet; but mares at this period distil a certain white moisture.

Females, likewise, are subject to the purifications of the menses, yet these purifications are in no female animal so abundant as in women. In sheep and goats, therefore, at the season of copulation, a certain
573 menstrual moisture is seen, which, also, flows for some time after copulation, and then ceases till they are about to bring forth. But then it again begins to flow, and shepherds know by this token that the time of their parturition is at hand. After delivery, however, an abundant purification follows, which at first is not very sanguineous, but is so afterwards; and the cow, the she-ass, and the mare emit a greater quantity of this menstrual moisture, on account of their magnitude, yet less in proportion to their bulk than sheep and goats. The cow, therefore, when stimulated to copulation has but a small purification, since she only emits as much menstrual moisture as will fill the half of a cup called *cotylus*,† or a little more. But the time in which cows are most adapted to copulation is when they are purified by the emission of the menstrual moisture.

A mare, however, brings forth the most easily of all quadrupeds, is the most free from menstrual purgations, and emits the least flux of blood, in proportion to the magnitude of her body.

But the menses are especially an indication of pregnancy in cows and mares, and they cease for the space of two, four, or six months; yet the failure of the menstrual flux cannot be known by those who are not very much accustomed to observe them, on which account some are of opinion that mares and cows are deprived of this purgation. Again, the females of mules have not this menstrual moisture, but the urine of the

† This cup contained nine ounces.

females is thicker than that of the males. In short, therefore, the excrement of the bladder is thicker in quadrupeds than men; and the females of sheep and goats discharge a urine thicker than that of the males. But, on the contrary, the urine of all she-asses is thinner than that of the males; and the urine of the ox is more acrid than that of the cow. All quadrupeds, however, after they have brought forth their young emit urine thicker than others that have a more abundant menstrual purgation. Animals, also, that begin to have milk at the time of coition, at that period emit a milk which resembles pus, and which is useless; but their milk becomes useful after they have brought forth. Female sheep and goats, likewise, and in a similar manner cows and all other quadrupeds, are more fat, and eat a greater quantity of food when they are pregnant. Moreover, for the most part, animals are more impelled to copulation in the spring. All of them, however, do not copulate at the same time of the spring, but in times adapted to the nourishment of their offspring.

Domestic female swine, therefore, are pregnant during four months, but the greatest number of offspring which they bring forth is twenty; though when they bring forth many, they are not able to nourish all of them. When they are old, also, they in a similar manner bring forth, but they copulate more slowly. They, likewise, become pregnant from 573
one copulation; but repeatedly copulate, because after copulation they emit that moisture which is called by some *subation*. This, therefore, happens to all female quadrupeds; but some females emit together with this seed. But during the gestation of the sow, any one of her offspring that happens to be vitiated and mutilated in the magnitude of its body is called *melachoiron*; and this may happen in any part of the womb. When the sow, also, has brought forth, she gives her first nipple to her first-born.

When sows, likewise, begin to be impelled to coition, they must not be suffered to copulate till their ears are pendant; for if they are not, they copulate in vain, and do not become pregnant. But sometimes a sow when she copulates, impelled by venereal desire, becomes pregnant, as we have before observed, from one copulation. It is, also, useful to the sow when she is about to copulate with the male swine, to give her barley, and to give her boiled barley after she has brought forth. The first progeny, likewise, of some sows are fine pigs; but other sows as they are more advanced in age generate a good progeny and fine pigs. But some say that a sow, for the most part, soon dies, if she loses one of her eyes. Most sows, however, live about fifteen years, and the life of some is but a little short of twenty years.

19 But sheep, and, in a similar manner, she-goats, become pregnant, indeed, in three or four copulations; yet if it happens to rain after they have copulated, they bring forth immaturely. Most of them, however, bring forth two, sometimes three, and sometimes even four offspring. Both sheep and goats, likewise, are pregnant for five months, and hence in some places that are warm, and mild, and abound with pasture, they bring forth twice [in the same year]. But the goat lives about eight, and the sheep about ten, years. For the most part, however, they do not live so long as this, the leaders of the flock excepted; for these live fifteen years. For in every flock of sheep there is a leader of the males, who, when the shepherd calls him by his name, precedes the flock. But the shepherds accustom him to do this when he is very young. The sheep, however, in the vicinity of Ethiopia live twelve or thirteen years; and the goats ten or eleven years.

Both the sheep and the goat, likewise, copulate as long as they live. And the females of these species of animals bring forth twins, in consequence of being well fed, and when either the male or female sheep or goat happens to be a twin. They, also, are generated either males or
574 females from the water which they drink; for some waters are prolific of females, and others of males. And this, likewise, arises from their copulations. In a similar manner, also, if sheep and goats copulate when the north wind blows they generate males; but females if the south wind blows. Such, indeed, is the power of the wind, that it causes those sheep or goats which were naturally accustomed to bring forth females to bring forth males, if they copulate when the north wind blows. It is necessary, however, that the females when they copulate should look to the north. But the sheep which were accustomed to copulate in the morning will not suffer the rams to copulate with them in the evening.

The lambs, also, are white or black, if there are white or black veins under the tongue of the ram; they are white, indeed, if these veins are white, but black if they are black; both, if the veins are both black and white; and red if they are red. But the sheep which drink salter water copulate prior to others that do not. It is necessary, however, that sheep should drink salt-water, not only before they copulate, but, also, before and after they have brought forth, and again in the spring.

But shepherds are not accustomed to appoint a leader of goats [as they do of sheep], because goats are naturally unstable, and are quick and alert. With respect to sheep, however, if the elder are impelled to copulation at the regular season, the shepherds say it is a sign that it will be a good year for sheep; but if the younger are thus impelled, that it will be a bad year for sheep.

20 Of dogs, however, there are many genera. But the Laconic dogs copulate in the eighth month of their age; and about this time, some of the males of this species begin to lift their leg when they make water. A bitch, also, becomes pregnant from one copulation; and this especially becomes evident in those bitches that copulate furtively; for having once copulated they become pregnant.

The Laconic bitch, however, is pregnant during a sixth part of the year; sixty-one or two, or three days at most. And her puppies when they are born are blind for twelve days. Having brought forth, also, she again copulates in the sixth month, and not before. Some bitches, however, are pregnant the fifth part of the year; *i.e.* seventy-two days. But the puppies of these bitches are blind during fourteen days. And there are some bitches that are pregnant a fourth part of the year; *i.e.* three whole months; but the puppies of these are blind for seventeen days. Bitches, also, for an equal space of time appear to possess an ardent desire of copulation. They have the menses, likewise, at this period for seven days; and at the same time it happens that their genital parts swell. At this time, however, they do not suffer the male to copulate with them, but in the seven following days. For they appear 574
for the most part to possess an ardent desire of copulation during the whole fourteen days; and some are under the influence of this passion for seventeen days. A purification, also, proceeds from the puppies, and the bitches that brought them forth, at one and the same time. But this purification is thick and phlegmatic; and bitches after delivery are purified in a quantity less than is analogous to the size of their body.

Bitches, also, have milk for the most part five days before they bring forth; though some have milk seven or four days prior to this period. The milk, likewise, is useful [for the purposes of nutrition] immediately after they are delivered of their young. But the Laconic bitch begins to have milk in thirty days after copulation. The milk, therefore, at first is thick, but in process of time becomes thinner. The milk, also, of bitches is thicker than that of other animals, except that of the sow and the hare.

There is, likewise, an indication in bitches when they are of an age adapted to copulation; for in these, in the same manner as in woman, there is a swelling and a cartilaginous hardness in the mammæ. Yet because these indications are obscure, they cannot easily be detected, except by such as are skilful; for the indication possesses no magnitude. This, therefore, happens to the females among dogs, but none of these indications happen to the male. The males, also, for the most part, elevate their leg in making water when they are six months old; but some of them do this when they are eight months, or even before they are six months old. For, in short, they do this when the begin to be robust. All bitches, however, make water sitting; though some have been seen to raise the leg when they made water. Bitches, also, bring forth at most twelve puppies; but generally five or six; and a bitch has been known to bring forth one only. Laconic bitches, however, for the most part bring forth eight. The females and males, likewise, of this species of animals copulate as long as they live. But it happens to be peculiar to Laconic dogs, that when they are exercised with labours they are more able to copulate than when they are idle.

Of Laconic dogs, likewise, the male lives ten years; but the female about twelve years. But of other dogs, the greater part live fourteen or fifteen years; and some live twenty years. Hence some think that Homer acted rightly in making the dog of Ulysses to die in his twentieth year. Among Laconic dogs, therefore, because the males labour more than the females, the latter live longer than the former; but in other dogs this is not very manifest; though, at the same time, the males live longer than the females.

Dogs, also, do not lose their teeth, except those that are called the canine teeth; and these both the females and males lose when they are four months old. But in consequence of dogs losing these teeth only, a difference of opinion has arisen among certain persons. For, because they lose only two teeth, some say that they do not lose any; for it is difficult to meet with these. But others, when they see these teeth, are, in short, of opinion that they, also, lose other teeth. The ages of dogs, also, are known from their teeth. For those that are young have white and sharp teeth, but those that are older have black and obtuse teeth.

21 The bull, also, renders the cow pregnant from one copulation, and copulates with so much violence, as to oblige the cow to bend herself. But if the impulse of the bull should be inefficacious, the cow after an interval of twenty days again admits copulation. Older bulls, therefore, do not cover the same cow frequently in the same day, but with some intermission. But younger bulls, in consequence of their vigour,

copulate with the same cow frequently, and also with many cows. The bull, however, is less lascivious than other male animals. He, also, copulates after having vanquished other bulls; and it frequently happens, that the vanquished bull again attacks the victor now debilitated by coition, and subdues him. Bulls and cows, also, begin to copulate, so as to generate in the first year of their age; but it rarely happens that bulls generate when they are one year, or eight months old. It is, however, acknowledged that they generate when they are two years old. Cows, also, are pregnant during nine months, and bring forth in the tenth month. But some contend that a cow is pregnant ten months, a few days excepted. Whatever fœtus, likewise, is produced before the above-mentioned time is abortive, and does not live long; for the horns of its feet are soft and imperfect. For the most part, also, the cow brings forth but one calf, and rarely two.

She, likewise, copulates and brings forth as long as she lives. But she lives for the most part about fifteen years; and bulls live for the same number of years, if they are castrated. Some cows, however, live twenty or more years, if they have a good habit of body. The keepers of oxen, also, appoint castrated bulls to be the leaders of the herd, in the same manner as shepherds appoint leaders of a flock of sheep; and these leaders of the herd live longer than the rest, because they are exercised by labour, and have better food. Oxen, likewise, are most vigorous when they are five years old; and hence some consider Homer as speaking rightly when he says,

> The male was five, the ox was nine years old;

for both are capable of effecting the same thing.

But the ox loses its teeth in the second year of its age, yet not all of them, but those only which the horse loses. And when they are attacked with the gout, they do not lose the horn of their feet, but the feet only very much swell.

Cows, likewise, when they bring forth have useful milk; but prior to this they are without milk. The milk, also, which is first produced, when it is congealed, becomes as hard as a stone; but this happens when it is not mingled with water. Heifers, likewise, that are not a year old do not copulate, unless something prodigious takes place; yet some have been found to copulate when they were but four months old. For the most part, also, they begin to copulate in the months of April and May; though some neither copulate nor are pregnant before autumn. But when many cows are pregnant, and very much admit copulation, it is considered as a sign of tempest and rain. Cows, likewise, have menstrual purgations, in the same manner as mares, but in a less quantity.

22 With respect to the horse, however, the male begins to copulate, and, also, the female, at the age of two years; yet it rarely happens that they copulate at this period, and when they do their progeny are smaller and more imbecile. For the most part, however, both male and female begin to copulate at the age of three years; and their offspring continually become better, till they are twenty years of age. The mare, also, is pregnant eleven months, but brings forth in the twelfth month. The horse, likewise, fills the mare, not in any fixed number of days, but sometimes in one, or two, or three days, and sometimes in a greater number of days. But the ass more rapidly fills the mare than the horse; and the copulation of horses is not laborious, like that of bulls. Both the male and female horse, likewise, are the most lascivious of all animals, after man. Younger horses, however, when their nutriment is good, and in abundance, copulate before the usual age. A mare,
576 therefore, is for the most part uniparous; nevertheless, she sometimes brings forth two, but never more. A certain mule, also, brought forth two, which was considered as a prodigy. A horse, therefore, copulates in the thirtieth month of his age; yet so that he alone generates colts of a good quality when he has ceased to lose his teeth. Some horses, however, as they say, have been known to fill mares while losing their teeth; such horses not being naturally unprolific.

The horse, therefore, has forty teeth, but loses the four primary teeth in the thirtieth month of his age, *viz.* two from the upper and two from the lower part. But when another year has elapsed, [*i.e.* in the forty-second month], he loses four other teeth, after the same manner, two from the upper, and two from the lower part; and again at the expiration of another year [*i.e.* about the fifty-fourth month] he loses four others, after the same manner. But when four years and six months have elapsed, he no longer loses any teeth. It has, however, happened, that a horse when he has lost his first teeth has at the same time lost all of them; and that he has lost all his teeth at the same time that he has lost his last teeth. But things of this kind happen but rarely; so that, for the most part, a horse is especially useful for the purposes of generation when he is four years and six months old.

Older horses, however, are more prolific, the older than the younger males, and the older than the younger females. Horses, likewise, cover their mothers, and, also, their daughters; and then the stud appears to be perfect, when they copulate with their own offspring. But the Scythians ride the pregnant mares, when the embryo more rapidly turns itself in the womb, and they say that this occasions their delivery to be more easy. Other quadrupeds, therefore, bring forth lying down, and on this account the embryos of all of them come forth on the side; but a mare, when the time of her delivery is near, erects herself, and sends her offspring into the light standing.

Most horses, likewise, live about eighteen or twenty years, and some live five-and-twenty or thirty years; and if any one pays attention to them, their life may be extended to fifty years. But the longest life of horses is, for the most part, thirty years; and the female generally lives five-and-twenty years; though some mares have been known to live forty 576
years. The males, however [in general] live a shorter time than the females, on account of copulation [by which the males are debilitated]; and those horses that are nurtured privately do not live so long as those that are in a stud. The female, therefore, in five years arrives at perfection in length and height; but the male in six years. But in the six following years they increase in bulk, and make a proficiency till the twentieth year of their age. The females, however, arrive at perfection prior to the males; but in the womb the males are more rapidly perfected than the females, in the same manner as in the human species. The very same thing, also, happens in other animals that are multiparous.

They, also, relate, that when six months have elapsed the mule no longer sucks from the mammæ of the mare, because if it should suck after this period it would tear and thus produce pain in the teats of the mother; but they say that a horse sucks his mother for a longer time. The horse and the mule, likewise, are vigorous after the loss of their teeth; but when they have lost all their teeth, it is not easy to know their age. Hence they say that the age of horses may be known before the loss of their teeth, but not after the loss of them. In short, however, their age may be especially known after the loss of their teeth by the canine tooth. For in those that are seven years old, this tooth is small in consequence of attrition from the bridle; but in those that are not seven years old, this tooth is large, indeed, but more obtuse, and in young horses it is sharp and small.

The male, also, copulates in every season of the year, and as long as he lives; and the female, likewise, copulates as long as she lives, and in every season of the year, unless some bond, or any other necessity, prevents her. There is not, indeed, any definite time of the copulation of the

male and female; yet they are not able from any casual copulation to rear the offspring which they have generated. In Opus, therefore, [a city of Locris], a stallion belonging to the stud copulated when he was forty years old; but it was necessary to assist him by raising his fore legs. Mares, also, begin to copulate in the spring. But a mare, after she has brought forth, is not immediately filled with a new offspring, but some time after; and if three or four years intervene, she is delivered of a better offspring. After delivery, however, it is necessary that she should
577 abstain from coition for at least one year, and act as if she were young in the business of copulation. A mare, therefore, brings forth by intervals, as we have before observed; but a mule continually. Of mares, likewise, some are entirely without offspring, but others conceive, indeed, yet cannot carry in the womb what they have conceived, so as to educe it with life into light. But they say that an indication of things of this kind is obtained from the embryo; in which, if it is dissected, other substances resembling the kidneys are found about the kidneys, so as to cause the appearance of four kidneys.

When a mare, likewise, brings forth, she immediately devours the chorion, and, also, eats the excrescence which grows on the forehead of the colt, and which is called *hippomanes*. But the magnitude of this excrescence is less than that of a small fig, and it is of a broad form, round and black. This *hippomanes* if any one plucks from the colt, the mother excited by the odour of it rises, and is rendered furious by the smell. And hence sorceresses search for and collect this hippomanes. If, however, an ass copulates with a mare, the mare corrupts the inherent embryo.

The keepers of studs of horses, likewise, do not constitute any one horse the leader of the rest, after the manner of a herd of oxen; because the nature of horses is not stable, but ardent and alert.

23 Both the male and female ass, also, copulate in the thirtieth month of their age, and at that time lose their first teeth; but they lose their second teeth, and, also, their third and fourth teeth, in the sixth month; which fourth teeth are called *gnoma* or *gnomons*. An ass, however, has been known to be pregnant when but one year old, and to bring forth and rear her offspring. After copulation, likewise, she discharges the seed she had received from the male, by making water, unless she is prevented; and hence the keepers of asses strike and pursue the females immediately after they have copulated. The ass, also, brings forth in the twelfth month; and, for the most part, she is delivered of only one

fœtus; for she is naturally uniparous; but sometimes she brings forth two. An ass, therefore, by copulating with a mare corrupts, as we have before observed, the seed which she received from the horse; but the horse does not corrupt the seed of the ass, when a mare has previously had connection with an ass. A pregnant ass, likewise, begins to have milk in the tenth month; and on the seventh day after her delivery again suffers coition, and on this day especially is filled from the male, and conceives. She is, also, filled, if she copulates after the seventh day. But if she does not bring forth before she has lost the teeth called *gnomons*, she no longer becomes filled from the male, nor is she pregnant during the remainder of her life. The ass, likewise, is not willing to bring forth in the sight of man, nor in the light; and, therefore, the keepers of asses lead them into dark places when they are about to bring forth. She, also, brings forth during the whole of her life, if she brings forth before she has lost the *gnomons*.

But the ass lives more than thirty years, and the female lives longer than the male. When, however, a horse copulates with an ass, or an ass with a horse, they are much more subject to abortion, than when the males and females of the same species mingle with each other, as for instance, a horse with a mare, and a male with a female ass.

The time of gestation, also, when a horse and an ass are mingled together is regulated by the male. My meaning is, that the mare is pregnant for the same length of time as the nature of the ass requires to its perfection. But the magnitude, form, and strength of the body are more assimilated to the female than the male. If, also, the males and females of different species continue to copulate without interruption, the females become rapidly unprolific. Hence the keepers of these animals do not suffer those of a different species to be mingled with each other without intermission, but after a certain time has elapsed. The mare, likewise, and the she-ass, do not willingly admit a male of another species, unless he has been nourished with the same milk; and hence their keepers take care, that some of the progeny of the ass may be suckled by the mare, and some of the progeny of the mare by the ass. But asses copulate with mares by violently covering them in the place where they feed, in the same manner as horses.

24 The mule, however, covers and copulates with the female after the loss of his first teeth; and when he is seven years old he, also, fills the female. When he covers a mare, also, he begets a young mule. But afterwards he no longer covers her. The female mule, likewise,

sometimes becomes pregnant, though she is unable to bring the fœtus she has conceived to perfection. The she-mules, however, in Syria, above Phœnicia, both copulate and bring forth; but this genus of mules is different from those of which we are speaking, though it resembles them. What are, likewise, called *ginni* are generated from a mare when she is diseased during her pregnancy, in the same manner as dwarfs are produced in women, and that imperfect fœtus which is called *metachoiron* in sows. The ginni, also, in the same manner as dwarfs, have large genital parts.

But the mule lives many years. For a mule has been known to live eighty years; as was the case with a mule at Athens, when the temple was building. This mule, though exempt from servitude on account of his age, yet being yoked to a car, used to assist in drawing it; so that it
578 was decreed, that no one should drive him away when he approached to any heap of corn. The female mule, however, grows old more slowly than the male. Some, also, say that the female mule is purified by the discharge of urine, and that the male by the smell of urine becomes more rapidly old.

25 Such, therefore, are the generations of these animals. But those who pay attention to quadrupeds are able to distinguish the young from the old. For if the skin when stripped from the lip is quickly restored, the quadruped is young; but if it remains for a long time shrivelled, the quadruped is old.

26 But the female camel, indeed, is pregnant ten months, and always brings forth only one offspring; for this animal is uniparous. The keepers, also, of camels separate from them an offspring that is one year old. The camel, likewise, lives a long time; for it lives more than fifty years. But the female brings forth in the spring, and then has milk till she again becomes pregnant. Both the flesh, also, and the milk of the camel, are more agreeable to the taste than those of other animals. But the milk of the camel is drank by mingling with it a double or triple quantity of water.

27 With respect to the elephant, however, both the male and female begin to copulate before the twentieth year of their age. But when the female has copulated, she is pregnant, as some say, for a year and six months; but according to others, for three years. The cause, however, of this difference of opinion, with respect to the time of her pregnancy,

arises from not surveying the time in which she copulates. But the female is delivered, lowering herself in the posterior part of her body, and with manifest signs of pain. And her young one when it is born sucks with its mouth, and not with its proboscis. It, also, walks and sees as soon as it is born.

28 Female wild boars, also, copulate at the beginning of winter. But they bring forth in the spring, retiring for this purpose into the most inaccessible places, and such as are particularly steep, narrow, and umbrageous. But the male for the most part, associates with the female thirty days. The number, however, of the offspring of the female boar, and the time of her pregnancy, are the same as those of domestic swine. To the voice of these, also, their voice is similar, except that the voice of the female boar is more frequently exerted than that of the male; for the male employs his voice but rarely. Wild boars, likewise, when they are castrated become larger and more fierce, as Homer, also, says:

> A gelded boar in dreadful woods she rear'd,
> The top resembling of some woody mount,
> With umbrage dark; nor like a corn-fed brute.

Boars, also, while they are young, after a manner castrate themselves; for labouring under an itching of the testicles, they destroy their testicles by rubbing them against trees.

29 But the female stag copulates, for the most part, retiring under the male, as we have before observed; for she frequently cannot support the efforts of the male, on account of their vigour. Sometimes, however, the females suffer the males to copulate with them after the manner of sheep; and when they are impelled with the desire of coition, they decline associating with each other. The male, likewise, is not constant to one female, but after a short time leaves her, and associates with other females. The time, also, in which they copulate is after arcturus, in the months of August and September.

Moreover, the female is pregnant eight months; but she becomes pregnant in a few days, and many females are filled from one male. For the most part, also, they bring forth but one fawn; though some few have been seen to bring forth two. They, likewise, bring forth near the public roads, through the fear of wild beasts. But the increase of the fawns is rapid. The menstrual purgations, also, do not happen to the female stags at other times; but when they are delivered, they have a phlegmatic purification. They, likewise, accustom the fawns to thickets. But this is a place to which stags retire, and is a steep rock which has only one entrance; in which place, also, they are accustomed to defend themselves, and repel invaders. With respect, however, to the life of the stag, it is fabulously reported to be long, but there does not appear to be any clear evidence in support of this assertion. For neither the gestation of the females, nor the increase of the fawns, affords any indications of a long life.[†] But in a certain mountain of Asia, which is called Elaphus, in the island Arginussa, where Alcibiades died, all the female stags have a cut ear; so that they are known by this circumstance, if they happen to migrate to another place. The embryo, also, of such stags in the womb have immediately this mark. The female stag, likewise, has four papillæ, in the same manner as the cow.

When, however, the females are filled, the males are separated from
579 them; and each from the impulse of venereal desire lives by himself alone, digs cavities in the earth, and produces a fetid smell resembling that of a goat. Their face, also, from the sweat that runs down it at that time, becomes black, like the face of a goat. And in this manner they live till the rainy season. But after this they return to their accustomed pasture. These things, however, happen to the stag, because he is naturally a lascivious and gross animal; and hence through the heat of lust, their fat is dissolved into that fetid sweat, through which their face becomes black. But they become excessively fat in summer, so that they are not able to run, and are, therefore, in the second or third course caught by those who pursue them on foot. They, likewise, fly to the water, through the heat, and the difficulty of respiration. But at the time in which they copulate their flesh is bad, and has a fetid smell, in the same manner as the flesh of goats. In the winter, therefore, they are thin and feeble; but in the spring they are especially vigorous in the course. When they fly from their pursuers, also, they sometimes rest, and stop till he who pursues them is near; and then they again betake themselves to flight. But they appear to do this in consequence of labouring under an imbecility in the intestines; for they have an intestine so thin and feeble, that it is burst by even gently striking them on the belly, the skin at the same time remaining entire.

† For animals which are not a long time in the womb, and which rapidly increase, do not generally live long. But the period of gestation of female stags is short, and the increase of their young is rapid.

30 Bears, however, copulate, as was before observed, not by covering, but lying on the ground. But the female bear is pregnant thirty days. She, also, brings forth one or two offspring, but never more than five. The embryo, likewise, which she brings forth is very small in proportion to the magnitude of the body of the mother; for it is less than a weasel, but larger than a rat. It is, also, when born, naked and blind, and its legs and most of its members are nearly rude and unformed. Bears, likewise, copulate in the month of December, and the females bring forth at the time in which these animals hide themselves. About this time, therefore, both the female and the male become very fat. But when they have nurtured their cubs for three months, they make their appearance in the spring.

The porcupine, also, hides itself, and is pregnant for the same number of days as the bear; and other particulars respecting it are the same as those which pertain to the bear. The she-bear, likewise, is with difficulty taken when she is pregnant.

31 We have already observed that the lion discharges his urine from behind, and, also, copulates with his buttocks against the buttocks of the female. This animal, however, does not copulate and bring forth at every season of the year; but it copulates, indeed, every year, and brings forth in the spring. Generally, likewise, the lioness brings forth two 579
whelps, but the most she brings forth are six, and sometimes she brings forth but one. That, however, which is reported of the lioness, that when she brings forth she at the same time throws away the womb, is a delirious fable, which arose from lions being rare, and the author of the fable being dubious of the cause why they are so. For the genus of lions is rare, and they are not produced in many places; but they are produced in the whole of Europe which is intercepted between the rivers Achelous and Nessus. The lioness, also, brings forth whelps so very small, that they can scarcely walk when they are two months old. The lionesses, however, in Syria bring forth five times during the course of their life. And at first they are delivered of five whelps, and afterwards always of one less than the preceding number; but after they have been delivered of these, they no longer bring forth any offspring, but remain unprolific. The lioness, also, has no mane, but the lion has. The lion, likewise, loses those four teeth which are called canine, two from the upper, and two from the lower part; but he loses them when he is six months old.

32 The hyæna, indeed, is of the colour of the wolf, but is more hairy, and has hair which extends the whole length of the spine of the back. That, however, which is related of this animal [by Callisthenes the poet, and others] that the male and female have the same genital parts is false. But the genital parts of the male, indeed, resemble those of the wolf and the dog; and the reason why they appear to be the same as those of the female, is because the male has under its tail something similar in figure to the private parts of the female; yet it has not any avenue. Under this, however, is the passage of the excrement. But the female hyæna has under her tail, in the same manner as the male, something similar to what is said to be the vulva, and this, also, has no avenue. After this is the avenue of the excrement; and under this is the true vulva. The female hyæna, also, has a womb, in the same manner as other animals who are naturally formed to have this part. The female hyæna, likewise, is rarely caught; for a certain hunter said that he had only found one female out of eleven hyænas that he had taken.

33 With respect to hares, however, they copulate from behind, *viz.* with buttock against buttock; for they discharge their urine backward. The females, also, copulate and bring forth at all seasons, and after delivery again become pregnant. They are, likewise, subject to superfetation, and bring forth every month. They are not, however, delivered of all their offspring at once, but successively with the
580 interposition of certain days. The female, likewise, has milk before she brings forth; and immediately after she is delivered copulates, and becomes again pregnant even while she suckles. But her milk in thickness resembles that of the sow; and her offspring are born blind, in the same manner as the offspring of most animals that have multifidous feet.

34 The male fox, however, copulates by getting on the female; and the female brings forth her young in the same manner as the bear, and even still more rude and unformed. But when she is about to bring forth, she so changes the place of her usual abode, that a pregnant fox is rarely caught. When, however, she is delivered, she licks her offspring with her tongue, and warms and brings it to perfection. But she never brings forth more than four.

35 The she-wolf, also, is pregnant and brings forth after the manner of a bitch. For she accords with a bitch in the age in which she

copulates, and in the number of the offspring which she brings forth; and her young, like those of the bitch, are born blind. Wolves, however, copulate at a certain time of the year, and the female is delivered at the commencement of the spring. There is, likewise, a report concerning the delivery of the she-wolf which harmonizes with a certain fable. For it is said that all wolves bring forth in twelve days of the year. But they fabulously assign the cause of this to be, that Latona, when she was pregnant from Jupiter with Apollo and Diana, being concealed for twelve days under the form of a wolf, through the fear of Juno, was brought from the Hyperborean mountains to the island of Delos. Whether, however, this is the time of their bringing forth, or not, is not yet determined, but is nothing more than an assertion, which does not appear to be true; as neither is the assertion true, that wolves only bring forth once in their life.

But the cat and ichneumon bring forth the same number of times as bitches, and are nourished with the same food. They, also, live about six years.

The panther, likewise, brings forth blind offspring, in the same manner as the she-wolf; and she never brings forth more than four.

The female thos, also, is pregnant in a manner similar to the bitch, and brings forth blind offspring. She is, likewise, delivered of two, three, or four in number. But the body of the thos is long towards the tail, and its height is not great. And though its legs are short, it runs remarkably swift. Because, likewise, its body is flexible, it can leap to a great distance.

36 There are, also, in Syria what are called mules, though they are of a different genus from the offspring of an horse and an ass. These mules, likewise, resemble our mules, in the same manner as wild resemble tame asses. But as wild asses, so, also, the Syrian mules run with remarkable swiftness. These mules, also, are generated from each other; of which this is an indication, that there yet remain from among the number of the Syrian mules some which were brought to Phrygia, in the time of Pharnaces, who was the father of Pharnabazus. Three of them, therefore, now remain; but formerly, as they say, there were nine.

37 The generation, however, of rats, both with respect to their multitude and the celerity with which they are generated, is more admirable than that of other animals. For a female rat having been shut in a vessel containing millet, on opening the vessel a short time after,

there were found in it a hundred and twenty rats. The generation, also, and conception of the rats which cover the surface of the fields are admirable. For in many places such an innumerable multitude of field-rats are wont to be generated, that they nearly consume the whole of the corn. The destruction, likewise, occasioned by these animals is so rapid, that some husbandmen, whose agricultural possessions were not very large, having on the preceding day determined to reap, on the following day, when they came with the reapers, have found all the corn consumed by the rats.

The disappearing, likewise, of rats is unaccountable. For in a few days they become entirely invisible, though prior to this they baffled all the attempts of men to destroy them, whether by fumigation, or subverting their retreats, or hunting them, or sending swine among them; for swine destroy the holes in which they hide themselves. Foxes, also, hunt them, and they are destroyed by wild weasels more than by any thing else, yet these do not prevent their fertility, and their rapid propagation; nor does any thing else, except copious showers of rain; and then they rapidly disappear. But in a certain part of Persia, some embryos of rats being cut open, pregnant female rats were seen in them. Some, also,
581 strenuously affirm that female rats by licking salt become pregnant without copulation.

The Egyptian rats, however, have hard hairs, nearly resembling those of the land hedge-hog. There are, also, other rats who walk upon two feet; for their fore legs are short, but their hind legs large. These rats, likewise, are very numerous. There are, also, many other genera of rats.

BOOK VII

1 It is now requisite to consider the generation of man, both his first formation in the female, and his posterior generation from his nativity to old age. We have already, indeed, discussed the difference of the male with respect to the female, and the parts pertaining to each.

The male, therefore, for the most part, begins to have seed when twice seven years have elapsed from his nativity. At the same time, also, hairs begin to grow about the organs of generation; just as plants, as Alcmæon of Crotona says, are first efflorescent when they are about to bear seed. About the same time, likewise, the voice begins to change into a rougher and more unequal tone, no longer being acute, nor in a certain respect grave, nor wholly equable, but resembling the slackened rough chords of a musical instrument; which they call a goatish sound. But this mutation of voice is greater in those who endeavour to gratify [at this time] venereal desires; for the voice of those who have a propensity to such gratifications changes into the voice of a man, but the contrary happens to those who abstain from venery. If, however, attention is paid to the mode of living, in the same manner as it is paid by those who are studious of music, the voice will be longer preserved, and will sustain a very small mutation. The mammæ, also, and the private parts become enlarged, not only in magnitude, but, likewise, in form. And about this time those who endeavour to procure an emission of the seed by friction, not only experience pleasure, but, also, pain from the egress of the seed.

About the same time, likewise, the mammæ swell in females, and what 581
are called the menses begin to flow; *i.e.* blood, which resembles the blood of animals recently slain. But what are called the whites happen to very young girls, and in a greater degree, if they use moist nutriment. They, also, prevent the increase, and extenuate the bodies of girls. The menses, however, in most females begin to flow when the breasts are raised to the altitude of two fingers. About this time, likewise, the voice of girls changes to a graver tone. For, in short, a woman has an acuter voice than a man, but the voice of young is more acute than that of older women, just as the voice of boys is more acute than that of men. The voice, however, of girls is more acute than that of boys; and the wind-pipe of a virgin emits a sharper sound than the wind-pipe of a boy.

Boys and girls, likewise, about this time, require that particular attention should be paid to them. For at the time of puberty they are especially impelled to the use of venery; so that if care is taken at this period that they experience no more excitation than what the mutation of the body occasions, and entirely refrain from venery, they are wont to be temperate in this respect during the rest of their life. For girls, if they have connection with a man when they are very young, become more intemperate in venery; and this is, also, the case with boys, whether they have connection with each other, or with those of the other sex. For the avenues become enlarged, and cause the humours to flow in greater abundance; and at the same time, the remembrance of the attendant pleasure produces the desire of its repetition.

Some males, however, are from their birth without hairs in the private parts, and are unprolific, in consequence of being injured about the place of fecundity. And this, likewise, happens to women from their birth.

Both males and females, likewise, change their habits, with respect to being more healthy and more diseased, and with respect to being more or less corpulent, and in a good condition of body. For, after puberty, some lads from being lean become corpulent, and more healthy; but the contrary happens to others. Virgins, also, experience a similar change. For such boys and such girls as prior to the age of puberty abound with excrementitious humours, being purified from these after puberty, the former by the emission of seed, and the latter by the menses, become more healthy, and in a better condition of body, in consequence of
582 being freed from the impediments to their health and nutrition; but such as do not abound with these humours become through these emissions more lean and diseased. For the secretion in this latter case is from nature and parts that are in a good condition; from boys in the seed, and from girls in the menses.

Farther still, there is, also, a remarkable difference in the mammæ of virgins, compared with each other. For some have very large, but others very small mammæ. For the most part, however, this happens to such girls as abound with excrements; since, in consequence of not being yet purified by the menses from a superabundance of moisture, the humours are necessarily elevated till they burst, and then the mammæ become enlarged, and afterwards preserve the magnitude which they have acquired. Males, also, that are of a moist temperament, that are smooth and have not large veins, whether they are younger or older, have mammæ more apparent and more feminine; and this is more particularly the case with men of a dark complexion, than with such as are white.

The seed, therefore, till thrice seven years is at first unprolific; but afterwards, it is prolific, indeed, yet the youth of both sexes generate small and imperfect offspring; which is, likewise, the case with most other animals. Young females, therefore, conceive more rapidly, but when they have conceived, bring forth with difficulty. Their bodies, likewise, become more imperfect, for the most part. Men, also, that are given to venery, and women that bring forth frequently, grow old more rapidly; for it seems that women, after they have brought forth thrice, admit of no further increase. Such women, however, as are intemperate in venery, become more temperate in this respect, by frequent pregnancy. After thrice seven years, also, women are properly adapted to the propagation of children; but men receive an accession of generative power after this period.

Their seed, however, is unprolific; but the seed is prolific which resembles hail, and males are generated from it more than females; and thin seed, and which is without globules, is prolific of females.

About this age, also, [*i.e.* in the twenty-first year] males begin to have a beard.

2 The menstrual purgation of women, however, takes place at the
decrease of the moon; on which account, it is sophistically said by some 582
that the moon† is of the female sex, because at one and the same time purification happens to women and decrease to the moon; and, also, repletion to both, after purification and decrease. And some women, indeed, have the menses every third month, though this is rare, but most have them every month. Such women, therefore, as have the menses but for a short time, *viz.* for two or three days, are liberated from them with ease; but such as have them for many days, with greater difficulty; for during these days they are in pain. For in some the purification is effected at once, but in others gradually, and the body of all females labours under an oppression as long as the menses flow. Many, likewise, when the menses begin to flow, experience a certain suffocation and tumult in the womb, as long as this effluxion continues. It is natural, therefore, to women to conceive after the discharge of the menses; and such as do not, live, for the most part, without offspring. Some women, however, conceive, though they have no menstrual purgation; and these are such as collect as much of a humid vapour as remains in women that have the menses after their purgations, though it is not in such abundance as to flow forth externally. Some, also, even conceive during the menses, but not afterwards; because immediately after the emission of the menses the womb in such as these is closed. In some women, likewise, after they have conceived, the menses still continue to flow, till

† For the reason why the moon was said by ancient wise men to be of the female sex, see note c, p. 143 of the *Organon* [TTS vol. XX].

the time of delivery. It happens, however, to these, that they bring forth an imperfect offspring; which either cannot be preserved to its increase, or becomes very imbecile. Frequently, also, in some women the womb descends, through the want of copulation, either on account of their juvenile age or in consequence of having abstained from venery for a long time; and the menses frequently flow thrice in a month, till they conceive. But then the womb again ascends to its proper place. And sometimes, though the womb in other respects is in a good condition, yet happening to be moist, it rejects the more humid part of the seed.

Women, also, as we have before observed, have the most abundant menstrual purgation of all female animals. For animals, indeed, which are not viviparous are entirely deprived of this purgation, because this excrement is in them converted to the nutriment of the body. For some females are larger than the males; and, farther still, some convert the superfluous moisture into scales of different species, others into a
583 multitude of feathers; but pedestrious and viviparous animals convert it into hairs, and the whole body; for man is the only animal that is smooth; and they, also, convert it into urine. For they emit thicker urine, and in greater abundance; but in women the excrement, instead of being converted into these, is converted into the menstrual purification. The like, also, takes place in males; for man, in proportion to the size of his body, emits the most seed of all animals. And of males, such as are naturally more moist, and not very fleshy, emit a greater quantity of seed, and those that are more white than those that are of a dark complexion. This, likewise, happens in women after the same manner; for in such of these as are corpulent much of the menstrual secretion is converted into the nutriment of the body; and in venereal connections, those that are naturally more white evaporate more of this excrement than those that are of a dark complexion. Such connection, also, is in a greater degree promoted by moist and sharp food.

3 An indication, however, of conception in women is when the place becomes dry immediately after connection. But if the lips of the vulva are smooth, the conception is difficult; for the seed glides away. The conception, also, is difficult, if the lips of the vulva are thick. But if when touched by the finger they feel rougher and adherent, and are, likewise, thin, then such women are adapted to conception. It is necessary, therefore, in order to conception to be furnished with such

wombs; but with the contrary, for the purpose of not conceiving. For women the lips of whose vulva are smooth do not conceive; on which account some women, in order that the seed may fall into the matrix, anoint the vulva prior to coition with oil of cedar, or with a composition of ceruse, or with frankincense diluted with oil. And if the seed remains during seven days it is an indication that conception has taken place; for what are called effluxions take place in this number of days. Some women, therefore, after conception emit certain menstrual purgations; which, if they have conceived a female, continue for thirty days, but if a male, for about forty days. After delivery, also, the purgations continue for the same number of days; though the time of their continuance is not accurately the same in all females. But after conception, and the above-mentioned number of days, the purgations cease, and the menstrual moisture is converted to the mammæ, and milk is produced from it. The first indication, also, of the milk is a certain substance appearing in the mammæ, in a small quantity, and of the consistence of a spider's web.

But women, when they have conceived, perceive that they are pregnant by a swelling in the iliac region and in the groin; for in some these parts become immediately more; and this happens more manifestly in women that are thin. And when they are pregnant with males, indeed, a motion is for the most part produced about the fortieth day in the right side of the womb; but when they are pregnant with females, there is a motion in the left side about the ninetieth day. Nothing, however, can be accurately affirmed in this respect; for many women that are pregnant with a female feel that the fœtus is moved in the right side of the womb, and that it is moved in the left side when they are pregnant with a male. These, however, and all such particulars differ, for the most part, according to the more and the less. The fœtus, also, about this time is distinguished by members, for prior to this it was a formless mass of flesh.

The conceptions, also, of the embryo which happen within the seventh day [after conception] are called effluxions; but when the embryo is conceived after the seventh, and within the fortieth day, such conception is called an abortion; and most embryos are conceived in these days [*i.e.* within the seventh, and within the fortieth day]. If any one, therefore, puts the embryo which is ejected in the fortieth day into any other fluid than cold water, it is dissolved and disappears; but if he puts it into cold water, it acquires a consistence as if it were in a membrane. And if this membrane is divided, the embryo will be seen within it of the size of a large ant. All the other members of it, likewise, will be manifest, together with the pudendum; and the eyes, in the same manner as in other animals, will be found to be proportionally very large. The female embryo, however, if it is ejected within three months, has not, for the most part, the members distinctly formed; but in the fourth month it is rapidly formed, and acquires the distinction of the members. The female, therefore, receives the whole perfection of the parts more slowly than the male, and is more frequently born in the tenth month than the male. The female, also, when born, more rapidly acquires youth, acme, and old age, than the male; and this is more particularly the case with those that have a numerous offspring, as we before observed.

4 When, however, the womb has received the seed, it immediately closes in most women, and remains closed till the seventh month; but in the eighth month it is opened, and the embryo, if prolific [*i.e.* if it is alive] descends together with the womb. Embryos, however, that are not prolific, but dead, neither fall into the inferior part in the eighth month, nor do the wombs in this space of time open: and it is an
584 indication that it is not prolific, when in the eighth month the womb is not opened, and the embryo does not descend. After conception, also, women experience a heaviness in the whole of their body, a dimness of sight, and pains in their head. Some females are thus affected more rapidly, and nearly on the tenth day after conception, but others more slowly, according as they happen to have a greater or less quantity of excrementitious humours. Most, likewise, experience a nausea of food, and sickness, and especially those whose menstrual purgations stop, and are not yet converted into the mammæ. Some women, therefore, are thus afflicted in the beginning, but others afterwards, when the embryo is more increased; and many frequently experience a stranguary at the last. For the most part, therefore, women that are pregnant with males suffer less inconvenience, and have a better colour; but the contrary takes place when they are pregnant with females, for these for the most part are pale, and oppressed with heaviness. Many pregnant women, likewise, have tumours in the legs, and a swelling of the flesh. Yet some when they are pregnant experience the contraries of these maladies.

Pregnant women, also, are subject to all-various desires, and to a rapid change of their desires; which some call *longing*. But their desires are more acute when they are pregnant with females, though when the object of their desire is present they are less able to enjoy it. To some few women, however, it happens that their body is in a better condition when they are pregnant. But, for the most part, they are in a worse condition when the fœtus begins to have hairs. The connascent hairs, likewise, of pregnant women become fewer, and fall off; but those parts which were not accustomed to have hairs are then covered with hair.

A male fœtus, also, is for the most part more frequently moved in the womb than a female fœtus, and is more rapidly brought forth; but a female is brought forth more slowly. The pain, likewise, in the delivery of females is continued and more slow; but in the delivery of males is rapid, indeed, but is more difficult to bear. Those females, also, that have connection with men prior to their delivery bring forth more readily. But women sometimes appear to have the pains of parturition, though they are not in a parturient state. This, however, originates from the embryo turning its head in the womb.

The delivery, therefore, of other animals receives its perfection in one way only, for all of them have one definite time of delivery; but woman is the only animal that has many times. For women are delivered in the seventh, eighth, and ninth month, and frequently in the tenth month; but some bring forth even in the eleventh month. Infants, therefore, that are born prior to the seventh month are never able to live; but those that are born in the seventh month are at first alive, but, for the most part, are imbecile; on which account it is usual to wrap them in wool. Many infants, likewise, that are born in the seven month have the avenues of their ears and nostrils closed, but as they grow these avenues are developed, and many such infants live. But infants are born in the eighth month in Egypt, and in certain places, where the women are more prolific, and can easily be pregnant with and bring forth many. The offspring, also, of such women are able to live, though they are monsters. In these places, therefore, infants born in the eighth month live, and are reared; but in Greece and its vicinity very few of those that are born in the eighth month live, but the greater part perish. And in consequence of this opinion, if an infant is preserved, they do not think that it was born in the eighth month, but that the mother was ignorant of the time of her conception. Women, also, especially labour under some inconvenience in the fourth and eighth months of their pregnancy, and if the fœtus perishes in the fourth or eighth month, the mothers, also, for the most part perish; so that not only infants born in the eighth month do not live, but when they perish, the life of their mothers, also, is in danger. After the same manner, it appears, when infants are born after the eleventh month, their mothers are ignorant of the time of their conception. For it frequently happens that women, who prior to their being pregnant have been troubled with flatulencies in the womb, when after this they have had connection with a man, and have conceived, have fancied they then conceived when they experienced these flatulencies.

Such, therefore, is the difference in the time of conception and delivery in women and other animals; and as some animals are uniparous, but others multiparous, the human species accords with each of these. For most women, and in most places, bring forth only one infant; yet frequently, and in many places, as in Egypt, they are delivered of twins. But in some places they bring forth three or four infants, as was observed before: and the greatest number of which they are delivered is five; for this has been seen to happen to many women. One woman, also, in four lyings-in was delivered of twenty infants; for she was delivered of five each time, and reared many of them. In other animals, therefore, though the twins should be a male and a female, yet they are no less reared than if both were males, or both females; but in the human species, few twins live, if the one is a male and the other a female.

A woman, likewise, and a mare admit of copulation when they are pregnant, the most of all animals; but other animals when they have conceived avoid the males, *viz.* such as do not admit of superfetation, as, for instance, hares. But a mare, indeed, when she is once pregnant does not admit of superfetation, but, for the most part, only brings forth one colt. Instances, however, of superfetation in women are rare, though they sometimes happen. If, therefore, the second fœtus is conceived a long time after the first, it does not arrive at perfection, but causes pain and corrupts the pre-existent fœtus. For it has happened, that an abortion taking place, twelve fœtus which were conceived from superfetation were ejected. But if the posterior fœtus was conceived a little after the first, then the second fœtus may be perfected and delivered together with the first, in the same manner as twins are born. Thus, also, Iphicles and Hercules are fabled to have been born together.† The following, also, is a manifest instance of superfetation in a woman: A certain adulteress brought forth two infants at the same time, one of which resembled her husband, and the other the adulterer. A certain woman, also, when pregnant with twins, conceived a third infant; and when the time of her delivery came she brought forth the twins perfect;

† *Viz.* So that Hercules was the son of Jupiter, but Iphicles was the son of Amphitrion.

but she was delivered of the third at the end of five months, and this died immediately. It happened, likewise, to another pregnant woman, that she was delivered of one infant at the end of seven months, and afterwards brought forth twins at the end of nine months; and of these, the former died, but the twins lived. Some women, also, that have conceived from superfetation produce an abortion of the prior fœtus, but are properly delivered of the second.

Most women, however, that copulate after the eighth month when they are pregnant, bring forth the infant covered with a mucous viscosity. And it frequently happens that the infant is born covered with the food which the mother has eaten. If the mother, likewise, has used salt with her food in a greater quantity than usual, the infant is born without nails.

5 With respect to milk, that which is produced prior to the seventh month is useless; but at the same time that the infant has life, the milk is useful. The first milk, however, is salt, in the same manner as the first milk of sheep.

Most pregnant women, likewise, are sensibly affected by wine; for if they drink it, they become relaxed and debilitated.

The commencement, however, of becoming pregnant to women, and of generative power to men, is the emission of seed by the latter and the efflux of the menses by the former; except that neither of these are immediately prolific at the commencement of the emission of the seed and the menses, nor when the former flows in a small quantity, and the latter is imbecile. We have, however, already shown at what age the menses and the seed originate. But in most women the menses cease about the fortieth year of their age; but those women that have them beyond this period continue to have them till they are fifty years old, and some at this age have been delivered of a child.

6 With respect to men, however, most are capable of generating till they are sixty years old, and if they are capable after this period they possess a generative power till they are seventy years old; and some have actually engendered at this period.

It, likewise, frequently happens to many men and women, who are not able to beget children from copulation with each other, that they are able when dissociated [from copulating with others]. The like, also, takes place in the procreation of males and females. For sometimes men and women, while they continue to copulate with each other, beget females alone, or males alone; but the contrary takes place when they are dissociated. They, likewise, change according to age; for some who when young engendered with each other females, when they become older engender males; but with others the contrary takes place. The same thing, also, happens with respect to engendering; for some who when young had no generative power, are able to generate when they are older. And others generate at first, but afterwards lose the power of generating. There are, likewise, some women who are scarcely able to conceive, but if they once become pregnant are delivered; but others, on the contrary, easily conceive, indeed, yet are not able to bring forth the fœtus with which they are pregnant. There are, also, men that beget females alone, and women that bring forth males alone. Thus, for instance, it is fabulously reported of Hercules, that of seventy-two children which he begat, only one of them was a daughter. But women that are [naturally] unable to conceive, if they happen to become pregnant, either through the measures adopted to remove their sterility, or from any casual circumstance, are for the most part rather pregnant with females than with males. It, also, happens to many men, that at first they are able to generate, but afterwards are not able, and that they again recover their former power.

Mutilated children, also, are generated from mutilated parents; as, for instance, lame children from lame parents, and blind from such as are blind. And, in short, parents that have preternatural marks, such as warts and moles, frequently beget children with similar marks. Sometimes, likewise, it happens that parents do not transmit marks of this kind to their children, but to their grandchildren. Thus, for instance, the son of a person that had a mark in his arm was not born with a similar mark, but the grandson of this person had a confused black spot in the same place of the arm. Things of this kind, therefore, seldom happen; but frequently children that have their members entire
586 are born from mutilated parents, and there is no certain rule for such events. Children, likewise, resemble their parents, or their ancestors, and sometimes they resemble neither. Sometimes, also, it happens that after many generations a child is born resembling some one of its ancestors. Thus, in Sicily, a woman who had committed adultery with an Ethiopian brought forth a daughter that was white, but this daughter had a son that was black. And, for the most part, indeed, females more resemble their mother, but males their father; though the contrary, also, takes place, for sometimes the females more resemble their father, and the males their mother. Children, also, partially resemble their parents,

so as to resemble their father in one part of their body, but their mother in another part. Twins, likewise, have been born not resembling each other; though, for the most part, they resemble each other. A certain woman, also, who had connection with a man the seventh day after her delivery, and became pregnant, brought forth an infant as much resembling the former, as if it had been a twin. But there are women who beget children resembling themselves; and there are others who beget such as resemble their husbands; in the same manner as a Pharsalian mare, that was called Dicæa, always brought forth horses resembling the horse by whom she was pregnant.

7 In the egress of the seed, however, a breath or hot air (πνευμα) is in the first place the leader; but the egress itself manifests that it is produced by a hot air; for nothing is thrown to a distance without pneumatic force. When, also, the seed is received into the womb, and has remained there for some time, it is enveloped with a membrane. For when it comes forth before it is distinctly formed, it has the appearance of an egg, enclosed in a membrane, when the shell is taken away. This membrane, likewise, is full of veins. All animals, however, that swim, or are pedestrious, or winged, whether they are viviparous, or oviparous, are generated in a similar manner; except that such as are viviparous have the navel adhering to the matrix; but those that are oviparous have it adhering to the egg; and others have it adhering in both ways, as is the case with a certain genus of fishes. And some animals, indeed, are enveloped as it were in membranes, but others, in what is called the chorion. The animal, also, is immediately involved in the interior chorion, but the interior chorion is involved in another membrane, the greater part of which adheres to the matrix; but the remaining part is separated from it, and contains water. Between these membranes, also, there is an aqueous and sanguineous moisture, which is called by women the first-fruits.

8 All animals, however, that have a navel are increased [in the womb] through the navel. But if the matrix has certain small cavities, the fœtus adheres to these through the navel; and if the matrix is smooth, the fœtus adheres to the matrix by the medium of a vein. The situation, also, of the fœtus in the womb is different in different animals. For all
quadrupeds are extended through the length of the womb; and those that 586
are without feet, such as fishes, are situated on the side of it; but the fœtus of bipeds, such as birds, are inflected and contracted; and man is

so convolved in himself, that he has his nose between his knees, his eyes above his knees, and his ears external to them. In a similar manner, all animals have the head at first in the superior part; but as they increase, and approach to the period of their egress from the womb, the head becomes situated in the inferior part. All animals, likewise, are naturally so born that the head is the first part that comes forth; and, therefore, that which is so born as to come forth inflected, or with the feet first, is born preternaturally. Quadrupeds, also, when they are perfect in the womb have excrements both moist and dry; the latter, indeed, in the extremity of the intestines, but the former, *i.e.* the urine, in the bladder. But if there are small cavities in the matrix, as the embryo increases, these cavities are diminished, and at length entirely disappear.

With respect, likewise, to the navel, it is as it were a shell surrounding the veins, which originate from the womb [through which the infant derives nutriment from the mother]. And in those wombs that have small cavities, such veins originate from these cavities, but in those that are without these cavities, they originate from another vein. But in the embryos of larger animals, as of oxen, there are four veins in the navel, and in smaller animals two. In such animals, however, as are very small, as in birds, there is only one vein. Of the four veins, also, of larger animals, two of them extend through the liver, in the parts of it which are called gates, and which are near to the great vein; but the other two are near to the aorta, in that part in which it is divided, and from one becomes two branches. Each pair of veins, likewise, is surrounded with membranes; and the navel surrounds the membranes like a shell. But as the embryo increases, the umbilical veins always subside. And when the embryo has arrived at maturity, it descends to the inferior concavity of the womb, there becomes manifestly moved, and sometimes is rolled about the private parts of its mother.

9 When women, also, are pregnant, pains fix themselves in many parts of their body, and especially in one or other of their thighs. But such of them as suffer very great pain in the belly, are delivered the most rapidly, and such of them as experience a pain in the loins, are delivered with difficulty; but the delivery is speedy of such as experience a pain in the lower region of the belly. When, therefore, they bring forth a male infant, a clear and pale lymph precedes the delivery; but when they bring forth a female infant, a sanguineous lymph precedes, but which, also, is clear. Neither of these lymphs, however, flow from some women at the time of their delivery.

The delivery, therefore, of the females of other animals is not laborious, but they manifestly suffer less pain from parturition than women. Women, however, experience at this time the most violent pains, and especially those that lead a sedentary life, and such as have not strong sides, and are unable to hold their breath. For their delivery is more difficult, if in the intermediate time, being compelled by their breath, they respire. In the first place, therefore, the water which is intercepted between the interior and exterior secundine comes forth from the motion of the embryo, and the bursting of the membranes, then the embryo follows, the secundines returning back, and the inward parts of them becoming external. 587

10 It is, likewise, the province of the midwife to cut off the umbilical chord; and the execution of this requires intelligence and skill. For it is not only requisite to be able to give assistance to women skilfully in the difficulties of labour, but, also, to be sagacious with respect to accidents which may occur, and in tying the umbilical chord of the infant. For if the secundine should fall out together with the infant, then the umbilical chord, which adheres in one part to the infant, and in the other to the secundines, must be tied with a woollen thread, and cut when it adheres to the secundines. But the part where it is tied coalesces, and the remainder falls. If, however, the ligature is untied, the embryo [*i.e.* the infant] dies, in consequence of the efflux of blood. If, also, the secundines do not immediately come forth together with the infant, but remain in the womb after the infant is born, then the umbilical chord must be cut, so that the part which is tied may adhere to the belly of the infant, and the other part may remain conjoined to the secundines. The infant, likewise, frequently appears to be dead, when from its being weak, the blood flows into and about the navel, before the chord is tied. Certain skilful midwives, however, cause the blood from the navel to return into the infant, and immediately the infant, who before was in an exanguious state, again revives.

To other animals, also, as we have before observed, it is natural to be born with the head foremost; but infants are, also, born with their hands extended along their sides. They, likewise, cry as soon as they are born, and lift their hands to their mouth. Some, also, immediately, and others after a short time, emit excrements, but all of them within the space of a day. And this excrement, as to its quantity, exceeds the proportion of the size of the infant, and is called by women *meconium*. But the colour of this excrement resembles that of blood, and is very black and pitchy.

After this, however, it is of a milky colour. For the infant immediately begins to suck.

The infant, however, does not cry prior to its egress from the womb, not even though, when it is born with difficulty, its head is out of but all the rest of its body is within the womb.

587 But those women with whom what is called the flooding precedes their delivery bring forth the infant with greater difficulty. If, however, the flooding is moderate after the delivery, and not in a greater quantity than before it, and if, likewise, it does not continue longer than forty days, in this case, the women become more strong, and more readily conceive.

Infants, also, when they are born, for forty days neither laugh nor weep, while they are awake, but at night [when they are asleep] they sometimes do both; and, for the most part, if they are tickled, they do not perceive it. For the greater part of the time, likewise, they sleep; but as they increase they always change to a more wakeful condition; and it is evident, indeed, that they dream, but it is late before they begin to remember their dreams.

In other animals, therefore, there is no difference of bones, but all of them are from the first equally perfect; but in infants the sinciput is soft, and it is late before it ossifies. And some animals, indeed, are born with teeth; but infants do not begin to have teeth till the seventh month. The anterior teeth, also, grow the first; and the upper teeth shoot forth first in some children, but the lower in others. In all such infants, however, whose nurses have a warmer milk, the teeth grow more rapidly.

11 After delivery, however, and the purifications [*i.e.* the floodings] have taken place, milk is produced in women in great abundance; and in some it flows not only from the nipples, but, also, from many parts of the mammæ, and in others even from the armpits. When the moisture, also, which ought to be converted into milk is not concocted, certain hard globules are formed which are called *strangalides*. For the whole of the mammæ is so spongy, that if a woman in drinking should swallow a hair, a pain would be produced in her mammæ which is called *pilare malum*; and this disease continues, till either the hair comes forth spontaneously, or is sucked out together with the milk. But women have milk till they again become pregnant, and then the milk ceases and is extinguished; not only in women, but, in a similar manner, in other viviparous quadrupeds. When the milk flows, likewise, there are, for the

most part, no menstrual purgations; though this purification has taken place in some women while they were suckling. For it does not happen that the moisture is at one and the same time impelled to many places. Thus, for instance, in those women that have haemorrhages, the menstrual purifications suffer a deterioration; and the same thing happens to others, when the moisture from the loins is secreted through the thighs, before it enters into the womb. Such women, also, as have not the menstrual purifications at this period, suffer no injury if they 588
happen at the same time to vomit.

12 Infants, also, that are more corpulent, that suck more abundantly, and are suckled by more corpulent nurses, are for the most part subject to convulsions. Wine, likewise, is hurtful in this disease, red wine more than white; and, also, wine unmingled with water is more noxious than diluted wine. This too is the case with most nutriment that contains a great quantity of air, and if there is a constipation of the belly. Most infants, however, perish before the seventh day; and hence a name is given to them on this day, because greater hopes are then entertained of their conservation. During the full moon, also, infants are more subject to convulsions; and their life is in danger if the convulsions begin from the back.

BOOK VIII

1 And thus much concerning the generation, and whatever else pertains to the nature of animals; but their actions and their lives differ according to their manners and their nutriment. For in most other animals there are, also, vestiges of the matters of the [rational] soul, which in men possess more apparent diversities. For in these animals there are mildness and fierceness, gentleness and roughness, fortitude and cowardice, fear and confidence, anger and craft, and in many of them there are similitudes of dianoetic intelligence, as we have, also, observed, in discussing the parts of animals; since some of them differ in the more and the less with respect to man, and man differs in the more and the less with respect to many animals. For some of the above-mentioned manners are more inherent in man, and some of them are more inherent in other animals; but others differ in the analogous. For as in man there are art and wisdom, and intelligence, thus, also, in some animals there is another such like physical power. But a thing of this kind will appear to be most evident to those who look to the puerile age. For in these the vestiges, as it were, and seeds may be seen of their future habits; but at this period the human soul, as I may say, in no respect differs from the soul of brutes. Hence it is not at all irrational to admit, that the manners of men are partly the same, partly similar, and partly analogous to those of brutes.

Nature, however, passes so gradually from inanimate natures to animals, that through the continuity, the confine and the medium of these are latent. For after the genus of inanimate natures, that of plants is the first; and these differ from each other in participating of life in a greater [or less] degree. The whole genus of plants, however, compared with other bodies, appears to be nearly as it were animated; but when compared with an animal, to be inanimate. But the transition from plants to animals is, as we have before observed, continued. For with respect to some of the bodies in the sea, it may be doubted whether they are animals or plants. For they adhere to other bodies; and many of such like substances, when separated from the bodies to which they adhere, perish. Thus, for instance, the pinnæ adhere to other bodies, but the *solenes* when divulsed are unable to live. In short, the whole genus of testaceous animals resembles plants, when compared with animals that have a progressive motion. And with respect to sense, some of them exhibit no indications of the possession of it; but others indicate it obscurely. The nature, also, of the body of some of them is fleshy; as in what are called *tethya*, and the genus of sea-nettles; but the sponge perfectly resembles plants.

Always, however, the difference is small, according to which some bodies appear to have more of life and motion than others, and this is, also, the case with the actions of life. For the work of such plants as are produced through seed, appears to be nothing else than that of making another being like themselves. The like, also, takes place in animals; for of some of these, it is not possible to assign any other work than that of generation; on which account, such actions as these are common to all animals. The lives, however, of those animals, differ with respect to bringing forth and nurturing their offspring, who have a larger portion of sense, and who copulate on account of the pleasure with which the act of copulation is attended. Some, therefore, as plants, accomplish at stated times the generation appropriate to their nature; but others labour 589
in the nurture of their offspring, and when their offspring have arrived at perfection separate themselves from and have no longer any communion with them; and others, who are more intelligent, and participate more of memory, conduct themselves towards their offspring in a manner more sociable and political. One part, therefore, of the life of animals consists in the actions which they exercise in the procreation of offspring; but another part consists in procuring nutriment. For about these two all their attention and life are employed. All kinds of nutriment, however, principally differ according to the matter from which the beings that are nourished are composed; for the natural increase of every animal is from the same matter as that from which it consists. But that which is according to nature is pleasant; and all animals pursue the pleasure which is adapted to their nature.

Animals, also, are divided according to places; for some of them are pedestrious, [*i.e.* terrestrial] but others are aquatic. But this difference being predicated in a twofold respect, *viz.* either from receiving air or water, some are called pedestrious, and others aquatic. And again, of those animals which do not receive either air or water sufficiently, but are naturally adapted to the temperament of refrigeration, some are called pedestrious, but others aquatic, though they neither respire nor receive the water, but derive their nutriment from and live in either of these elements. For many animals, that receive the air and bring forth their young on the land, derive their nutriment from aquatic places, and live the greater part of their time in the water; these animals alone, as it seems, partaking of an aquatic and terrestrial nature, and may, therefore, be considered as both pedestrious and aquatic.

2 No animal, however, which receives the water is either pedestrious, or derives its nutriment from the earth; but the pedestrious animals which receive the air are many. And some, indeed, receive the air in such a way, as not to be able to live when separated from the nature of water; such as the animals which are called marine tortoises, crocodiles, the hippopotamus, and the phocæ; and likewise smaller animals, such as the *emys*, and the genus of frogs. For all these, unless they respire for a certain time, are suffocated; and they are accustomed to bring forth and nurture their young on the dry land, but they live in the water.

The dolphin, however, is the most extraordinary of all animals; and this is, also, the case with every other animal of a similar nature, whether it be in the cetaceous or aquatic genus, as the whale, and such other animals among these as have a pipe. For it is not easy to rank each of these either among aquatic or terrestrial animals, if those are to be considered as terrestrial which receive the air, and those as naturally aquatic which receive the water. For each of these participates both of the aquatic and terrestrial nature; since they receive the sea and return it through a pipe, and they, likewise, receive the air by their lungs; for they have this part, and respire through it. Hence the dolphin, when it is caught in nets, is soon suffocated, in consequence of not respiring; though out of the water it lives a long time, murmuring and making the same noise as other animals that respire. The dolphin, likewise, sleeps with his nose raised above the water, in order that he may respire. It is, however, absurd, since the divisions are contrary, to arrange these animals in both of them; but it seems that the aquatic genus must be still further divided.

For some of these receive and return the water from the same cause through which animals that respire receive the air, *viz.* for the sake of refrigeration; but others on account of nutriment. For it is necessary, since they receive their food in water, that they should together with it receive the water, and receiving have an organ by which they may return it. Those animals, therefore, which use water in a manner analogous to respiration have gills; but those sanguineous animals that use water on account on food have a pipe. The like, also, takes place in the mollia, and in crustaceous animals; for these receive water on account of nutriment. After another manner, however, those animals are aquatic, from the temperament of their body and their life, that receive air, indeed, yet live in water; but those are pedestrious that receive water and have gills, yet betake themselves to the dry land and derive their nutriment from it. Nevertheless only one animal of this kind is at present known, and this is called the *cordylus*; for this animal has no lungs, but has gills, and goes on the dry land, and thence derives its nutriment. It is, however, a quadruped, as being naturally adapted to walk.

Moreover the nature of all these [ambiguous] animals is, as it were, distorted, just as some males participate of female and some females of male characteristics. For animals receiving a difference in the small parts of their body, are seen to differ largely according to the nature of the whole body. But this is evident in castrated animals. For in consequence of a mutilation in a small part the animal is changed into a female; so that it is evident that in the original formation of animals, from something of the smallest magnitude being changed, if it has but a principal subsistence, one animal becomes a female, but another a male. And, in short, the members being taken away which cause the distinction of the sexes, that which is produced is neither male nor female. In a similar manner, therefore, on account of a small corporeal mutation and difference, some animals are terrestrial, and others aquatic. And some, indeed, do not communicate with both natures, but others do communicate, in consequence of participating something of the matter from which they derive their nutriment in their original formation. For to every animal that which is according to nature is friendly, as we have before observed.

But animals being divided into the aquatic and pedestrious triply, *viz.* by receiving air or water, and by the temperament of their bodies, but, in the third place, by their food, their lives are conformable to these divisions. For some are aquatic or pedestrious according to their temperament and food, and according to their receiving water or air; but others from their temperament or their life alone.

Of testaceous animals, therefore, those which do not change their place are nourished with potable water. For the sea-water is percolated through dense bodies, and the more attenuated part of it is concocted, and receives a generation as it were from the beginning. But that sea-water by percolation may become potable is evident; since this has been ascertained by experiment. For if a vessel made of thin wax, to which a string is fastened, is dropped into the sea empty, and remains there for the space of a day and a night, there will be found within it a considerable quantity of potable water.

Sea-nettles, however, are nourished with any small fishes they may happen to meet with. But they have a mouth in the middle of their body, which is very apparent in large sea-nettles. They have, likewise, in the same manner as oysters, a passage in the superior part of their body, through which they discharge their excrements. For the sea-nettle, in the same manner as the oyster, appears to have a fleshy substance internally, and to use a rock as a shell. But the lepas frees itself from the place to which it adheres, and goes in search of 590 nutriment. Of Such testaceous animals, however, as are motive, those, indeed, that are carnivorous, as, for instance, the purple, feed on small fishes; for this animal is carnivorous, and on this account it is caught by things of this kind; but others feed on plants that grow in the sea.

The marine tortoises, likewise, feed on shells; for they have a body the most robust of all animals. For whatever they take into their mouth, whether it be a stone, or any thing else, they break and devour; and, going on shore, they feed on grass. But they labour and frequently perish, when swimming on the surface of the sea, they are dried by the sun; for they cannot again easily plunge themselves in the depths of the sea.

Crustaceous, also, are nourished after the same manner as testaceous animals. Thus, for instance, crabs of the rock feed on stones, wood, mud, sea-weed, and dung; and they are, also, carnivorous. But lobsters vanquish, indeed, large fishes, and a remarkable reciprocation happens to some of these. For the polypi subdue the lobsters; so that lobsters when they perceive the polypi near them in the same net die through fear. Lobsters, however, vanquish the congers; for the latter cannot escape on account of the roughness of the former; and the congers eat the polypi; for the former, on account of their smoothness, cannot be caught and retained by the latter. All the mollia, likewise, are carnivorous. Lobsters, however, feed on the little fishes, which they hunt after near their retreats. For their habitations are in such places of the sea as are rough and stony. And whatever they catch they carry to their mouth with a forcipated foot, in the same manner as crabs. Naturally, likewise, when they are not terrified, their progressive motion is to the anterior part, letting fall their horns on their side; but when they are terrified, they make their retreat backward, and extend their horns. They, also, fight with each other, in the same manner as rams, with their horns, raising these and striking with them. They are, likewise, frequently seen with each other in a great number, as if they were a herd.

Crustaceous animals, therefore, live after this manner; but of the 591 mollia, the small calamary and the cuttle-fish vanquish large fishes. But the polypi principally collect shell-fish, and extracting the fleshy part feed on it; on which account, those who hunt for them know their retreats by the shells [which they find in them]. That, however, which some assert, that the polypus eats itself, is false; but some of them have their arms gnawed by the congers.

All fishes, however, at the time in which they bring forth are nourished with their own eggs; but with respect to other nutriment, all of them do not use the same. For some of them are carnivorous only, such as the selache, the congers, cockles, the tunny, the wolf, the dentals, the amia, the *orphi*, and lamprey; but barbels are nourished with sea-weed, oysters, and mud, and, also, with flesh. But the *dascillus* feeds on mud and dung; the *scarus* and *melanurus* feed on sea-weeds; and the stock-fish on dung and sea-weeds. This last, also, eats the *prasium*; and is the only fish that is hunted with the *colycynthe* or gourd. All fishes, likewise, devour each other, except the mullet, and this is especially the case with the congers; but the *cephalus* and the mullet are, in short, the only fishes that do not feed on flesh; of which this is an indication, that nothing fleshy is ever found in the belly of these fishes, nor do fishermen use the flesh of animals as a bait in catching them, but bread alone. Every mullet, however, feeds on sea weed and sand. And the *cephalus*, indeed, with some call *cheilon*, or *the thick-lip*, lives near shores; but the other kind of cephalus which is called *peræas* does not. The *peræas*, however, is fed with its own mucus; on which account it is always empty. But the cephali feed on mud, on which account they are heavy and mucous; but they do not eat any fishes. In consequence, however, of living in mud, they frequently plunge themselves in water, that they may wash off the mud with which they are covered. But none of the large fishes devour the young of the cephali, on which account their progeny are very numerous. When, however, their young are increased in size, then they are devoured by other fishes, and especially [591] by the archanus. But the mullet is the most gluttonous and insatiable of all fishes; on which account its belly is distended, and when it is not hungry it is in a bad condition. When it is terrified, also, it hides its head, as if by so doing the whole of its body was concealed. The dental, likewise, is carnivorous, and devours the mollia; and frequently when this fish and the cockles pursue smaller fishes, they throw their belly out of their mouth, because the belly of fishes is near their mouth, and they have not a stomach.

Some fishes, therefore, as we have before observed, are only carnivorous, such as the dolphin, the dental, the goldfish, the selache, and the mollia; but others, for the most part, feed on mud, sea-weed, moss, and what is called *caulium*, and other marine plants; and these fishes are such as the *phycis*, the gudgeon, and those that live in rocks. But the *phycis* does not eat any other flesh than that of the squill. Fishes, likewise, frequently devour each other, as we have before observed, and the larger devour the smaller. This, however, is an indication, that they are carnivorous, that they are caught with baits made from the flesh of animals. And the amia, the tunny, and the wolf, are for the most part carnivorous, but they, also, eat sea-weed. But the sargus feeds on the barbel; and when the barbel, having stirred the mud, comes forth, (for it is able to excavate the mud), the sargus descends into the mud, there feeds and prevents other fishes weaker than himself from descending together with him. What is called the scarus, also, appears to be the only fish that ruminates after the manner of quadrupeds.

Other fishes, therefore, hunt such as are smaller than themselves with their mouths opposite to their prey, in the same manner as they naturally swim; but the selache, dolphins, and all of the cetaceous kind, seize their prey in a supine position. For they have the mouth beneath; on which account smaller fishes more easily elude them; since otherwise very few fishes would escape. For the celerity of the dolphin in pursuing its prey, and the voraciousness with which it devours them, are wonderful.

592 Some few eels, likewise, in certain places are nourished with mud, and with such eatables as any one may happen to throw to them; nevertheless the greater part of them is nourished with potable water. And those that nurture eels are particularly careful that the water in which they are kept is pure, and that there is a perpetual flux and reflux of it on ground covered with gravel. For they are soon suffocated unless the water is pure; since they have small gills; and hence those that fish for them disturb the water in which they are contained. And eels, indeed, are caught in the river Strymon about the rising of the pleiades [*i.e.* in the month of November], for then the water and the mud are disturbed by contrary winds; but without such winds, it is better not to attempt to fish for eels. Moveover, eels, when they are dead to not swim on the surface of the water, nor are borne upward, like most other fishes. For they have a small belly,† and but few of them are fat. When taken out of the water, likewise, they live five or six days; and they live

† As eels have a small belly, they contain but little air, and, therefore, are specifically heavier than water. Hence when they are dead they do not rise to the surface of the water.

longer when the north than when the south wind blows. If, also, they are taken out of lakes, and put into fish-ponds in summer, they die; but not if they are taken out in winter. They, likewise, cannot sustain violent mutations. Thus, in summer, if they are plunged into cold water, a great number of them frequently perish. But they are suffocated, if they are nourished with a small quantity of water. The very same thing, also, happens to other fishes; for they are suffocated, if they are always kept in the same small quantity of water, in the same manner as animals that respire perish if they are surrounded with but little air. Some eels, likewise, live seven or eight years.

River fishes, however, feed on each other, and, likewise, on herbs and roots, and other substances which they find in the mud. They, also, feed more by night than by day, and withdraw themselves in the day into deep water. Such, therefore, are the particulars pertaining to the nutriment of fishes.

3 Of birds, however, all such as have crooked claws are carnivorous; but they cannot swallow corn, even if it is put into their mouth; as, for 592
instance, all the genera of eagles, kites, and the two species of hawks, *viz.* the hawk which pursues the ring-dove, and that which attacks the chaffinch and the buzzard. These two species of hawks, also, differ very much from each other in magnitude. But the buzzard is as large as the kite, and is seen at all times. Among carnivorous birds, also, are the osprey and the vulture; but the osprey is larger than the eagle, and its colour resembles that of ashes. Of vultures, however, there are two species; the one small, and more white, but the other large, and more of an ashen colour. Farther still, of birds of the night some have crooked claws, as the night-raven, the common owl, and the great owl. But the great owl in form resembles the common owl, but in magnitude is not less than the eagle. Among birds of the night, likewise, are the screech-owl, the horned owl, and the small owl. But of these, the horned owl is larger than a cock, and the screech owl is of a similar size. Both of them, also, pursue the magpie; but the small owl is less than the common owl. All these, however, which are three, are similar to the view, and all of them are carnivorous. Some birds, likewise, that have not crooked claws are carnivorous, as, for instance, the swallow.

Some birds, also, feed on worms, as the chaffinch, the sparrow, the *bathis*, the greenfinch, and the titmouse. There are, however, three species of the titmouse. One, indeed, and that the largest, is the titmouse-chaffinch; for it is of the size of the chaffinch. The second is

the titmouse of the mountain, because it dwells in mountains, and has a long tail. But the third is similar to the preceding two, yet differs in magnitude; for it is the smallest of the three. The becafige, likewise, the black-head, the bullfinch, the robin-redbreast, the *epilais*, the asilus, and the bird called *tyrannus*, or the tyrant, feed on worms. But the tyrant is a little larger than a cricket, has a purple crest, and is an elegant and well proportioned bird. Next to these are the bird called *anthos* or *the flower*, which is as large as the chaffinch; and the chaffinch of the mountain, which resembles the common chaffinch, and is nearly of the same size with it, except that it has an azure neck, and dwells in the mountains. And farther still, among the birds that feed on worms, is the wrenreaper. These birds, therefore, and such as these, either entirely or for the most part, feed on worms. But birds of the following kind feed
593 on thorns, *viz.* the thistle-finch, the *thraupis*, and besides these, what is called the gold-cap. For all these feed on thorns, but none of them on worms, or any thing animated. These, also, sleep as well as feed on thorns.

Other birds, however, feed and principally live on gnats, which they hunt; such as both the large and small woodpecker. But some call both these *treepiercers*. They resemble, however, each other, and have a similar voice, except that the voice of the larger woodpecker is louder; and both these fly to trees on which they feed. The green woodpecker, also, lives on gnats. But this bird is of the same size as the turtle-dove, and its colour is entirely green. It, likewise, digs forcibly into trees, upon which it usually feeds, and has a loud voice. This bird, too, is principally found about Peloponnesus. With these, also, may be enumerated another bird, called a wagtail, which is not larger than the thistle-finch, and it is of an ashy colour and speckled. Its voice, likewise, is weak, and it digs into trees.

Certain other birds, however, feed on fruits; such as the large and the small ring-dove, the pigeon, the vinous pigeon, and the turtle-dove. The large and the small ring-dove, therefore, and the pigeon are always seen; but the turtle-dove is seen in the summer; for in winter it disappears, since it then hides itself. The vinous pigeon, however, is principally seen and caught in autumn; and this bird is in magnitude larger than the common pigeon, but less than the ring-dove. This bird, also, is principally caught while it is drinking water. And those birds that feed on fruits come to these parts [*viz.* to Greece] when they have young. But all other birds which visit our country in summer, build their nests in it, and nourish their young with animals, except those which belong to the species of pigeons.

With respect to all birds, however, as I may say, some derive their nourishment from the earth, others from rivers and lakes, and others from the sea. But those which obtain their nutriment in water may be distinguished into two kinds, *viz.* into the *palmipedes*, or flat-footed, and the *fissipedes*, or cloven-footed. The palmipedes, therefore, for the most part, live in water; but the fissipedes about water. And of these, such as 593
are not carnivorous feed on the fruits of the earth. But the heron and the white heron live about lakes and rivers. And the white is of a less size than the other heron, and has a broad and long bill. To these, also, may be added the stork and the seacob, the latter of which is of an ashy colour, and, likewise, the red sparrow, the *cincle*, and the *tryngas*. But the tryngas is the largest of these small birds; for it is of the size of the thrush. All these birds, likewise, move their tail. Farther still, with these the *calidris* may be numerated. But this bird has a variegated plumage, and the whole of it is of an ashy colour. The genus of the halcyons, also, is aquatic. But of these there are two species; one of which is vocal, and perches on reeds, and the other is mute, but is larger than the vocal halcyon. Both of them, also, have an azure back. The wren, likewise, lives near water. But the halcyon and the *cerylus* live near the sea; and crows frequent shores, and feed on animals thrown up by the sea; for the crow is omnivorous. The white sea-gull, also, the coot, the plungeon, and the hawk of the rock, dwell near the sea. Of the palmipedes, however, those that are more heavy, such as the swan, the duck, the *phalaris*, and the didapper, dwell about rivers and shores. This is, also, the case with the teal, which resembles the duck, but is not so large a bird; and, likewise, with what is called the raven, which in magnitude is equal to the stork, except that its legs are less. But the raven is flat-footed, swims well, and its colour is black. This bird, also, perches on trees, and is the only bird of this kind which builds its nest in trees. To these may be added the goose, the small goose which goes in a flock, the goose-fox, the goat, and the *penelops*. But the eagle of the sea dwells near the sea, and frequents lakes.

Many birds, however, are omnivorous. But those that have crooked claws, not only feed on such other terrestrial and aquatic animals as they are able to vanquish, but, also, on other birds; except that they do not 594
feed on birds of the same species, and in this differ from fishes, who frequently devour those of the same species with themselves. The whole genus of birds, likewise, drinks but little; but those that have crooked

claws entirely abstain from drinking, a few excepted, and these drink but seldom. This is particularly the case with the kestrel. The kite, also, has been seen to drink, though rarely.

4 Animals, also, which are covered with a scaly skin, such as the lizard and other quadrupeds, and, likewise, serpents, are omnivorous. For they feed both on flesh and grass. But serpents are the most dainty of animals. These, therefore, drink but little, and such other animals as have spongy lungs. But all animals have spongy lungs that have but little blood, and that are oviparous. Serpents, however, are immoderately fond of wine; and on this account, some in order to catch vipers place wine in shells on the hedges; for they are caught when they are intoxicated. But serpents being carnivorous, whatever animal they seize, when they have sucked out the juice of it, they emit the remaining body of the animal by the passage through which they discharge their excrement. Other such like animals, also, as, for instance, spiders, do the same; but spiders, when they suck the juice of any animal keep the animal external to themselves, and serpents have the animal which they suck in their belly. A serpent, therefore, seizes and eats whatever he may happen to meet with; for he eats birds and beasts, and devours their eggs. But seizing his prey, he raises himself till he is in an upright position; and afterwards he so gradually collects and contracts himself, that when he becomes extended that which he devours descends. This endeavour of serpents in devouring their prey is necessary, because their stomach is long and thin. The spiders called phalangia, also, and serpents can live a long time without food. But this is evident from those serpents and phalangia which are nourished by druggists.

5 With respect, however, to quadrupeds and viviparous animals, all such as are savage and have serrated teeth are carnivorous; except wolves, who, as they say, are the only savage animals that eat when they are hungry a certain earth. At another time, however, they do not touch grass; but when they are ill, in the same manner as dogs, they eat certain herbs, which they vomit, and are purged. But those wolves that search for prey apart from other wolves more frequently devour men than those that are hunters. The animal, however, which some call glanus, but others an hyæna, is in magnitude not less than the wolf, but it has
594 a mane like a horse, though its hairs are harder and thicker, and are extended through the whole of its spine. This animal, also, insidiously attacks and hunts after men; and, vomiting in the same manner as men,

it seizes and devours dogs. It, likewise, digs into sepulchres, through a desire of eating human flesh.

But the bear is omnivorous; for it eats the fruits of trees, which, through the pliability of its body, it climbs. It, also, eats leguminous fruits. Destroying, likewise, the hives of bees, it eats their honey, and feeds on crabs and ants; and is carnivorous. For on account of its strength, this animal not only attacks stags, but, also, wild-boars, if it can invade them latently. It, likewise, attacks bulls. For attacking the bull in front, he falls on his back, and while the bull endeavours to strike him [as he lies in this supine position] the bear throws his arms round the horns of the bull, bites his shoulder, and lays him prostrate. For a short time, likewise, the bear walks erect on two feet, and eats every kind of flesh, previously macerating it.

The lion, also, is carnivorous, in the same manner as other animals that are savage, and have serrated teeth. But he eats voraciously, and swallows a great quantity of flesh without dividing it. Afterwards, for two or three days he abstains from food; for he is able to do so on account of his excessive repletion. He drinks, however, but little; and rarely emits excrement; for he does not emit it till the third day, or just as it may happen. But his excrements are dry and hard, and resemble the excrements of the dog. He, likewise, emits a very pungent flatus, and his urine has a feted smell. Hence dogs [that hunt the lion] smell the trees; for the lion makes water against them, lifting up his leg in the same manner as dogs. His breath, also, produces a strong smell in the food which he eats. For when a lion is dissected, the interior parts of his body produce a strong exhalation.

Some savage quadrupeds, however, procure their nutriment about lakes and rivers; but no quadruped except the sea-calf searches for food about the sea. Among savage quadrupeds, also, are what is called castor, or the beaver, the *satherion*, the *satyrion*, the otter, and what is called the *latax*. But this animal is broader than the otter, and has very strong teeth; for 595 frequently going out by night, it cuts off with its teeth the bushes that grow on the banks of rivers. The otter, also, bites men, and does not, as they say, leave them till he hears the sound of their bones. But the hair of the *latax* is hard, and the form of the hair is a medium between that of the sea-calf and the stag.

6 Animals, also, that have serrated teeth drink lapping; and this is, likewise, the case with some that have not serrated teeth, as wild rats. But animals which have equal and continued teeth, as horses and oxen,

drink by attraction. The bear, however, neither attracts nor laps, but drinks by biting the water. All birds, likewise, except those that have long necks, drink by attraction; for those that have long necks drink by intervals and raising their head. But the porphyrio alone bites the water. All horned animals, however, both the tame and the wild, and such as have not serrated teeth, except the dog, feed on fruits and grass, unless they are very hungry; for the dog never eats fruits and grass. But the hog is the greatest eater of roots of all animals, because his snout is naturally adapted to this operation; and he, also, accommodates himself the most of all animals to every kind of food. He, likewise, becomes most rapidly bulky in proportion to the size of his body; for he becomes fat in the space of six days. The progress, however, which he makes in becoming fat is known by those who pay attention to things of this kind; who, in order that he may become fat, first attenuate his body by fasting. But he becomes fat after having previously fasted for three days; and nearly all other animals become fat through previous fasting. After three days, however, have elapsed, those that fatten swine feed them abundantly. But the Thracians when they fatten swine, on the first day give them drink; afterwards, omitting one day, they again give them drink; after which, they omit two, and then three and four days, and as far as to seven days before they give them drink. This animal, also, is fattened with barley, millet, figs, acorns, pears, and cucumbers. Quiet, however, especially contributes to fatten these animals, and others that have a good stomach; and it is advantageous to the fattening of swine for them to roll themselves in mud. Swine, also, feed gregariously according to their age [*i.e.* so that the younger feed with the younger, and the elder with the elder].

The hog, likewise, fights with the wolf; and when dead loses a sixth part of the weight which he had when living, this sixth part consisting in bristles, blood, and things of this kind. And sows, as well as all other female animals, become thinner when they suckle. Such, therefore, are the particulars respecting these animals.

7 Oxen, however, feed both on grain and grass. But they are fattened with grain which contains a considerable quantity of air, such as with the pulse called orobus, beans that are ground, and the leaves of beans; and if any one cutting the skin of the oldest oxen blows through the incision, and afterwards gives them food, they will become fat. They are, likewise, fattened by eating barley, whether whole or triturated; and, also, by eating sweet things, such as figs, grapes, wine, and the leaves of

the elm tree. But they are especially fattened by the sun, and the use of warm baths. With respect, likewise, to the horns of calves, if they are anointed with warm wax, they become supple, and may be easily bent as any one pleases; and calves suffer less pain in their feet, if their horns are anointed with wax, or pitch, or oil. Herds of oxen, likewise, suffer more inconvenience from frost than from snow. They, also, become larger, if they abstain for many years from venery. Hence the inhabitants of Epirus preserve what are called Pyrrhic[†] cows nine years without copulation; and they call them when they are increased *apotauroi*, or maidens. But of these cows they say that four hundred are the property of the kings of Epirus; and that it has been found by experience that they cannot live in any other climate.

8 Horses, however, mules, and asses feed, indeed, on grain and grass, but are especially fattened by drinking water. For beasts of burden enjoy their food in proportion to the quantity of water which they drink; and by how much less they refuse to drink, by so much the more easily they become fat. Forage, however, when it is full of juice causes their hair to be smooth; but it is not good for them when it is dry and hard. The first crop of lucern, also, is bad, as, likewise, is grass when it is watered with fetid water. Oxen, also, endeavour to drink pure water; but horses drink the same water as camels. And the camel drinks turbulent and thick water; so that he does not drink the water of rivers till he has first disturbed it. Camels, likewise, can live four days without 596
drink, and after this period they drink a great quantity of water.

9 The greatest quantity, also, of barley that the elephant eats at one time is nine Macedonian medimni.[‡] It is dangerous, however, to supply him with this quantity of food; but it is usual to give him six or seven medimni. Of meal, also, they give him five medimni; and of wine five mares. But a maris contains six cotylæ.[§] An elephant, likewise, has been known to drink at one time fourteen Macedonian measures of water, and on the evening of the same day eight measures more. Most camels, however, live about thirty years, and some live a much longer

† So called from Pyrrhus, the king.

‡ Almost two of our bushels.

§ A cotyle is the third part of a chœnix, and a chœnix is a measure containing about a quarter of a peck.

time. For some live a hundred years. But with respect to the elephant, some say that he lives three hundred, and others two hundred years.

10 Sheep and goats, however, feed on grass; and sheep, indeed, feed stably fixed to a certain place, but goats rapidly change their places while they feed, and only crop the top of the grass. Drink, likewise, especially fattens sheep; and hence, in summer, shepherds give a medimnus of salt every five days to a hundred sheep; for the leader of the flock becomes by this mean more healthy and fat. Frequently, also, they mingle a great quantity of salt with the straw which they give to sheep, in order that being thirsty they may drink more abundantly; and in autumn they give them a gourd sprinkled with salt; for this produces in them a greater quantity of milk. By walking, likewise, in mid-day, sheep drink more abundantly in the evening. And those sheep that have had salt in their nutriment, have their breasts when they bring forth more distended. The leaves of the olive, also, both cultivated and wild, the herb swine-snout, straw, and grass, fatten sheep; but all these fatten them in a greater degree when they are sprinkled with salt. These, likewise, fatten them still more, if prior to eating them they have fasted three days. In autumn, also, northern is better for sheep than southern water, and it is profitable to feed them in the evening. But journeys and labour render them thin. Shepherds, likewise, know strong sheep from those that are weak by this, that the former in winter have frost upon them, but the
596 latter have not; for sheep that are not strong, by the motion and agitation of their body [in consequence of the cold] shake off the frost. The flesh, also, of every quadruped that feeds in marshy places is inferior to the flesh of those quadrupeds that feed in more elevated places. Again, sheep with broad tails more easily bear the cold than those with long tails, and those that have short wool, than those that have wool in great abundance. Sheep, also, with curled wool bear the cold with more difficulty than others. Sheep, therefore, are more healthy than goats; but goats are stronger than sheep. But the skin and wool of sheep which have been devoured by wolves, and, also, the garments made from such wool, are much more subject to vermin than others.

11 With respect to insects, however, such as have teeth are omnivorous, but those that have only a tongue, are nourished by moisture which they suck with their tongue from a variety of substances. And of these latter, some, indeed, are omnivorous; for they taste all kinds of juices, as flies; but others are nourished with blood, as

the ox-fly, and the fly asilus; and others live on the juices of plants and fruits. The bee, however, is the only insect that sucks the juice of nothing putrid, and it employs as food no other substance than that which has a sweet juice. Bees, also, drink the most sweet water, where it runs limpid and pure. Such, therefore, is the food of the genera of animals.

12 All the actions, however, of animals are conversant about copulation, the nurture of their young, the means of procuring sufficient nutriment, and providing a habitation which may defend them from the inconveniences arising from heat and cold, and the mutations of the seasons. For all animals have a connascent sense of the mutation according to heat and cold. And as of men, some in winter betake themselves into houses, but others, who have dominion over a large tract of land, spend the summer in cold and the winter in warm places; the like, also, happens to animals that are able to change the places of their abode. And some, indeed, find the assistance of which they are in want in the places to which they are accustomed; but others for this purpose migrate; betaking themselves after the autumnal equinox from Pontus 597
and other cold places, that they may avoid the approaching winter. After the vernal equinox, however, they migrate from warm to cold places, fearing the heat; some, indeed, departing from neighbouring places, but others, as cranes, from one extremity of the earth to another. For they migrate from the Scythian plains, to the marshes† above Egypt, whence the Nile flows. But these places [to which the cranes migrate] are inhabited by the Pygmies. For this is no fable, but both the men that inhabit these places and the horses are, as they are said to be, a small race; and the men dwell in caverns.

Pelicans, likewise, migrate and fly from Strymon [a river of Thrace] to the Ister [or Danube], and there beget offspring. But they fly in a great troop, those that are before waiting for those behind, because when they fly above mountains, those that are behind cannot be seen by those before.

After the same manner, also, fishes [migrate from one place to another], some, indeed, from Pontus, but others to Pontus. And some, indeed, in the winter migrate from the sea to the land, in search of warmth; but in the summer migrate from the shore to the sea, avoiding the heat. More imbecil birds, likewise, [which cannot migrate from one

† These marshes are in Ethiopia, and near the equator.

region to another], in winter descend from elevated places to the plains, in search of heat; but in summer ascend to the mountains, in order to avoid the heat. More imbecile animals, however, migrate prior to others, both in excessive heat and cold. Thus, for instance, mackerel migrate prior to the tunny, and quails prior to cranes. For mackerel and quails begin to migrate in the month of August, but the tunny and cranes in the month of September. But all animals are fatter, when they migrate from cold than when they migrate from hot places. Thus, for instance, quails are fatter in autumn than in spring. But it happens that at one and the same time they migrate from cold places and from the season of heat. Animals, also, are more impelled to copulation in the spring season, and when they migrate from warm places.

Of birds, therefore, cranes, as we have before observed, migrate from
597 one extremity of the earth to another; but they fly in the direction of the wind. What is asserted, however, respecting a stone is false. For it is said, that they carry a stone in order to balance them as they fly; which stone, when it falls, is useful for the purpose of trying gold.

The large ring-dove, also, and the stock-dove migrate, and do not stay here during the winter; and this is, likewise, the case with the swallow and the turtle-dove; but pigeons stay with us. In a similar manner quails migrate, except some few quails, and, also, turtle-doves, which winter in places exposed to the sun. But the large ring-dove and the turtle-dove fly in troops, both when they approach to us, and again when the season arrives of their departure. Quails, however, when they fly, if the air is serene or the wind northern, assemble together, and accomplish their journey prosperously. But if the wind is southern, their journey is difficult, because they are then impeded in their flight; for this wind is moist and heavy. Hence those that hunt quails endeavour to catch them when the south wind blows. These birds, likewise, fly with difficulty on account of their weight; for their body is large; and on this account they make a noise when they fly, for flying is laborious to them. When, therefore, they come to Greece, they are without leaders, but when they depart from it, they have leaders from birds of a different species, *viz.* the *glottis*, the quail-mother, the great owl, and the *cychramus*, the last of which recalls the quails to their journey by night. Hence fowlers, when they hear the voice of the *cychramus*, know that the quails are about to depart. But the quail-mother is in form similar to the birds that live near lakes; and the glottis has a tongue [from whence it is denominated], which extends to a great length out of its bill. The great owl, however, resembles the horned owl, and has little wings about its ears. But some call it the raven of the night. This bird, also, is loquacious and an imitator, and is caught by imitating the leaping of one of the fowlers, and is thus surrounded by another of them and taken, in the same manner as the horned owl. In short, all birds that have crooked claws have a short neck, and a broad tongue, and are imitative. For the Indian bird, the parrot, which is called *human tongue*, is a bird of this kind, and becomes more loquacious when he has drank wine. The crane, however, the swan, the pelican, and the small goose, are gregarious birds.

13 Of fishes, however, some, as we have before observed, migrate from the sea to the land, and from the land to the sea, avoiding the 598
excesses of heat and cold. But fishes that live near the shore are better than those that live in the deep sea; for they have more abundant and better nutriment. For the productions of the place which the sun irradiates with his beams are more abundant, better, and more delicate, as is evident in gardens; and the black algula grows near the shore, but the other species of algula resembles a wild plant. Farther still, places near the sea-shore are well tempered with heat and cold; on which account the flesh of such like fishes has more consistence; but the flesh of fishes that live in the deep sea is moist and loose. But the fishes that live near shores are, the dental, the cantharus, the orphus, the goldfish, the mullet, the barbel, the thrush, the sea-dragon, the callionymus, the gudgeon, and all those that live in rocks. Those, also, that live in the deep sea are, the pastinaca, the selache, the white congers, cockles, the roach, and the *glaucus*. But the pagrus, the scorpion, the black conger, the lamprey, and the cuculus or gurnard, live near shores, and, also, in the deep sea. There are, however, differences of these according to places. Thus, for instance, about Crete, gudgeons and all fishes that live in rocks are fat. But the tunny is better after the rising of arcturus, because at that time it is no longer infested by the marine asilus or gadbee; for in consequence of the molestation which it suffers from this animal in summer, it is not so good. Many fishes, also, are produced in maritime lakes, such as the stock-fish, the goldfish, the barbel, and nearly most other fishes. The amiæ, likewise, are produced in Alopeconnesus; and the greater part of the genera of fishes are to be found in the lake Biston.

The greater part, however, of the fishes called *collias* do not enter the Euxine Pontus, but in summer live in Propontis, and there bring forth, but they winter in the Ægean sea. The tunny, however, the pilcher,

and the amiæ, migrate into the Euxine Pontus in spring, and continue there during the summer. But this is, also, the case with nearly most of the fishes called *ruades*, and most of the gregarious fishes; and most fishes are gregarious, and those that are so have a leader. All those, however, migrate into the Euxine Pontus, on account of nutriment; for the food in that sea is better, and more abundant on account of its admixture with potable water; and, also, because this sea is less infested with large voracious fishes. For except the *phocene* and the dolphin, there is no
598 voracious fish in the Euxine Pontus; and the dolphin in this sea is small; but in the sea which immediately joins with Pontus there are large dolphins. The tunny, therefore, and other fishes migrate into the Pontus on account of nutriment, and for the sake of their offspring. For there are places in this sea adapted to the delivery of fishes; and the potable and more sweet water which it contains nourishes the little fishes. When the fishes, however, are delivered, and their progeny have acquired a larger size, they again swim out of the Pontus after the rising of the pleiades [*i.e.* in November]. And, indeed, if the winter is southern [*i.e.* if the southern winds blow during the winter], they migrate from this sea more slowly; but if the winter is northern, they migrate more rapidly; because the northern winds are favourable to their egress from the Pontus, and at the time of their egress, the offspring of these fishes, which are then small, are caught about Byzantium, in consequence of not continuing long in the Pontus. Other fishes, therefore, are seen both when they enter into and when they depart from the Pontus, but the *trichiæ* alone, are, indeed, caught when they enter into the Pontus, but are not seen when they depart from it. And if it happens that the Byzantine fishermen catch one of these fishes at the time in which they depart from the Pontus, they purify their nets, because it is not usual for these fishes to be seen when they leave the Euxine Pontus. The reason, however, why the trichiæ alone are not seen to depart from the Pontus, is because after they have entered into it, they are the only fishes that enter into the Danube; and afterwards swim through the Danube, where it is divided into the Adriatic sea; of which this is an indication, that what happens there is contrary to what happens in the Pontus, for when they swim into the Adriatic they are not caught, but when they swim out of it they are caught. But the tunny swim into the Pontus on the right hand, keeping close to the land; and they swim out of the Pontus on the left hand. They do this, however, as some assert, because they naturally see more acutely with the right eye, but with the left eye they do not see acutely.

The ruades, therefore, make their voyage in the day, but rest and feed by night, if the moon does not shine; for if it does, they continue their voyage, and take no rest. But it is said by some of those that dwell near the sea, that during the winter solstice these fishes do not move, but continue quiet in the place where they may happen to be found, till the vernal equinox. The colias, therefore, is caught when it enters into the Pontus; but is not caught during its egress from that sea. Those, also, of them that are in the Propontis are the best prior to their delivery. But the other ruades are more frequently caught on their egress from the Pontus, and they are then the best. When, however, they enter the Pontus very near the Ægean sea, and are caught, they are then the fattest; but the more they ascend towards the Pontus the thinner they 599
become. Frequently, likewise, when the south wind blows in a contrary direction, they swim in company with the colias and mackerel, and are more frequently caught below than about Byzantium.

After this manner, therefore, the migrations of fishes are accomplished. The very same thing, also, takes place in terrestrial animals, with respect to concealing themselves. For in winter they are naturally impelled to hide themselves, but quit their retreats in the warmer season of the year. There are, also, animals that hide themselves as a defence both against heat and cold. And of some animals, the whole genus hides itself; but with respect to others, one part of the species hides itself, and the other does not. For all testaceous animals hide themselves; as, for instance, in the sea, the purples and buccina, and every other such like genus of animals. The retreat, however, of animals which are not fixed to one place is more manifest, for some of these at certain times withdraw to their retreats, as cockles; but others conceal themselves under a covering with which they are invested, as land-snails. The mutation, however, of those that are fixed to a place is immanifest. All testaceous animals, likewise, do not hide themselves at the same season of the year; but snails, indeed, hide themselves in the winter, and the purple and the buccina, during the time of the dog-star, hide themselves for thirty days. Cockles, also, hide themselves about the same time. Most of them, however, hide themselves in very cold and very warm weather.

14 Nearly all insects, likewise, hide themselves, except those that dwell in houses together with men, and such as perish before the end of the year. But insects hide themselves in the winter, and some, indeed, conceal themselves for many days, but others, as bees, only during more winterly days; for bees, also, hide themselves. That bees, however, hide

themselves is indicated by this, that [at certain times] they do not appear to taste any of the food placed before them, and if any bee then creeps out of its retreat, it is seen to be transparent, and manifestly contains nothing in its belly. Their inactivity, also, continues from the rising of the pleiades [*i.e.* from the month of November] to the spring. Animals, however, form their retreats and conceal themselves in warm situations, and in places where they were accustomed to dwell.

15 Many sanguineous animals, likewise, hide themselves; as, for instance, those that are covered with scales, serpents, and lizards, stellions, and river crocodiles. For these animals conceal themselves during the four more winterly months, and do not then eat any thing.
599 Other serpents, therefore, hide themselves in the earth; but vipers conceal themselves under stones.

Many fishes, also, hide themselves; but most manifestly the hippurus[†] and the coracinus[‡], in winter. For these fishes alone are never caught, except at certain regular times, and always at the same times. But nearly all other fishes hide themselves; as, for instance, the lamprey, the *orphus*, and the conger. Fishes, likewise, that live in rocks conceal themselves in couples, the males with the females, in the same manner as they build their nests; as, for instance, the thrush, the merula[§], and the perch. The tunnys, also, conceal themselves in winter in the depths of the sea, and become very fat after their concealment. The fishing for them, likewise, commences from the rising of the pleiades, and continues till the setting of arcturus; but at every other time they are tranquil and concealed. Some of them, however, are caught about the time of their concealment, as is, also, the case with some other fishes that hide themselves, at the time when they betake themselves to warm places, or when the fine weather is extended to an unusual length; for they sometimes come forth from their retreats, for nutriment, and especially during the full moon. Many fishes, also, that hide themselves are most pleasant to the taste. The *primadiæ*, however, conceal themselves in mud; of which this is an indication, that at that time they are not caught, but are seen to have a great quantity of mud on their back, and their fins ruffled. They begin to move, however, in the spring, and come forth on the land, for the

† A kind of lobster.

‡ A blackish sea-fish, having a head shining like gold.

§ A fish resembling a perch.

sake of copulating and bringing forth. They are, also, caught when pregnant, and then appear to be seasonable; but the autumnal and winter primadiæ are not so good. At the same time, likewise, the males are seen to be full of genital seed. While the fœtus, therefore, of the primadiæ are yet small, it is difficult to catch them, but when they are larger they are caught in great abundance, on account of the gad-flies with which they are infested. Some fishes, likewise, hide themselves in sand, but others in mud, having no other part uncovered but their mouth. Most fishes, therefore, hide themselves in winter; but such as are crustaceous, those that live in rocks, the thornback, and the selache, conceal themselves only during more winterly days; but this is evident from their not being caught when it is cold. Some fishes, however, as, for instance, the *glaucus*, conceal themselves in summer. For this fish hides itself in summer for about sixty days. The fish called the ass, likewise, and the gold-fish, hide themselves. But this appears to be an indication that the ass hides itself for a long time, *viz.* that for a long time it is not caught. And that some fishes hide themselves in summer appears to be indicated by this, that they are caught during the rising of certain stars, and especially of the dog-star. For then the sea is subverted; which is particularly the case in the Bosphorus. For the mud ascends to the surface, and the fishes are carried up with it. It is, likewise, said, that if the bottom of the sea is frequently scraped [with a drag-net], more fishes will be caught in the same net at the second throw than at the first. But when great rains take place, many animals are seen [in the sea], which before were either not seen at all, or were not frequently seen.

16 Many birds, likewise, hide themselves; nor do all of them, as some fancy, migrate to warmer climates; but some, indeed, being near to places of this kind in which they perpetually dwell, as, for instance, kites and swallows, migrate to them; and others being more remote from places of this kind, do not migrate, but conceal themselves. For many swallows have been seen in hollow places entirely stripped of their feathers, and kites have been seen flying from such like places, when they first appear. Those birds, however, that have crooked, and those that have straight claws, equally hide themselves. For the stork, the blackbird, the turtle-dove, and the lark conceal themselves. And that the turtle-dove, indeed, hides itself is universally acknowledged. For no one, as I may say, is reported to have seen the turtle-dove any where in winter. But this bird when it begins to hide itself is very fat, and

though it loses its feathers during its retreat, yet it still continues to be fat. Of ring-doves, however, some hide themselves, but others do not, but migrate together with swallows. The thrush and the starling, also, hide themselves, and for a few days, among birds that have crooked claws, the kite and the owl.

17 Among viviparous quadrupeds, also, porcupines and bears hide themselves. That savage bears, therefore, hide themselves is evident; but it is dubious whether it is from cold, or some other cause. For both the males and females become about this time very fat, so as to be unwieldy. The female, also, brings forth at this season, and conceals herself till the period arrives of leading forth the young bears from her retreat. But she does this in the spring, and in the third month after the winter solstice. The bear, likewise, is concealed for about forty days at least. And it is said, that for fourteen of these days it does not move at all, but in most of the days after these it is concealed, indeed, yet is in a vigilant state, and in motion. A pregnant bear, however, has either never been caught by any one, or by very few. It is evident, also, that the bear during the time of its concealment does not eat any thing; for it does not come forth from its retreat: but when it is caught, both its belly and intestines are found to be empty. It is, likewise, said, that the intestine of the bear at that time is nearly adherent, and that on this account, as soon as it quits its retreat, it eats a certain herb which is called *arus*, for the purpose of relaxing and separating the intestine. The dormouse, also, conceals itself in trees, and at that time becomes very fat; and the white Pontic rat, likewise, hides itself.

But of those that hide themselves, some put off what is called their old age. This, however, is their exterior skin, and the covering which they received in their formation. With respect to pedestrious and viviparous animals, therefore, it is dubious why among these the bear hides itself, as we have before observed; but nearly the greater part of those that are covered with scales hide themselves; and such as have a skin soft, and not of a testaceous nature, as the tortoise, put off their old age. For the tortoise and emys [or aquatic rat] rank among animals that have a scaly skin. Thus, for instance, the stellio and the lizard, and especially serpents, divest themselves of their old age in spring, and again in autumn. Vipers, also, put off their old age in spring and autumn; for that which is asserted by some is false, *viz.* that vipers are the only serpents that do not divest themselves of their old age. When, however, serpents begin to divest themselves of their old age, they say that it is first drawn off from their eyes, so that they appear to be blind to those who do not perceive this circumstance; but afterwards, it is drawn off from the head, and then from the rest of the body. In a night and a day, also, nearly the whole of their old age is drawn off, beginning from the head, and ending at the tail. But the inward part of the skin of their old age, when it is drawn off, becomes outward. For it is put off in the same manner as an embryo from the chorion, in which it is enveloped.

Such insects, likewise, as put off their old age are divested of it after the same manner; as, for instance, the silk-worm, the gnat, and those whose wings are enclosed in a sheath, as the beetle. None of them, however, put off their skin, till a new one is formed under it. For as the fœtus of viviparous animals is enveloped in the chorion, thus, also, the fœtus which originate from worms are enveloped in a shell, similarly with bees and locusts. Grasshoppers, also, when they come forth from their shell sit upon olive-trees and reeds; but the shell bursting they come forth, leaving but little moisture within it; and in no long time afterwards they fly and sing.

Of marine animals, also, locusts and lobsters put off the skin of their old age, sometimes, indeed, in the spring, and sometimes in the autumn after the birth of their young. For certain locusts have been caught having the parts about the thorax soft, because the shell was on all sides burst; but the parts beneath their thorax were hard, because the shell in those parts was not yet burst. For they do not divest themselves of the skin of their old age in the same manner as serpents. Locusts, also, hide themselves about five months. Crabs, likewise, put off their old age; confessedly, indeed, such of them as have a softer shell; and it is said that this is, also, the case with such of them as have a harder shell, as, for instance, with the crabs called *maiæ*. But when they have put off their skin, a very soft shell is formed [in the place of the old]; and crabs are not very well able to walk. Animals, likewise, of this kind do not put off their skin once, but frequently. And thus much concerning animals that hide themselves and put off their old age.

18 Animals, however, do not live prosperously at the same seasons, nor similarly happy in all excesses of heat and cold. Farther still, health and disease vary according to seasons in animals of different genera, and on the whole are not the same in all animals. Hot and dry seasons, therefore, are advantageous to birds, both with respect to their health and delivery, and especially to ring-doves; but rain is advantageous to fishes, a very few excepted. On the contrary, rainy seasons are

disadvantageous to birds, but hot and dry seasons to fishes. For, in short, it is not advantageous towards to drink much. Birds, therefore,
601 with crooked claws, as we have before observed, live entirely without drink. Hesiod, however, was ignorant of this; for in his narration of the siege of Ninus, he represents the eagle the assessor of the divination drinking. Other birds, indeed, drink, yet do not drink much. In a similar manner, neither does any other animal drink largely that has spongy lungs, or is oviparous. When birds, however, are afflicted with any malady, it is indicated by their feathers; for they are disordered, and are not disposed in the same manner as when they are well.

19 But of fishes, the greater part, as we have before observed, are in a prosperous condition in rainy seasons. For then they not only have more food, but, in short, rain-water is advantageous to them, in the same manner as it is to the productions of the earth. For leguminous substances, though they are watered, yet make a greater proficiency when irrigated by rain. The very same thing, also, happens to reeds which grow in lakes; for none of them, as I may say, are increased without rain. That fishes, however, are delighted with sweet water is indicated by this, that most of them migrate into the Euxine Pontus, for the purpose of staying there during the summer; for on account of the multitude of rivers which flow into the Pontus, the water of it is more sweet, and the nutriment it contains more abundant. Farther still, many fishes, such as the amia and the mullet, swim into rivers, and in rivers and lakes are in a prosperous condition. But gudgeons become fat in rivers; and, in short, places which abound with lakes have the best fishes. With respect to waters themselves, however, summer rains are more advantageous to most fishes, and, also, spring and autumnal rains; and fishes are benefited by a serene winter. In short, those seasons which are favourable to man, are, also, favourable to the greater part of fishes.

But fishes do not prosper in cold places. And those that have a stone in their head, as the *chromis*, the wolf, the umbra, and the pagus suffer most in the winter; for on account of the stone they are congealed by the cold, and fall, deprived of life. To most fishes, therefore, rain is
602 profitable; but, on the contrary, it is noxious to the mullet, the cephalus, and the fish which some call *murin*. For many of them become rapidly blind by rain-water, if it falls in great abundance. And the cephalus is more accustomed to this disease of blindness in winter. For the eyes of this fish become white, and at that time it is lean when it is caught; and at length its eyes are entirely destroyed. It seems, however, that they are

not so much affected in this manner by excessive rain, as by cold. In other places, therefore, and, also, about the Argive Nauplia, near to Tenagos, many of the cephali are caught blind in very cold weather. Many, also, have been caught whose eyes were white. The gold-fish, likewise, suffers in winter; but the *acharnas* in summer, at which time it becomes thin. The *coracinus*, however, as I may say, is the only fish that is benefited by more dry seasons, and this because it happens that in dry seasons there is more heat.

Such places, likewise, as are naturally near shores are more favourable to the prosperous condition of fishes than those that are in the deep sea; but such fishes as are amphibious prosper either near shores or in the deep sea. There are, however, peculiar places which are favourable to particular fishes; and, in short, those places are advantageous to them which abound with weeds. Fishes, therefore, which feed in all various places, are fatter when they are caught in weedy places. For those that feed on weeds have an abundance of nutriment, and those that feed on flesh find a greater number of fishes for their food in weedy places.

It is, also, of consequence to fishes, whether places are northern or southern. For long fishes are in a better condition in northern places, and, therefore, in summer, more long fishes are caught than broad fishes in northern situations in the same region. But the tunny and the sword-fish are tormented with the gad-fly about the time when the dog-star rises; for then both of these fishes have under their fins a little worm as it were, which is called *oistros*, and which resembles, indeed, a scorpion, but in magnitude is equal to a spider. These fishes, however, suffer so much from this insect, that the sword-fish sometimes through pain leaps no less than the dolphin; on which account it frequently falls into ships. The tunny, likewise, delights in heat the most of all fishes, and for the sake of heat betakes itself to the sand on the shore, and swims on the 602
surface of the sea that it may become warm. Small fishes, also, are preserved from the tunny and sword-fish because they escape their notice; for large fishes pursue such as are large. A great part, however, of the eggs and spawn of fishes perish on account of heat [*i.e.* in consequence of fishes going in search of warmth], for through this they injure whatever they touch.

Fishes, likewise, are principally caught before the rising and after the setting of the sun; and, in short, about the rising and setting of the sun; for the castings of the nets are then said to be seasonable. Hence fishermen at this time draw up their nets; for at these periods the sight of fishes is especially deceived. For they rest at night, and see more clearly in proportion as there is more light.

Fishes, also, do not appear to be subject to any contagious maladies, such as frequently happen to men, and, among viviparous quadrupeds, to horses and oxen, and to some other animals, both tame and wild. Fishes, however, appear to be sometimes ill, of which this is conceived by fishermen to be an indication, that some fishes when caught are thin, and resemble animals that are unwell. Their colour, also, is found to differ from that of many others that are caught at the same time, that are fat, and of the same genus. And thus much concerning marine fishes.

20 With respect, however, to fishes that live in rivers and lakes, these, also, are not subject to contagious maladies; but some of them are afflicted with peculiar diseases. Thus, for instance, the glanis in consequence of swimming on the surface of the water is injured by the deadly influence of the dog-star, and is laid asleep by very loud thunder. Sometimes, likewise, the carp is affected in this manner, but in a less degree. But the glanis perishes when struck in shallow water by the serpent called a dragon. In the *ballerus* and the *tillon*, however, a worm is engendered under the dog-star which causes them to rise to the top of the water, and renders them weak. And when they are raised to the top of the water, they are destroyed by the heat. But the *chalcis* is afflicted with a very violent disease; for it is destroyed by a great quantity of lice produced under its gills. No other fish, however, is subject to a disease of this kind. Fishes, likewise, die by eating the herb long-wort; and hence some fishermen that they may catch fishes infect rivers and lakes
603 with the poison of this herb; but the Phœnicians after this manner catch fishes in the sea. Some, also, adopt two other modes of catching fishes. For because fishes in winter avoid the profundities of rivers (and potable or sweet is colder than salt water), hence fishermen dig a canal in a river, and cover it with herbs and stones, thus forming as it were a cavern, having its egress from the rivers; and when it is frosty weather from this canal they catch fishes in a bow-net. Another similar mode, also, of catching them is employed by fishermen, both in summer and winter. For in the middle of the river they make an enclosure with small branches of trees and stones, so as to leave an avenue, and in this placing a bow-net, the stones being moved, the fishes are caught [while they run to the avenue].

Rainy seasons, however, are advantageous to testaceous animals, except the purples; of which this is an indication, that if the purples taste of the potable water which runs into the sea, they die on the same day in which they taste it. The purple, likewise, lives after it is caught about fifty days. But these fishes are nourished by each other; for a certain weed as it were and moss grow on their shells. Fishermen, therefore, throw food to the purples, in order that they may weigh more and sell for a greater price. To other testaceous animals, however, dry seasons are disadvantageous, for in these seasons they are of a less size, and of an inferior quality; and cockles then become more red. In the Pyrrhæan Euripus, however, cockles sometimes fail, not only on account of the instrument with which they are caught by fishermen, but, also, on account of dry weather. But rainy seasons are advantageous to other testaceous animals, because the sea-water then becomes sweeter. Testaceous animals, however, are neither produced in the Euxine Pontus nor in rivers, on account of the cold, except a few of the bivalves; but the univalves are especially congealed by cold. After this manner, therefore, aquatic animals subsist.

21 Of quadrupeds, however, swine are subject to three diseases; one of which, indeed, is called a quincy; in which disease, the jaws and throat are principally inflamed. This disease, also, attacks any other part of the body; for it frequently attacks the foot, and sometimes the ear. But the diseased part becomes immediately putrid, and then the part next to this, till the ulcer arrives at the lungs; and then the animal dies. This disease, likewise, is rapidly increased; and swine when they are attacked by it in any part of the body eat nothing. Swine-herds, however, cure this disease when they perceive it is but small in no other way than by cutting off the whole of the infected part.

Swine, also, are subject to two other diseases, both which are called the evil; of which one is a pain and heaviness of the head, with which most swine are afflicted; but the other is a flux of the belly, and this appears to be incurable. The former malady, however, is alleviated by applying wine to the nostrils, and washing them with it. Swine, also, with difficulty escape from this disease; for it destroys them in three or four days. But swine principally become subject to a sore throat, when the summer supplies them with abundance of food, and they are very fat. This malady, however, is remedied by giving them the fruit of the mulberry-tree, by washing them abundantly in a hot bath, and by opening a vein which is under the tongue.

Swine, likewise, whose flesh is humid are afflicted with a malady resembling hail-stones in their legs, neck, and shoulders, in which parts many hail-stones are formed. And the flesh, indeed, of those swine that have but few hail-stones is sweet; but the flesh of those that have many becomes very humid and insipid. It may, also, be clearly seen when swine have this disease. For they have hail-stones in the inferior part of the tongue; their hairs when plucked from their back appear bloody; and, farther still, they cannot stand on their hind feet. While they suck, however, they are not afflicted with this malady. And they are liberated from this disease by that kind of corn which is called *tiphe*, which is useful for food. But in order to fatten them peas and figs are the best nutriment; and, in short, various and not simple food is best calculated for this purpose. For, like other animals, they are delighted with change of food; and at the same time they say, that by this mean they become partly inflated, partly fleshy, and partly fat. They, also, add, that they 604 willingly eat acorns, but that this nutriment renders their flesh humid, and that pregnant sows if they eat abundantly of acorns become abortive, in the same manner as sheep; for sheep are more manifestly thus affected by eating acorns. Swine, also, are the only animals with which we are acquainted that are afflicted with the malady of hail-stones.

22 Dogs, also, are subject to three diseases; and these are denominated, rage, a sore throat, and gout. But of these, rage produces madness; and all animals that are bit by dogs when they have this disease become furious, except man. This disease, also, destroys dogs, and all animals that are bitten by mad dogs, except man.† A sort throat, likewise, destroys dogs; and but few of them escape death when attacked by the gout. Camels, also, are subject to rage. But it is said that elephants are free from other diseases, and that they are afflicted with flatulency.

23 Moreover, gregarious oxen are subject to two diseases; of which, one is the gout, but the other is called the evil. In the gout, therefore, the feet of oxen swell, but they neither die, nor lose their hoofs; and they are relieved, by anointing their horns with warm pitch. But when they have the evil, their respiration is hot and frequent; and what the fever is in men, that the evil is in oxen. Indications, also, that oxen are unwell are, the hanging their ears, and an incapacity of eating. But they die rapidly, and when dissected their lungs are found to be putrid.

† As this is not the case at present, perhaps, in the days of Aristotle, a method was known of curing the bite of a mad dog in the human species.

24 Of horses, however, those that are gregarious, indeed, are not subject to any other disease than the gout. But with this disease they are afflicted, and sometimes through it lose their hoofs; though when they lose them, others immediately grow in their place; for the loss of one hoof is accompanied with the production of another. The indications, also, of disease in a horse are, the leaping of the right testicle, and a certain hollow and wrinkled substance being formed a little below the nostrils. But horses which feed in stables are subject to many diseases. For [in the first place] they are afflicted with the colick; which is indicated when they lift their hind to their fore legs, so as that they knock against each other. But if after they have fasted a few days they become furious, they are relieved by opening a vein, and taking away some blood. They are, likewise, subject to a disease called *tetanos* [or an equable and perpetual convulsion of the whole body]; which is indicated by a tension of all the veins, and of the head and neck, and by walking with their legs stiff. Horses, also, are troubled with purulency. And they are afflicted with another malady which is called barley, and which is indicated by the palate becoming soft, and the respiration fervid. But this disease is incurable, unless the horse that is afflicted with it becomes spontaneously convalescent. Horses, likewise, are subject to another malady, by which they are said to be lymphatic; in which disorder it happens that they stand still at the sound of a flute, and bend their head; and when any one gets on their back, they run with great rapidity till they are stopped. A horse, also, if he is seized with rage, is always sorrowful. But the indications of this malady are the inclining his ears to his mane, and again extending them, fainting away, and breathing short. The following maladies, also, to which horses are subject are incurable. A pain in the heart, an indication of which is a subsidence of the flank. Secondly, a displacing of the bladder, indications of which are the inability of making water, and drawing up the hoofs and the hip. Thirdly, the swallowing the insect *staphulinus*, which is about the size of a *sphondule* [or book-worm]. The bite, also, of a shrew-mouse is pernicious both to horses and to other cattle; for it produces pustules. But the bite is more pernicious, if the shrew-mouse happens to be pregnant, for the pustules caused by the bite burst; but this is not the case if the shrew-mouse is not pregnant. The bite, also, of an animal which some call *chalcis*, and others *zygnis*, either kills horses, or causes them to suffer great pain. But this animal resembles a small lizard, and its colour is that of the blind serpent. In short, those that are skilful say that horses and sheep are nearly subject to the same diseases as men. A

horse, likewise, and every beast of burden, are destroyed by the poison of red arsenic. But this poison is given percolated in water. The smell, also, of an extinguished lamp causes a pregnant mare to become abortive.
605 And this, likewise, happens to be the case with some pregnant women.

After this manner, therefore, the diseases of horses subsist. But what is called the hippomanes grows, indeed, as it is said, on the forehead of colts; and mares, licking and wiping off the hippomanes, devour it. Other particulars, however, respecting the hippomanes may be considered as the fables of women and enchanters. But it is acknowledged that mares prior to their delivery emit a liquor which is called *polion*.

Horses, likewise, on hearing, know the voice of the horses with which they happen to fight. But horses delight in meadows and marshes; for they drink turbid water; and if the water happens to be pure and clear, they disturb it with their hoofs, and after they have drank wash themselves in it. For, in short, the horse is an animal which loves to wash itself, and is delighted with water. Hence the nature of the river horse is such that it cannot live without water. The ox, however, is in this respect contrary to the horse; for unless the water is pure, and cold, and clear, he is unwilling to drink.

25 Moreover, asses are principally subject to one disease, which is called *melida*. But this malady is first produced in the head, and causes a thick and red phlegm to flow from the nostrils. And if this malady descends to the lungs, the ass dies; but when it is confined to the head, it is not mortal. The ass, likewise, endures cold with very great difficulty; and on this account, there are no asses about the Euxine Pontus and Scythia.

26 But elephants are subject to flatulent diseases, through which they are rendered incapable of ejecting either the moist excrement [*i.e.* urine] or the excrement of the belly. They are, also, diseased if they feed on earth, unless they eat it very frequently; for then they sustain no injury; and sometimes they even swallow stones. Elephants, likewise, are subject to a flux of the belly; but when they are afflicted with this malady, their keepers cure them by giving them hot water to drink, and hay steeped in honey to eat, and each of these stops the complaint. When they are weary, likewise, from want of sleep, they are restored to health by rubbing their shoulders with salt and oil and warm water.
605 When, also, they have a pain in their shoulders, their keepers give them

the roasted flesh of swine, and by this mean afford them relief. But with respect to oil, some elephants drink it, and others do not. If, likewise, it should happen that a dart remains in the body of an elephant, it is said that it is ejected if he drinks oil; but to those elephants that do not drink oil, they give a root boiled in oil. And thus much concerning the diseases of quadrupeds.

27 Of insects, however, the greater part are in a prosperous condition when the temperature of the year resembles the season in which they were born, *viz.* when it resembles the spring, being moist and tepid. But in the hives of bees certain animals are produced which injure the hives; and, also, a worm of the nature of a spider, which weaves a web, and is detrimental to the hives. This worm is called by some *cleros*, and by others *purauste*; and it produces in the hive an animal resembling itself, *viz.* having the form of a spider, and causes the swarm of bees to be diseased. Another insect, also, resembling the butterfly which flies about lamps, is noxious to bees. This insect brings forth a certain substance full of dust, and is not stung by the bees, but is alone expelled by smoke. Caterpillars, also, are produced in hives which they call augers, and which the bees do not attack. But bees are principally subject to diseases when the flowers of trees are spoiled with rust, and in dry seasons. All insects, likewise, perish when rubbed with oil; but most rapidly when, their head being rubbed with oil, they are exposed to the sun.

28 In short, animals differ according to places. For as in some places certain animals are not at all produced, so in others they are produced, indeed, yet they are smaller, live a shorter time, and are subject to a greater number of maladies. Sometimes, also, there is a difference of this kind in places near to each other. Thus, for instance, there are certain places in Milesia, which border on each other, in some of which grasshoppers are produced, and in others not. In Cephalenia, likewise, a river separates certain places, in some of which there are grasshoppers, and in others there are not. In Pordoselene, also, there is a road between two places, in one of which there are weasels, but there are none in the other. And in Bœotia, there are many moles about Orchomenus; but in the Lebadic vicinity there are none; and even if 606
they are brought to this place they are unwilling to dig into the earth. But in Ithaca, the hares which are brought thither from other places cannot live, but they are seen dead turned towards the sea through which they entered the island.

And in Sicily, indeed, there are no hippomyrmeces, or horse-ants; but in Cyrene, there were formerly no croaking frogs. In the whole of Libya, likewise, there is neither a wild boar, nor a stag, nor a wild goat. But in India, as Ctesias says, though he is not worthy of belief, there is neither a tame nor a wild hog; but sanguineous and all such animals as hide themselves [at certain times of the year] are there very large. And in Pontus, indeed, there are neither the mollia, nor testaceous animals, except in some few places; but in the sea of Erythra all the testaceous animals are excessively large. In Syria, also, the sheep have tails a cubit in breadth; but the goats have ears of the length of one palm and four breadths of a finger, and some have their ears hanging towards the earth. The Syrian oxen, likewise, and the camels have manes on the top of their shoulders. And in Cilicia the goats are shorn in the same manner as sheep in other places. In Libya, also, the horned kind of rams have horns as soon as they are born, not only the males, as Homer says, but, likewise, the females; but the contrary takes place in Pontus, about Scythia, for the rams are without horns.

And in Egypt, some animals, as oxen and sheep, are larger than those in Greece; but others, as asses, wolves, hares, foxes, ravens, and hawks, are less; and others are of a similar size, as crows and goats. Nutriment, however, is assigned as the cause of this difference; because some animals are supplied with it abundantly, but others sparingly. Thus, for instance, wolves and hawks, which are carnivorous animals, are supplied with but little food; for the small birds on which they feed are rare. Hares, also, and such animals as are not carnivorous, have but little food, because neither the fruits which have a hard covering, nor those which are enveloped with a soft pulp, are of a long continuance. In many places, likewise, the temperature of the climate is the cause of this difference. Thus, for instance, in Illyria, Thrace, and Epirus, the asses are small; but in Scythia and Celtica there are no asses at all; for the cold in those climates is very severe. But in Arabia, lizards are more than a cubit in length; and the domestic rats are larger than the field rats. The fore legs, also, of these rats are one palm in length, and the hid legs are as long as the first joint of the finger. But in Libya, the magnitude of the serpents is said to be immense. And some persons that have sailed to Africa relate that they have seen the bones of many oxen, which it was evident to them had been devoured by serpents. They add, that on their leaving the coast, the serpents rapidly pursued their three-ranked gallies, some of which they attacked and over-turned.

Farther still, there are more lions in Europe, and in that part of Europe which is situated between Achelous and Nessus [than in Asia]. And there are panthers in Asia, but not in Europe. In short, wild beasts are more savage in Asia, but all of them are more courageous in Europe, and those in Libya have a greater variety of forms. Hence it is proverbially said, that Libya always produces something new. For in consequence of the want of rain in Libya, wild beasts which are not in of the same class assemble at the same waters; and those whose times of gestation are the same, and who do not much differ in magnitude from each other, bring forth at the places where they assemble. They are, also, on this occasion mild towards each other, in consequence of the want of drink. For, contrary to other animals, they are more in want of drink in winter than in summer. For it is unusual to them to drink, because in Libya there is no rain in summer; and rats in that country die when they 607
drink.

In other regions, also, certain animals are generated from the mixture of animals of a different species. Thus, in Cyrene, wolves mingle with dogs, and generate; and Laconic dogs are the progeny of a fox and a dog. They say, likewise, that Indian dogs are generated from a dog and a tiger; not, however, immediately, but in the third mixture. For it is said that the first progeny is of a savage nature. But leading the bitches into solitary places they chain them, and many of them are destroyed by the tigers, unless these savage animals happen to be impelled to venery.

29 Places, likewise, produce a difference in the manners of animals. Thus, for instance, the manners of animals in mountainous and rude places are different from those in plains and cultivated places. For their aspect is more ferocious, and they are more robust in the former than in the latter situation; as is evident in the swine in mount Athos. For the males of the valleys avoid the females of the mountain.

Places, also, produce a great difference in the effect of the bite of savage animals. Thus, for instance, about Pharos, and other places, scorpions are not dangerous; but in other places, and in Caria, many large and dangerous serpents are generated; and if they bite any man or beast, the man or beast dies, and they, likewise, kill swine, though swine are not injured by other virulent bites. Black swine, also, are more easily destroyed by these serpents; but they principally die, if when bit they betake themselves to water. There is a great difference, however, in the bite of serpents. For the asp is produced in Libya, from which serpent a putrefying substance is made, and the bite of this serpent is incurable.

In a shrub, likewise, called *silphion* a small serpent is produced, the bite of which is said to be cured by drinking wine in which a certain stone taken from the tomb of an ancient king has been steeped. But in certain parts of Italy, the bite of the stellio [an animal resembling a lizard] is deadly. The bites, however, of all venomous animals are more pernicious if they happen to have devoured each other. Thus the bite of the viper is more pernicious after it has devoured a scorpion. The saliva of man, also, is hostile to most venomous animals. There is, likewise, a certain small serpent which some persons called sacred, and which very large serpents avoid. This serpent is never longer than a cubit, and is hairy to the view. Whatever this serpent bites, likewise, becomes immediately circularly putrid. In India, also, there is a certain little serpent, for the bite of which alone the inhabitants have no remedy.

607 30 Animals, likewise, differ in being in a prosperous condition or the contrary, with respect to the times of their gestation. For testaceous animals, such as cockles, and all other shell-fish, and, likewise, those that have a soft shell, as, for instance, lobsters, are best when they are pregnant. The conception, indeed, of testaceous animals is spoken of; for those that have a soft shell have been seen copulating and bringing forth, but no crustaceous animals have been seen doing either. And the mollia, such as the small calamary, the cuttle-fish, and the polypus, are best when they are pregnant. Nearly all fishes, likewise, are good when they begin to be pregnant; but as they advance in their pregnancy, some are good, and others are not. The *mœnis*, therefore, is good when pregnant. But the form of the female is more round; and the male is longer and broader. It happens, also, that when the female begins to be pregnant, the male has a black and more variegated colour, and is then in the worst condition to be eaten. The males, likewise, at this time, are called by some goats. The fishes, also, which are called merulæ [and which resemble a perch], the thrush, and the squill, change their colour according to the seasons, in the same manner as some birds. For in spring they become black, and after the spring they again become white. The *phycis*, likewise, changes its colour; for at every other time it is white, but in the spring it is of a variegated colour. This, also, is the only marine fish, as they say, which makes a nest, and brings forth in a nest. The *mœnis*, too, as we have before observed, and the *smaris* change their colour, and from being more white are again changed in the

summer, and become black. But this is especially manifest in their fins and gills.

The coracinus, also, in the same manner as the *mœnis*, is best when pregnant. But the mullet, the wolf, and those fishes that are covered with scales, are nearly all of them bad when pregnant. A few, likewise, as the glaucus, are similarly good, both when they are pregnant and when they are not. Fishes, also, when old are bad; and the tunny when old are bad even for the purposes of salting, since much of their flesh is consumed by the salt. The like, also, happens in other fishes. But old age in fishes is manifestly indicated by the magnitude and hardness of their scales. An old tunny, likewise, was once caught which weighed fifteen talents; but the interval between the two parts of its bifurcated tail was five cubits and one palm.

The fishes, however, of rivers and lakes are best after they have brought forth, and after the emission of their genital seed. But some of them, as the *saperdis* are, indeed, good when pregnant, and some, as the glanis, are at that time bad. The males, therefore, of all other fishes are better than the females; but the female *glanis* is better than the male. Among eels, likewise, those which they call the females are better than the males; for those which they thus denominate are not females,[†] but only differ in their appearance from the males.

† This was shown by Aristotle in Book IV, Chap. xi.

BOOK IX

1 With respect to the manners of animals, of those, indeed, that are more obscure, and live for a shorter time, the manners are less manifest to us according to sense, but of those whose life is more extended, the manners are more apparent. For it seems that animals possess a certain natural power, about each of the passions of the soul, with respect to wisdom and folly, fortitude and timidity, mildness and asperity, and other habits of a similar kind. Some animals participate at one and the same time of a certain discipline, and erudition, partly from each other, and partly from man; and such are those animals which not only distinguish the differences of sound by the hearing, but, also, the different significations of words.

In all those genera of animals, however, in which there is the male and female, nature has nearly similarly distinguished the manners of the females from those of the males; but this is particularly evident in the human species, in the larger animals, and in viviparous quadrupeds. For the manners of the females are more soft, they more rapidly become mild, and are more tractable and docile. Thus, among the Laconic dogs, the natural disposition of the females is more excellent than that of the males; but with respect to the dogs of the Molossi, the hunting kind differ, indeed, in no respect from other dogs, but those that follow the sheep surpass the rest in the magnitude of their body, and in the courage which they display in attacking wild beasts. And the dogs which are the progeny of the Molossian and Laconic species excel both these in courage and the endurance of labour. All females, however, are less animated than the males, except the she-bear and the female panther; for
608 the female of these appears to be more courageous than the male; but in the other genera the females are softer, and more crafty, their manners are less simple, they are more petulant, and they are more attentive to the nurture of their young. On the contrary, the males are more animated, and ferocious; their manners are more simple, and they are less insidious.

Vestiges, indeed, of these manners are, as I may say, in all animals; but they are more apparent in those whose manners are more characteristic, and most of all in man. For man has the most perfect nature [of all animals], so that these habits are in him more apparent. Hence woman is more compassionate than man, and has a greater propensity to tears. She is, also, more envious, more querulous, more slanderous, and more contentious. Farther still, the female is more dispirited, more despondent, more impudent, and more given to falsehood, than the male. She is, likewise, more easily deceived, and more apt to remember;

and again, the female is more vigilant, less active, and in short, less disposed to motion, and receptive of less nutriment than the male. But the male, as we have before observed, is more disposed to give assistance in danger, and is more courageous than the female; since among the mollia, also, when the female cuttle-fish is struck with the harpoon, the male affords her assistance; but the female flies when the male is struck.

Those animals, therefore, fight with each other, that dwell in the same places, and derive their subsistence from the same nutriment. For if the nutriment is rare, animals of the same tribe fight with each other; since they say that the phocæ or sea-calves fight for the place in which they feed, the males with the males, and the females with the females, till one of them is either killed by or relinquishes the place to the other, and the offspring of the one contend in a similar manner with the offspring of the other. Farther still, all other animals fight with those that are carnivorous, and the carnivorous fight with others; for their nutriment is from animals. Hence, also, augurs from the concord or discord of carnivorous animals form their presages; from their discord, indeed, inferring war, but from their concord, peace.

It appears, also, that animals which are formidable and ferocious, if they are supplied with abundance of nutriment, conduct themselves mildly towards men, and in a similar manner towards each other. But this is rendered evident by the attention which in Egypt is paid to animals. For there the most ferocious animals live with each other, in consequence of not being in want of nutriment. For they are rendered mild by the good treatment they receive. Thus, for instance, in some 609
places crocodiles are mild towards the priest, from the care which he takes of their nutriment. The very same thing, also, is seen in other regions, and the parts of these.

The eagle, therefore, and the dragon are hostile to each other; for the eagle makes serpents his food. The ichneumon, also, and the phalanx are hostile; for the ichneumon hunts the phalangia. But among birds, those that are speckled, larks, the woodpecker, and the chloreus, fight with each other; for they eat the eggs of each other. The crow, also, and the owl are mutually hostile; for the crow at mid-day, seizes and eats the eggs of the owl, because the owl does not see acutely in the day; but the owl by night seizes and eats the eggs of the crow; for the one is more powerful by day, but the other by night. The owl, likewise, and the *orchilus* are hostile to each other; for the *orchilus*, also, eats the eggs of the owl. But by day, other birds fly round the owl, which is called admiring the owl; and flying to it, pluck out its feathers. Hence fowlers employ the owl for the purpose of catching all sorts of little birds. The bird, also, which is called *presbeus*, or the elder [and by some the king of birds], and, likewise, the weasel, are hostile to the owl; for they eat the eggs and the young of the owl. The turtle-dove, also, and the pyralis or spinturnix[†] are enemies to each other; for the place in which they feed, and their mode of living, are the same. The courier, likewise, and the libyus, the kite, and the raven, are enemies; for the kite takes from the raven whatever the raven has, because the kite is more powerful with his claws, and in flying; so that nutriment makes these, also, enemies to each other.

Farther still, the birds which derive their nutriment from the sea are, likewise, hostile to each other, such as the *brenthus*, the sea-gull, and the harpa [a species of vulture]; and this is, also, the case with the triorches [a kind of hawk], the toad, and the serpent; for the triorches eats these. The turtle-dove, likewise, and the chloreus are enemies to each other; for the chloreus kills the turtle-dove; and the crow kills the bird which is called typanus. The little owl, also, and other birds with crooked claws, eat the *colaris*, and hence the colaris and those birds fight with each other. The stellio, likewise, fights with the spider; for the stellio eats spiders. And the marine bird the *pipus* fights with the heron; for the former eats the eggs and the young of the latter. The little bird, likewise, called *ægithus* and the ass are hostile to each other, in consequence of the ass rubbing his ulcers against the brambles in which the *ægithus* builds its nest; on this account, therefore, and, likewise, if he brays, he throws down the eggs and the young of the ægithus; for the young ones fall through fear. Hence, through the injury she sustains, the ægithus flies to the ass, and tears his ulcers.

609 The wolf, also, is hostile to the ass, to the bull, and to the fox. For being a carnivorous animal, he attacks bulls, asses, and the fox [that he may feed on their flesh]; and from the same cause, the fox and the circus, or gentle falcon, are enemies. For the circus being a bird with crooked claws, and carnivorous, it attacks the fox, and torments him by striking him with his claws. The raven, also, is hostile to the bull and the ass, by flying to and striking them, and tearing out their eyes. The eagle and the heron, too, are enemies to each other; for the eagle having crooked claws attacks the heron, and the heron in defending itself dies. The merlin, likewise, is hostile to the *ægypius*; and the *crex* to the

† The spinturnix is a bird which was accustomed to fly to altars and carry away a burning coal, which was considered as an ill omen.

courier, the blackbird, and the *chlorion*, which is fabulously asserted by some to be generated from a funeral-pile; for the crex injures these birds, and their young. The sitta, also, or nut-jobber, and the wren are hostile to the eagle; for the sitta breaks the eggs of the eagle. But the eagle is hostile to the sitta, both on this account, and because the eagle is a carnivorous bird. The bird, likewise, called *anthos*, or the flower, is hostile to the horse; for it drives away the horse from his pasture, because the anthos feeds on grass. For the anthos has a dimness of sight, and does not see acutely. But it imitates the voice of the horse, and flying to him, terrifies him, and drives him away. The horse, however, when he catches the anthos kills him. The anthos, likewise, lives near rivers and marshes; but its colour is beautiful, and its food is easily produced. The ass, also, is hostile to the *colote* [an animal, which, according to some, is a species of lizard]; for it lies in the manger of the ass, and by entering into his nostrils, prevents him from eating.

Of herons, however, there are three genera, *viz.* the black, the white, and that which is called the star. Of these, the black heron copulates with difficulty, and, therefore, during the act of copulation it vociferates, and, as they say, emits blood from its eyes; and it, likewise, brings forth with difficulty and labour. It, also, fights with those animals that injure it, *viz.* with the eagle, of which bird it is the prey; with the fox, by whom it is destroyed during the night; and with the lark, for this bird steals its eggs. But the serpent is hostile to the weasel and the hog; to the weasel, indeed, on account of habitation, when both have the same place of abode: for they live on the same nutriment [in hollows of the earth], and the hog eats serpents. The merlin, also, is hostile to the fox; for this bird strikes and tears the fox, and destroys his offspring; for it has crooked claws. The raven, however, and the fox are friends to each other; for the raven fights with the merlin, and aids the fox when he is struck by the merlin. The *ægypius*, also, and the merlin are hostile to each other, for both have crooked claws. And the *ægypius* and the swan 610
fight with the eagle; but the swan is frequently victorious. Swans, likewise, the most of all birds, eat each other.

With respect to savage animals, however, some are always hostile to each other, but others, just as it may happen, in the same manner as among men.

But the ass and the *akanthides* are enemies to each other; for the subsistence of the latter is from thorns, but the ass eats thorns while they are yet tender. The anthos, likewise, or flower, the *akanthis*, and the *ægithus*, are enemies to each other. It is, also, said, that the blood of the ægithus and anthos will not mingle with each other.

The crow, however, and the heron are friends, and, also, the rush, the lark, the *lædus*, and the courier. For the courier lives near rivers and thickets; but the lædus, on rocks and mountains, and loves the place in which he dwells. The *piphex*, likewise, the harpa, and the kite are friends; and, also, the fox and the serpent: for both delight in caverns. The blackbird and the turtle-dove, too, are friends.

But the lion and the thos are hostile to each other; for being carnivorous they live on the same food [and, therefore, fight for it with each other].

Elephants, also, fight furiously with each other, and strike each other with their teeth; but the vanquished elephant is severely enslaved, and cannot endure the voice of his conqueror. Elephants, likewise, differ from each other in fortitude in a wonderful degree. But the Indians use in war both the male and female elephants. The females, however, are of a smaller size, and are far less animated than the males. The elephant, also, throws down walls by means of his large teeth. He, likewise, impels palm-trees with his forehead till they bend, and afterwards lays them prostrate on the earth with his feet. But the hunting of the elephant is as follows: Getting on the back of certain tame and courageous elephants, they pursue [those savage elephants they wish to take], and when they have caught them, they order the tame elephants to strike them till they are debilitated; but then the leader of the elephant leaping on the back of a savage elephant, manages him with his hook or scythe. And after this the elephant rapidly becomes tame and obedient to his governor. As long, therefore, as the rulers of the wild elephants sit on them, they are all of them mild; but when they descend from them, some of them return to their former ferocity, and others do not. But in order to tame the more ferocious elephants, they bind their fore-legs with chains, and thus render them quiet. The hunting, however, is of elephants that are large, and of those that are young.

2 Such, therefore, are the friendships and enmities of these animals, arising from their nutriment, and the life which they lead. With respect
610 to fishes, however, some herd with each other, and are friends; but others do not herd together, and are enemies. Some, therefore, herd together when they are pregnant; but others when they have brought forth. And, in short, the following fishes are gregarious; the tunnys, the *mœnides*, the gudgeons, the *bocæ*, the lacerti, the coracina, the dentices,

the barbels, the sphyræme, the *anthiæ*, the *elegini*, the *atherini*, the *sargini*, the needles, the calamaries, the *julides*, the pilchers, the mackerels, and the *colias*. But of these, some are not only gregarious, but there is, also, a union between the male and female; for with respect to the rest, all of them, indeed, copulate, but are only gregarious at certain times, *viz.* as we have before observed, either when they are pregnant, or when they have brought forth. But the wolf and the mullet, though they are most hostile to each other, yet herd together at certain times. For not only fishes of the same genus, but, also, those that have similar nutriment, if it is supplied in abundance, frequently herd together.

Mullets and congers, likewise, frequently live, though deprived of their tail, as far as to that part by which their excrement is emitted. But the mullet is devoured by the wolf, and the conger by the lamprey. The more powerful, also, are hostile to the less powerful fishes; for the former eat the latter. And thus much concerning marine fishes.

3 The manners, however, of animals, as we have before observed, differ according to timidity and mildness, courage and docility, intelligence and stupidity. For the manners of sheep, as it is proverbially said, are simple and stupid. For sheep are the most stupid of all animals; since they creep into desert places for no reason, frequently in winter leave their folds, and when they are caught in the snow are not willing to depart from it, unless the shepherd compels them, but remain in it and perish, unless the shepherds lead away the males; for then they follow. And with respect to she-goats, if any one lays hold of one goat by the extremity of a species of hair which depends from her chin, the other goats will stand still as if stupid, and look at him. But sheep are colder when they go to sleep than goats; for sheep sleep more quietly, and approach men more familiarly; but goats endure cold with more difficulty than sheep.

Shepherds, however, teach sheep to assemble together when a noise is made; for if any sheep is left by herself when it thunders, and does not
run with the rest, she produces an abortion if she happens to be 611
pregnant. Hence, if a noise is made in the house, they run together from custom. But he-goats are destroyed by wild beasts when they wander from the herd to which they belong. Sheep, also, and she-goats lie down according to alliance [*i.e.* by families], the one opposite to the other; but when the sun begins to descend, shepherds say that she-goats no longer lie looking at, but turned from each other.

4 But cows feed in company with those to whom they are accustomed; and if one cow wanders, the rest follows her; on which account, also, herdsmen, if they cannot find one cow, immediately search after the whole herd.

With respect to mares, however, that feed together, when any one dies the others nurture her foals, and, in short, the genus of horses appears to be naturally a lover of its offspring; of which this is an indication, that frequently barren mares pay attention to foals that are deprived of their mothers, though in consequence of not having milk, they cause the destruction of the foals.

5 Among wild quadrupeds, however, the hind appears to excel in prudence; in the first place, because she brings forth near public roads, which through fear of men are not frequented by savage animals. In the second place, after she has brought forth, she first devours the chorion in which the fœtus is enveloped, and afterwards she runs to eat a herb called *seselis*; and having eaten it, she returns to her offspring. And in the third place, she leads her offspring to the coverts; thus accustoming them to the place where they are to retreat [when in danger]. But this is a steep rock, having only one entrance; in which place they say the hind stays and defends herself.

Farther still, the male [*i.e.* the stag] when he becomes fat, and he becomes very fat in autumn, does not make himself apparent any where, but emigrates, as being easily captured on account of his fatness. Stags, also, cast off their horns in places which are difficult of access, and cannot easily be found; whence, also, the proverb is derived, *where the stags cast off their horns*. For they avoid being seen, as if they had thrown away their arms. It is, likewise, said, that no one has ever found the left horn of a stag; for the stag hides it, as possessing a certain medicinal power. Horns, therefore, do not grow in stags that are but one year old, except that there is a certain beginning of horns, as it were for the sake of an indication [of future horns]. But this indication or mark is short and hairy. Stags, however, that are two years old have horns, that at first are straight like a stake; on which account their horns are then called *stakes*. But in the third year stags have bifurcated horns [*i.e.* horns with two branches], and in the fourth year their horns are more rude and inelegant [and have three branches], and after this manner the number of branches always increases till they are six years old.
611 From this period, however, the horns always grow similar, so that the age of stags can no longer be known by their horns; but old stags are

principally known by two marks. For some of them have no teeth, and others only a few; and their *defenders* no longer grow. But those branches which project before the forehead of stags from the growing horns are called their *defenders*, because with these the stag defends himself. Old stags, however, have not these defenders, but their horns increase in a straight direction. Stags, also, cast off their horns every year, and they cast them off about the month of April. But when this happens they hide themselves in the day, as was before observed; and they conceal themselves in thickets in order to avoid flies. During this time, likewise, they feed by night in the thickets till they have new horns. The horns, however, grow at first enveloped as it were in skin, and are hairy; but when they begin to increase, they expose them to the sun, that the horn may be concocted and dried. When their horns, likewise, are so hard that they feel no pain by rubbing them against trees, then they leave these retreats, from the confidence they feel in now having the means of defence. A stag, also, was once caught with a considerable quantity of green ivy growing to his horns, which was [accidentally] implanted in them while they were yet tender, as in a green tree.

Stags, likewise, when they are bit by a phalangium, or any other similar insect, collect and eat crabs. It appears, also, that a drink made from crabs is good for man; but it is unpleasant.

But the females of stags when they bring forth immediately eat the chorion [in which the fœtus is enveloped in the womb], and on this account the chorion cannot be obtained by any one; for the hind seizes it before it falls to the ground. It, likewise, appears that this chorion contains a medicinal property.

Hinds, also, when they are hunted, are caught by playing on the pipe, and singing; for they are allured by the pleasure they receive. If, therefore, there are two hunters, the one openly sings or plays on the pipe, but the other strikes the stag from behind, when his companion gives the signal. Hence, if it happens that the ears of the hind are erect, she hears acutely, and it is not possible to be concealed from her; but if it happens that her ears hang down, the hunter may be concealed from her view [when he attacks her].

6 She-bears, however, when they fly, drive before them their young, and [when the young ones are fatigued] carry them. But when they are on the point of being caught, they leap into trees. When, likewise, they come forth from their retreats, they in the first place eat the herb *arum*,

612 as was before observed, and masticate wood, as if they were cutting their teeth.

Many other quadrupeds, also, act prudently with respect to procuring remedies for themselves; for they say that in Crete the wild she-goats, when wounded with a dart, search for the herb dictamnum, or garden ginger; for it seems that this herb has a power of expelling darts from the body. And bitches when they are ill produce for themselves an emetic by eating a certain grass. But the panther, when she has eaten the poison which is called *the death of the panther*, searches for human dung, which is a remedy against this poison. This poison, also, destroys lions; on which account hunters hang from a certain tree human dung in a vessel, in order that the panther may not depart to a great distance; for the panther leaping to the dung, and hoping to obtain it, dies. It is, likewise, said, that the panther in consequence of perceiving that wild beasts are delighted with her odour, hides herself, and then hunts them as they draw near, and that in this manner she catches hinds.

But the ichneumon of Egypt when it sees the serpent called the asp, does not attack it, till it has called other ichneumons to its assistance. And these cover themselves with mud as a defence against the blows and the bites of the asp; for they first wash themselves in water, and afterwards roll themselves in the earth.

At the time, also, when the crocodile opens its mouth, the [little water-fowl called the] *trochilus* flies into his mouth, and cleans his teeth; and the bird, indeed, by this mean obtains food, but the crocodile perceives that he is benefited, and does not injure the bird. The crocodile, likewise, when he wishes the bird to depart, moves his neck in order that he may not bite it.

But the tortoise when it has devoured a viper eats wild marjoram; and a certain person on perceiving a tortoise frequently going in search of a viper after eating wild marjoram, plucked the herb out of the ground; in consequence of which the tortoise died.

The weasel, also, when it fights with a serpent eats rue; for the odour of this herb is hostile to serpents.

But the dragon when he collects autumnal fruits, swallows, after he has eaten them, the juice of the bitter lettuce; and this he has been seen to do. Dogs, also, when they are troubled with worms eat the green shoots of corn. But storks and other birds apply wild marjoram to any wound they may have received in fighting.

Many persons, likewise, have seen the cricket when it fights with a
612 serpent seize the neck of the serpent. The weasel, also, appears to catch

birds sagaciously; for it strangles them in the same manner as wolves strangle sheep. But it principally fights with those serpents that hunt for rats, because this animal, also, hunts rats.

With respect to hedgehogs, likewise, it has happened that their sagacity has been noticed in many places. For in consequence of perceiving a change in the north or south winds, some of them have been seen to change their habitations in the cavities of the earth, and those that have been nurtured in houses have been known to pass from one wall to another. Hence they say, that a certain person in Byzantium, from this presentiment of the hedgehog, acquired the reputation of being able to predict the changes of the weather.

With respect to the pole-cat, it is an animal of the size of a small dog of Malta; but in the thickness of its hair, its form, and the whiteness of its belly, and, also, in the craftiness of its manners, it resembles the weasel; and becomes very mild and tame. This animal, however, is injurious to bee-hives, for it delights in honey. It, likewise, eats birds in the same manner as cats. But its genital parts are bony, as we have before observed; and the pudendum of the male appears to be a remedy against the stranguary; for which purpose the scrapings of it are given.

7 In short, with respect to the lives of other animals, many imitations may be surveyed in them of human life; and accuracy of thought may be more clearly seen in smaller than in larger animals. And in the first place, among birds, this is seen in the swallow, in the construction of her nest. For she mingles mud and straw together, in the same order in which they are mingled by man, uniting with the mud small pieces of straw; and if she is in want of mud, she wets herself, and then rolls with her wings in the dust. Farther still, in forming her bed, in the first place, in the same manner as man, she puts underneath harder substances [and afterwards softer], and makes it commensurate to the magnitude of her body. Both the male and female swallows, likewise, labour in the nurture of their young; and in giving them food they are careful in observing the young one that received it before, that he may not receive it twice. At first, also, they throw out of the nest the dung made by the little swallows; but when the young ones are grown larger, they teach them to turn themselves so as to emit their excrement out of the nest.

There are other particulars, also, respecting pigeons which have a theory of this kind. For they neither copulate with many females, nor does the male, leaving the female with which he was before conjoined, unite himself to another, unless by the death of the female he becomes

a widower. Again, the attention of the male pigeon to the female when she is parturient, and his sympathy with her pains, are very remarkable;
613 and if the female is averse to go to her nest on account of the pain attending her delivery, the male strikes her, and compels her to enter. But when the young are brought forth, the parents open the mouth of the little pigeons, and insert into it chewed salt earth in preference to any other earth: previously preparing it for their nourishment. When, however, the young ones [being adult], are about to leave the nest, the male copulates with all of them.

For the most part, therefore, pigeons after this manner [*i.e.* with this constancy], love each other; but sometimes certain pigeons commit adultery with the females belonging to other males. The pigeon, likewise, is a fighting animal; and they invade each other, and enter into each other's nest. This, however, is but seldom the case. And they fight, indeed, when at a distance from their nests less violently, but when near their nests their combats are attended with extreme violence.

It appears, also, to be the peculiarity of pigeons, of the small ring-doves, and of the turtle-doves, not to raise their head when they drink, unless they have drank sufficiently.

The turtle-dove, also, and the female wood-pigeon are content with one male, and do not admit any other. And both the male and the female incubate. But it is not easy to distinguish the male from the female, except by the internal parts.

Wood-pigeons, likewise, live a long time; for they live five-and-twenty or thirty years, and some live forty years. But as they become old their claws increase; but they are cut by those that nurture them. Old age, however, does not appear to injure them in any other respect.

Turtle-doves and pigeons, also, live eight years, after they are blinded by those that keep them. But partridges live about fifteen years. Wood-pigeons, likewise, and turtle-doves, always build their nests in the same places. And, in short, among animals the males generally live longer than the females; but in the genera of pigeons and doves, it is said that the males die prior to the females, which is inferred from those pigeons and doves that are nurtured at home in order that they may allure others into confinement.

It is, also, asserted by some that they live only one year; which they infer from this circumstance, that in the spring [when sparrows begin to be seen] no sparrow is found to have a black spot in its neck, which it has afterwards; which indicates that no sparrow of the former year had survived. But it is said that the female sparrows live longer; for these have been caught together with young male sparrows, and have been found to be manifestly older by the hardness of their bill. 613

Turtle-doves, likewise, in summer live in cold places, but in winter in warm places. On the contrary, chaffinches in summer live in warm, but in winter in cold places.

8 Heavy birds, however, such as quails, partridges, and others of the like kind, do not build nests; for as they do not fly high nests are of no advantage to them. But when they have made a hole in the dust in a smooth place (for they bring forth in no other place), they collect together certain thorns and twigs of trees, in order to conceal themselves from eagles and sparrow-hawks, and in these holes they bring forth and incubate. Afterwards, when the eggs are hatched they immediately take the young ones with them, because they are unable by flying to supply them with food. But quails and partridges, likewise, rest sometimes [in going from one place to another], and assemble their young under them, in the same manner as hens; and they do not lay their eggs and incubate in the same place, lest by continuing in the same place for a long time some one should discover it.

When any bird-catcher, also, approaches to the nest of a partridge, the mother rolls herself before him, as if she might now be taken, and allures him with the hope of catching her, till each of her young ones is fled; but afterwards, she flies to her young, and recalls them. The partridge, therefore, brings forth not less than ten, and frequently sixteen eggs.

This bird, however, as I have before observed, is malignant and very crafty. But in the spring the male partridges, together with the females with whom they are paired, separate themselves from the troops, singing and fighting. In consequence, however, of male partridges being extremely lascivious, if they happen to find any eggs, they roll them away and break them, that the female may not incubate them. But the female in order to prevent this, withdrawing herself from the male, brings forth [secretly]; though it frequently happens from the urgency of parturition, that she brings forth any where, even if the male is present; and that the eggs may be preserved, she does not go to them [lest by this mean being discovered by the male he should break them]. If, also, she happens to be seen by a man, she acts in the same manner by her eggs as by her young [which we have already mentioned]; for she presents herself before the feet of the man, till she has drawn him away from the eggs.

614 But when the female, having withdrawn from the male, incubates, the males make an outcry, and assembling together fight; and these partridges are called widowers. The partridge, also, who is conquered follows the conqueror, who alone has connection with the vanquished partridge. And if any partridge is vanquished by the partridge that was conquered before, or by any other, the victor has connection with him latently. This, however, does not take place always, but at a certain time of the year; and the same thing is found to happen among quails. This [preposterous venery], likewise, sometimes happens among cocks. For in temples where they are kept without hens, all the cocks have connection with the one that is dedicated. And of partridges, those males that are tame have connection with the males that are wild, and the former strike the latter, and insult them.

But the mode of catching partridges by such of them as are males is as follows. The partridge employed to catch them is put into a cage covered with twigs, the door of which is left open. Then the partridge who is the leader of the wild partridges, on viewing the partridge employed to catch him, enters the cage, that he may fight with him. But the leader being taken in the cage, another wild partridge approaches to fight, after the same manner. If, therefore, the partridge that catches them is a male, this method is adopted; but if a female partridge is employed for this purpose, when she is put into the cage she sings, and the leader of the wild partridges approaching to her, the rest of the troop assembling strike him, and drive him away from the female, because he came to her and not to them. On this account the leader of the wild partridges frequently approaches in silence to the female partridge employed to catch him, lest some other partridge on hearing his voice should come forth, and fight with him. And skilful bird-catchers say that the male when he approaches to the female causes her to be silent, lest the other males hearing her, he should be under the necessary of fighting with them.

The partridge, however, not only sings, but emits a stridulous sound and other voices. Frequently, also, the female while she is in the act of incubation rises from the eggs when she perceives the male approaching to the female partridge employed to catch him, and running to him suffers him to have connection with her, that she may draw him away from the other female. Both partridges, likewise, and quails are so violently stimulated with the desire of coition, that they fall into the hands of the bird-catchers, and frequently perch on their heads.

Such, therefore, are the particulars which take place respecting the copulation and the method of catching partridges, and, also, respecting the craftiness of their manners. But both quails and partridges, and some other birds which do not fly high, build, as we have before observed, their nests on the ground. And farther still, among birds of this kind, the lark, the woodcock, and the quail, do not perch on trees, but on the ground.

9 The woodpecker, however, does not sit on the ground; but it pierces oaks, in order that the worms and gnats which are latent in them may 614
come forth. For it catches them on their egress with its tongue, which is broad and large. The woodpecker, also, rapidly walks on trees, in every kind of position, and, likewise, in a supine posture, in the same manner as the stellio. But it has claws naturally stronger than those of the tame jackdaw, in order that it may creep along trees with security; for adhering by these it ascends the trunks of trees.

Of woodpeckers, likewise, one kind is less than the blackbird, and is distinguished by small reddish spots. But another kind is larger than the blackbird. And the third kind of woodpeckers is not much less than a hen. These birds, however, as we have before observed, build their nests in trees, and particularly in the olive-tree. And they feed on ants, and the worms which live in trees. It is, likewise, said, that the woodpeckers in their search after worms excavate the trees so excessively as to throw them down. A certain tame woodpecker, also, having inserted an almond in the cleft of a piece of wood in order that he might peck it firmly, with the third stroke of his bill broke the shell of the almond, and ate the nucleus.

10 Cranes, also, exhibit many indications of prudence. For they migrate to a considerable distance, and fly high, in order that they may see remote objects. And if they see clouds which indicate rain, they descend to the earth, and repose themselves. Farther still, [when they emigrate], they have a leader of their journey; and those cranes that are among the last of the troop give signals by their clamour, so as that their voice may be heard [by the whole troop]. When, however, they repose, the rest, indeed, sleep with their head under their wing, and alternately standing on one foot; but the leader observes what passes, with his head uncovered; and when he perceives any thing, gives a signal by a cry.

But pelicans that dwell near rivers swallow large and smooth shells, and when they have digested them in the place which precedes the stomach, they vomit them, in order that they may extract the flesh from the gaping shells, and eat it.

11 With respect to wild birds, however, their habitations are contrived with a view to the lives and preservation of their young. Some of them, also, are more prolific, and pay more attention to their offspring; but others are of an opposite character. And some of them, indeed, are more ingenious in procuring subsistence, but others are deprived of this ingenuity. With respect to their habitations, also, some construct them
615 near great floods, caverns, and rocks, as the bird which is called the bird of the rock. But this bird has both a bad colour, and a bad voice; and it makes its appearance by night, but conceals itself in the day. The sparrow-hawk, however, builds its nest in steep rocks; and though it is a carnivorous bird, yet it does not eat the heart of the birds which it subdues. Some persons have, likewise, observed the same thing of the quail, and the thrush, and others of other birds. There is a difference, also, with respect to catching the sparrow-hawk; for it is not caught in a similar manner in summer and in winter.

It is, also, reported of the vulture by certain persons, that no one has ever seen either its nest or its young; and on this account, Herodotus, the father of the sophist of Brysson, says that the vulture comes from another more elevated country. But he infers this because a great number of vultures suddenly make their appearance without its being manifest whence they arrive. The reason of this, however, is because the vulture brings forth in inaccessible rocks. The vulture, likewise, brings forth one egg, or two eggs at most.

Some wild birds, however, as, for instance, the lapwing and the *brinthus*, dwell in mountains and woods; but the *brinthus* is a bird that provides for itself good nutriment, and sings.

But the trochilus lives in groves and holes. This bird, also, is caught with difficulty, since it has a great propensity to flying, and its character is imbecil. It procures, however, good nutriment for itself, and is ingenious. It is, likewise, called the *senator*, and the king; on which account, also, it is said, that the eagle is hostile to the trochilus.

12 Some birds, also, live near the sea; as, for instance, the *cinclus*. But with respect to the manners of this bird, it is crafty, and difficult to be taken, and when caught becomes extremely tame. This bird, however, is mutilated; for its posterior parts are very debile. All the *palmipedes*, however, live near the sea, and rivers, and lakes; for nature herself

searches after that which is appropriate. Many of the *fissipedes*, likewise, live near water and marshes; for instance, the anthos, or flower, lives near rivers. But this bird is of a beautiful colour, and nourishes itself with good food.

The bird, also, which is called *cataractes* lives near the sea; but when it plunges into the depth of the sea, it remains there for no less a time than a man might pass over an acre of land; but this bird is less than a sparrow-hawk. Swans, likewise, are among the number of the palmipedes, and live near lakes and marshes. They, also, procure good nutriment for themselves, their manners are good, they are prolific, and are happy in their old age. And if the eagle attacks them, they oppose 615
and vanquish him. They never commence the combat, however, with the eagle themselves. Swans, likewise, sing, and especially at the time of their death. They, also, fly into the deep sea; and some persons who have sailed to Libya, have met with many swans in the sea singing with a mournful voice, and have seen some of them dying.

The bird, however, called *cymindis* is rarely seen; for it dwells in mountains. But this bird is black, and it is about the size of the sparrow-hawk called the *pigeon-killer*, and its form is long and thin. The Ionians, however call it *cymindis*, and it is mentioned by Homer in the Iliad, when he sings:

> *Chalcis* its name with those of heav'nly birth,
> But called *cymindis* by the sons of earth.
>
> *Iliad*, XIV, 291. Trans. Pope

The *hybris*, also, which some say is the same bird as the ptonx, is not seen by day, because its sight is dull; but it hunts by night in the same manner as the eagle. It, likewise, fights so violently with the eagle, that both of them are frequently caught alive by the shepherds. The *hybris*, therefore, brings forth two eggs; and builds its nest in rocks and caverns.

Cranes, likewise, fight so violently with each other, that they are caught fighting; for they suffer themselves to be taken rather than desist. The crane, also, brings forth two eggs.

13 The magpie, however, has a great variety of voices; for every day, as I may say, she emits a different voice; and she brings forth about nine eggs. She, also, constructs her nest in trees, from hair and wool. But when acorns fall, she collects them and conceals them in a repository.

Concerning storks, therefore, it is said by many persons that they are nourished by their young. Some, likewise, assert that the bird *merops* is nourished by its young, and this not only when it is old, but, also, immediately, as soon as they are able; and that the father and mother remain in the nest [while the young ones supply them with food]. The idea, also, of the plumage of this bird is as follows: the inferior parts of it are yellow; but the superior parts, in the same manner as those of the halcyon, are of an azure colour; and the extremities of its wings are red. Towards autumn, likewise, it brings forth about six or seven eggs in steep rocks, the earth of which is soft; and it penetrates into these about four cubits.

But the bird called the *chloris* in consequence of having its inferior parts of a yellowish colour, is about the size of the lark, and brings forth
616 four or five eggs. This bird, also, builds its nest from the herb *symphytum* or walwort plucked by the roots, and puts underneath it hair and wool. The blackbird and the magpie, likewise, do the very same thing, and the internal part of their nests is composed from the same materials.

The nest of the thistle-finch, also, is artificially constructed; for the parts are knit together like a ball of flax, and it has a small aperture for an entrance.

They say, likewise, that the bird called *cinnamomus* brings cinnamon from a certain place, and constructs her nest from it. But this bird builds its nest in lofty trees, and in the branches of trees. The inhabitants, however, discharge arrows furnished with lead at the nest, with which they throw it down, and thus collect the cinnamon from the ruins of the nest.

14 With respect to the halcyon, it is, indeed, not much larger than the sparrow; but its colour is a mixture of azure, green, and purple. Its whole body, likewise, its wings, and the parts about its neck, are distinguished by a mixture of these colours, but each of the colours is not separately disposed. Its beak, however, is greenish, long, and thin. Such, therefore, is its form. But its nest resembles marine balls, and what is called the foam of the sea, except that the colour of it is different; for it is of a reddish colour. The figure, however, of its nest is similar to that of a gourd which has a long neck; but the magnitude of it exceeds that of the largest sponge. Some of the nests, however, of the halcyon are less than this. All the nests, too, are internally distinguished by various enclosures, and are partly solid, and partly hollow [after the manner of sponges]. It is with difficulty, likewise, that they are perforated or cut with a sharp piece of iron; but they are easily

broken by a blow, and the force of the hand, in the same manner as the froth of the sea. The opening, also, of these nests is narrow, and only affords a small entrance; so that the sea when it is impelled to the shore [when the halcyon builds its nest] cannot enter into the nests. But the cavities of the nest resemble those of the sponge. It is, however, doubtful from what materials the halcyon constructs its nests; but it seems that it is principally constructed from the thorns of the needle fish; for the halcyon lives by eating fishes. This bird, also, ascends [from the sea] into rivers. For the most part, likewise, it brings forth about five eggs. But it is prolific as long as it lives, and begins to lay eggs in the fourth month of its age.

15 The lapwing, however, principally constructs its nest from human 616
dung. But the colour of its plumage changes in summer from what it was in winter; which is, also, the case with the greater part of other wild birds. The titmouse, likewise, as they say, brings forth the greatest number of eggs; but according to some persons, the bird called the *black-head*, brings forth the greatest number of eggs, after the Libyan ostrich; for it has been seen to lay seventeen eggs. It lays, however, more than twenty; and it is said that it always brings forth an odd number of eggs. The titmouse, also, builds its nest in trees, and feeds on worms. It is, likewise, peculiar to this bird and the nightingale, in contradistinction to other to other birds, not to have the point of the tongue.

But the *ægithus* procures good nutriment for itself, and is prolific. It is, however, lame in one of its feet. The *chlorion*, also, is very docile, and ingenious in contriving the means of subsistence; but it does not fly well, and has a bad colour.

16 The *elea*, likewise, in the same manner as another bird, feeds itself well, and in the summer exposes itself to the wind, and sits in the shade; but in winter it sits in sunny places, and perched on the tops of reeds surveys on all sides the marshes. It is, however, a small bird, and has a good voice.

The bird, also, called the *gnaphalus* has a good voice, a beautiful colour, is ingenious in procuring the means of subsistence, and is of a graceful form. It appears, however, to be a foreign bird; for it is seldom seen in places which are not appropriate to it.

17 With respect to the *crex*, however, its manners are hostile; but it is ingenious in procuring the means of subsistence, though, otherwise, it

is an unhappy bird. But the bird called *sitte* is, indeed, of hostile manners, yet is tractable, well formed, and feeds well. It is, also, said that this bird is a remedy [for certain maladies], because it possesses the knowledge of many things. It is, likewise, very prolific, has good offspring, and lives by piercing trees.

But the little owl feeds by night, and is seldom seen in the day. It, likewise, dwells in rocks and caverns. For it uses both marine and terrestrial food. It, also, lives well, and is ingenious in its contrivances.

There is, likewise, a little bird which is called *certhius*. This bird with respect to its manners is bold, lives near trees, and feeds on the worms which are produced from putrid wood. It is, also, ingenious in procuring subsistence, and has a clear voice. But the *acanthides* are birds that live ill, and have a bad colour, yet their voice is canorous.

18 Of the species of herons, however, the black, as we have before observed, copulates with difficulty; but is ingenious in its contrivances, is ravenous, and active in chasing other birds. But it gives chase in the day. It is, however, of a bad colour, and its belly is always wet. But of
617 the remaining two, for there are three species of herons, the white heron has, indeed, a beautiful colour, copulates without labour, and builds its nest, and lays its eggs in a becoming manner, in trees. It procures its nutriment, also, in marshes, lakes, plains, and meadows. But the starry heron, which is, likewise, called slothful, is fabulously said, indeed, to have been formerly changed from a slave into a bird; and it is, according to the appellation assigned to it, the most slothful of all the herons.

Herons, therefore, live after this manner; but the bird called *phoix* has a peculiarity which distinguishes it from other birds; for it principally attacks the eyes of birds. But it is hostile to the harpa [a species of vulture], because it is nourished with similar food.

19 Of blackbirds, however, there are two genera; one, indeed, being black, and to be found every where; but the other being white. The magnitude of the latter, also, is equal to that of the former, and its voice is similar. But the latter is produced in Cyllene in Arcadia, and no where else. To the genus of blackbirds, also, a certain bird may be referred, *viz.* the *laius*; which resembles, indeed, the black blackbird, but in magnitude it is a little less. This bird lives in rocks and tiles; but it has not, like the blackbird, a beak of a red colour.

20 Of thrushes, however, there are three species. And one of these, indeed, is called the mistletoe-eater, and feeds on nothing else than mistletoe and resin. This species of thrushes, also, is of the size of the magpie. Another species is called *trichas*; but its voice is acute, and in size it is equal to the blackbird. And the third species is that which some call *ilias*; but it is the least of the three species, and its plumage is less variegated.

21 There is, also, a certain bird which lives in rocks, and which is called the blue-bird. This bird principally dwells in Scyros; but it is less than the blackbird, and a little larger than the chaffinch. Its feet, likewise, are large, and it climbs rocks. But it is wholly of an azure colour, and its beak is thin and long. Its legs, also, are short, and resemble those of the bird called *pipus*.

22 The *chlorion*, however, is wholly of a green colour. This bird is not seen in winter; but principally becomes apparent about the summer solstice. It, likewise, changes its habitation when arcturus rises. But its magnitude is that of the turtle-dove.

The bird, however, called the *soft-skull* always dwells in the same places, and in these is caught. But with respect to the form of this bird, 617
its head is large and cartilaginous, but in size it is less than the small thrush. Its mouth, however, is strong, small, and round; but its colour is wholly ashy. Its feet, also, are good, but its wings are bad; and it is particularly the prey of the owl.

23 Farther still, the bird called *pardalus* is for the most part a gregarious bird, and is not to be seen by itself. But its colour is wholly ashy, and in magnitude it resembles the *chlorion*, and *soft-skull*. Its feet, also, are good, and its wings are not bad. But its voice is loud, and not flat.

The *collyrion*, also, feeds on the same substances as the blackbird; and its magnitude is the same as that of the preceding birds. But it is principally caught in winter. All the above enumerated birds, however, are seen at all seasons of the year. Again, the raven and the crow principally live in cities. For these birds are always apparent, and neither change the places of their habitation, nor conceal themselves.

24 But of the tame jackdaw there are three species; one, indeed, is the *coracias*, which is of the same size as the crow, and its beak is red. The

second species is called the *wolf*; and the third is small, and is jocose. There is, also, another species of jackdaws about Lydia and Phrygia; and these are planipedes, or have their toes joined together by a common membrane.

25 Of larks, however, there are two genera. And one of these, indeed, lives on the earth, and is tufted; but the other is gregarious, and does not live by itself like the former. Its colour, however, is similar to that of the former species, but its magnitude is less, and it is not tufted. It is, also, eaten.

26 The *ascalopas*, likewise, is caught in the enclosures of gardens. But in magnitude it is equal to a hen, its beak is long, and in colour it resembles the *attagas*. It, likewise, runs rapidly, and is very much attached to man.

But the plumage of the starling is variegated, and its magnitude is equal to that of the blackbird.

27 Of the Egyptian ibis, however, there are two kinds. For some of them, indeed, are white, but others black. The white, therefore, are produced in every part of Egypt, except Pelusium; but the black are not produced in any other part of Egypt than Pelusium.

28 With respect to the small horn-coots, some of them, indeed, are always seen at every season of the year, are, therefore, called *perpetual-*
618 *horn-coots*, and are not eaten, because they are not fit to eat. But others of them sometimes appear in autumn. They are, however, seen for one day, or two at most, are eatable, and they are very much approved of for this purpose. They, also, differ from the perpetual-horn-coots in nothing else, as I may say, than their fatness; and these, indeed, are not vocal, but the perpetual-horn-coots are. With respect to the generation of them, however, what it is has not yet been ascertained by any one, except that they are seen when the west-wind blows; for this is certain.

29 The cuckoo, however, as we have elsewhere observed,† does not construct a nest, but brings forth in the nest of other birds, and especially in the nests of the wood-pigeon, the hen-linnet, and the lark, on the ground; and, also, in the nest of the bird called *chloris*, which

† In Book VI, Chap. vii.

builds its nest in trees. The cuckoo, therefore, brings forth one egg, which she does not herself incubate; but the bird in whose nest she lays it hatches and nourishes it. And, as they say, when the young of the cuckoo is grown, the bird which nourished it ejects her own offspring, and thus they perish. But others assert that the bird which nourishes the young cuckoo kills her own offspring, and gives it to the young cuckoo to eat. For because the offspring of the cuckoo is beautiful, she despises her own. Some persons, therefore, confess that they have been spectators of most of these particulars; but all do not agree in explaining how the offspring of the bird which nurtures the young cuckoo perishes. For some say that the cuckoo, returning to the nest in which she had deposited her egg, eats the offspring of the bird that received it; but others assert that the offspring of the cuckoo, in consequence of surpassing in magnitude the offspring of its nurse, devours all the food that is brought into the nest, and thus causes them to die with hunger. And others again assert that the young of the cuckoo, as being stronger, kills the young of its nurse. But, however, this may be, the cuckoo appears to conduct herself prudently with respect to the procreation of her offspring. For, from a consciousness of her timidity, and that she is incapable of defending her young, she gives them to other birds to be nurtured and preserved. For this bird is excessively timid, since she suffers herself to be plucked by small birds, and flies from them.

30 With respect to the birds, however, called *apodes*, or *without feet*, and which some denominate cypselli or martlets, that they resemble, indeed, swallows, has been before observed by us; for they cannot easily be distinguished from swallows, except in this, that their legs are covered with feathers. These birds build their nests in long boxes fashioned from 618
clay, so as to have a sufficient opening for an entrance. They, also, construct their nests in narrow places under rocks and caverns, so as to escape the view both of men and beast.

But the bird called *caprimulgus*[†] is a bird of the mountain, and is a little larger than the blackbird, but is less than the cuckoo. This bird, therefore, brings forth two or three eggs at most, but it is characterized by indolent manners. It, also, flies to she-goats, and sucks them; whence it derives its name. But it is said that when it sucks the mamma of the

† A bird like a gull, that in the night sucks goats, and causes their udders to mortify.

goat, the mamma is extinguished, and the goat becomes blind. This bird, likewise, does not see acutely by day, but it sees better by night.

31 With respect to ravens, however, two only are found in regions of a small extent, and where there is not nutriment sufficient for many. As soon as their young, also, are able to fly, they first of all expel them from the nest, and afterwards drive them from the whole of the country. But the raven brings forth four or five eggs. About the time, however, in which the Median guests perished in Pharsalum, the ravens deserted the places about Athens and Peloponnesus, as having a certain sensation by which they were able to communicate knowledge with each other.

32 But of eagles there are many genera; one, indeed, which is called *pygargus*; and this species of eagle is found in plains and forests, and about cities; but some call this eagle the fawn-slayer. In consequence, likewise, of its daring courage it flies to mountains and woods; but the remaining genera of eagles are rarely seen in plains and forests. There is, also, another genus of eagle, which is called *plangus*, and is the second in magnitude and strength. But it dwells in thickets, valleys, and lakes, and is called the duck-destroyer and *morphnus*. It is, also, mentioned by Homer[†] in the departure of Priam [from Troy to redeem the body of Hector]. There is, likewise, another species of eagle of a black colour, and which in magnitude is the least, but is the strongest of all the species of eagles. This eagle dwells in mountains and woods, and is called the black eagle, and the hare-destroyer. But this is the only eagle that nurtures its young, and conducts them out of the nest. This eagle, also, flies rapidly, is elegantly formed, is unenvious and intrepid, warlike and auspicious, and is neither querulous nor clangous. Another species of eagle is *percnopterus*, or has its wings diversified with black spots, but its head is white, and in size it is the largest of all the eagles. But its wings are very short, and its tail long, and it resembles a vulture; it is called the mountain-stork, and the vulture-eagle. It, also, dwells in forests, and has
619 all the bad qualities of other eagles, but none of their good ones; since it is pursued and caught by ravens and other birds; for it is heavy, is badly nourished, and feeds on dead bodies. But it is always hungry, querulous, and clangous. There is, likewise, another genus of eagles, which are called eagles of the sea. But these eagles have a long and thick neck, curved wings, and a broad tail. They, also, dwell near the sea and

† *Iliad*, Book XXIV. v.316.

shores; and when they have seized their prey, and are unable to carry it, they are frequently plunged with it into the depths of the sea. There is, likewise, another genus of eagles, which are called genuine; for it is said that these are the only genuine eagles; since the other genera are mingled, and are adulterously begot by each other from eagles, hawks, and the smallest birds. This, however, is the largest of all the eagles, and is larger than the osprey. It, also, exceeds other eagles in a sesquialter ratio. But the colour of its plumage is yellow, and, in the same manner as the bird called *cymindis*, is rarely seen.

The time, however, of chasing the eagle, and in which it flies, is from breakfast, or the first repast, till the evening; for early in the morning it remains quiet till the time in which the forum is full. But when the eagle becomes old, the superior part of its beak increases, and always becomes more curved, and at length the eagle dies from hunger. It is, however, fabulously said, that the eagle suffers this punishment, because when he was formerly a man he had injured his guest. The eagle, also, reserves for its young the surplus of the nutriment which it does not itself consume. For, because it is not easy for it to supply its young with food every day, it sometimes has not any thing to give them. If, likewise, the eagle finds any one invading its nest, it strikes him with its wings, and lacerates him with its claws. But it builds its nest not in plains, but in lofty places, and especially in steep rocks, and, likewise, in trees. Eagles, also, nourish their young till they are able to fly; but then they drive them from the nest, and expel them from the whole of the region in the vicinity of the nest. For one pair of eagles occupies a great extent of territory, and on this account the pair do not suffer other eagles to dwell near them. Again, eagles hunt after their prey, not from places near to their nest, but from those that are far distant. But when they have hunted and caught their prey, they do not immediately carry it to the nest, but first make trial of its weight, and rest themselves. Hares, likewise, are not immediately seized by them, but they suffer 619
them to run into the plain, and they do not directly descend to the ground, but their descent is gradually less and less. The eagle, however, does both these in order that he may be more safe from stratagems. He, also, perches on elevated places, because he is slowly elevated from the earth. And he flies high that he may have an extended view; on which account men say, that he is the only bird that is divine. But all birds that have crooked claws perch but very little on rocks, because the hardness of the rock is an impediment to the curvature of their claws. The eagle, also, chases fawns, hares, foxes, and such other animals as he

is able to subdue. He, likewise, lives long; as is evident from the same nest of the eagle always remaining.

33 In Scythia, however, there is a genus of birds called bustards, or wild turkeys. This bird brings forth two young ones; yet does not incubate its eggs, but concealing them in the skin of the hare or the fox, leaves them. It observes them, however, from the top of a tree, when it does not happen to be hunting for food. And if any one climbs up the tree [towards its nest] it attacks and strikes him with its wings, in the same manner as the eagle.

34 But the owl, the night-raven, and such other birds as cannot see by day, hunt and procure themselves food by night. This, however, is not done by them during the whole of the night, but in the evening, and towards the morning. These birds, also, hunt after the rat, the lizard, the *sphondyle* [a kind of moth], and other such like small animals.

The bird, however, called the osprey is fortunate in its offspring, procures food for itself with facility in abundance, is placid, and nourishes both its own offspring and the young of the eagle. For when the eagle expels the young from the nest, the osprey receives and nurtures them; since the eagle expels her young prematurely, while they are yet in want of food, and incapable of flying. But it seems that the eagle ejects its young through envy; for it is naturally an envious and greedy bird, and, also, rapacious. When it eats, likewise, it devours a great quantity of food. It envies its young, therefore, when they have arrived at maturity, because they are then good to eat, and tears them with its claws. The young eagles, also, fight with each other for a place in the nest, and for food; but the mother expels and strikes them. And
620 the young eagles when expelled cry, and the osprey receives them. The sight of the osprey, however, is imperfect, and its eyes are injured.

The sight, however, of the eagle of the sea is most acute, and this bird compels its offspring while they are yet without feathers to look to the sun. She, also, strikes any one of her offspring that is unwilling to do this, and turns it to the sun. She, likewise, kills the young eagle that is the first that weeps, but she nurtures the other. But she dwells near the sea, and lives, as we have before observed, by hunting the birds that frequent the sea. She, also, attacks them severally, observing when they emerge from the sea. She, also, attacks them severally, observing when they emerge from the sea. When, however, the emerging bird sees the eagle of the sea, it again merges itself through fear, in order that it may be at some other time [more opportunely] emerge. But the eagle of the sea, perceiving the bird through the acuteness of its sight, continues perpetually flying, till either the bird is suffocated in the water, or is caught by the eagle of the sea as it ascends. For the eagle of the sea does not attack marine birds in a troop; since they repel it by sprinkling it with water from their wings.

35 & 36 Coots, however, are caught with the foam of the sea; for they peck it with avidity. Hence those that hunt them sprinkle them with foam. The flesh, also, of these birds has a pleasant odour, except in the posterior part of their body, which smells of mud. These birds, likewise, become fat.

But of sparrow-hawks, the strongest, indeed, is the buzzard; the second is the merlin; and the third is the circus, or gentle falcon. The goss-hawk, however, the *phassophonus*, or wood-pigeon-slayer, and the *pernes* are different from each other. But the broader sparrow-hawks are called sub-buzzards; and others are called speckled sparrow-hawks, and chaffinch sparrow-hawks. The smooth sparrow-hawk, however, and those that eat hedge-frogs, live with great facility, and fly low. Some persons also say that there are not less than ten genera of sparrow-hawks; but they differ from each other. For some of them attack and carry away the pigeon when she sits on the ground, but do not touch her when she is flying. But others seize the pigeon when she is perched on a tree, or on any thing else, and do not touch her when she is on the ground, or when she is flying. Others, again, neither meddle with the pigeon when she is on the ground, nor when she is perched on any thing, but endeavour to catch her when she is flying. It is, likewise, said that pigeons recognize each of these genera; so that if a flying sparrow-hawk is among the number of those that attack the pigeon when flying, the pigeon will remain in the place where she happens to be sitting; but if the sparrow-hawk is one of those that attack the pigeon when sitting, the pigeon does not remain where she was, but flies away. In that part of Thrace, however, which was once called Cedropolis, little birds are hunted in the marshes by men in conjunction with sparrow-hawks. For the men having sticks, beat with them the reeds and the trees, in order 620 that the little birds may fly away; but the sparrow-hawks, which are seen high in the air, pursue them. And the little birds being terrified again fly downward to the earth; but the men striking them with their sticks catch them, and make the sparrow-hawks partakers of the prey. About the Palus Mœotis, also, it is said that wolves are the associates of fishermen; and that when the fishermen do not make these partakers of the fish which they have caught, the wolves tear the nets while they are drying on the ground.

37 After this manner, therefore, the particulars respecting birds subsist. In marine animals, also, many indications of sagacity in the lives of each may be surveyed. For the reports which are circulated about the frog called the fisher and the cramp-fish are true. For this frog hides itself in sand or mud, and leaves out of the sand certain capillaments which hang before its eyes, and which are globular in the extremity, and are given to the animal as a bait. The little fishes, therefore, run to the globules of the capillaments [conceiving them to be food]; and then the frog gradually contracts the capillaments, till he has brought the globules together with the little fishes to his mouth. But the cramp-fish numbs the fishes which he intends to subdue, and thus catches and feeds on them. For this purpose, however, he conceals himself in sand and mud, and numbing the fishes that swim near him, he makes them his prey; and of this certain persons have been eye-witnesses. The *pastinaca*, also, hides itself, yet not in a manner similar to the cramp-fish. That the frog called the fisher and the cramp-fish, however, live after this manner is indicated by mullets being frequently found in their belly when they are caught, though the mullet is the swiftest, and they are the slowest of fishes. In the next place, the frog when he has no longer little fishes on the globules of his capillaments, is thinner when he is caught; and men are evidently numbed by the cramp-fish. But the fish called the ass, the frog, the plaice, and the file-fish hide themselves in sand; and when they have rendered themselves immanifest, they agitate what the fishermen call the rods which they have in their mouth; but the little fishes run to these as to pieces of sea-weed with which they are fed.

No voracious fish, also, is found where the anthias is seen. Hence those that search for sponges call the anthias a sacred fish. This, however, is merely a casual circumstance; just as it is entirely accidental
621 that where there are snails there is neither a swine nor a partridge. For swine and partridges eat all the snails they find.

But the marine serpent resembles in colour, indeed, and the form of its body, the conger, except that it is of a darker hue, and is more robust. If, however, after it is caught it is dismissed, it rapidly hides itself in the sand, which it perforates with its nose. Its mouth, also, is more acute than the mouth of land serpents.

With respect, however, to what is called the scolopendra, when it has swallowed the fish-hook, it turns its inward parts outward, till it has thrown up the hook, and afterward again turns them inward. But marine in the same manner as land scolopendræ are attracted by odour. They do not, therefore, bite with the mouth [those they infect with poison], but the touch of the whole of their body is the cause of the infection, in the same manner as the animals which are called the nettles of the sea.

But with respect to the fishes called foxes, when they perceive that they have swallowed the fish-hook, they procure themselves assistance in the same manner as the scolopendra. For they run to the line, and gnaw the greatest part of it; for in some places they are caught by a complication of hooks, in currents and deep water.

The amiæ, also, when they see a certain voracious fish, turn themselves, and the largest of them swim in a circle [round the less]. And if the voracious fish touches any one of the amiæ the rest defend it; for they have strong teeth. It has, also, been observed that other fishes and the *lamia* have been wounded by the amiæ, which they had attacked.

Of river-fishes, however, the male glanis pays great attention to its offspring. For the female, indeed, when she has brought forth leaves her eggs; but the male abides in that place which contains the greatest number of the eggs, and stays there as their guardian, affording no other assistance than that of driving away other little fishes, that they may not plunder the eggs; and he does this for the space of forty or fifty days, till the young ones are so much increased in size, as to be able to fly from other fishes. Fishermen, however, know where to find the glanis who is the guardian of the eggs; for while he drives away the little fishes, he leaps and produces a murmuring sound. So great, also, is his attachment to his offspring, that though the fishermen endeavour to bring the eggs from deep to shallow water, he does not leave the eggs, but follows them. But if the guardian glanis is young, he is soon caught by the hook, in consequence of seizing the little fishes that approach to the eggs; but if he is accustomed to danger, and eats the bait from the hook, 621 he does not desert his charge, but with the bite of his very hard teeth breaks the hook.

All aquatic animals, however, as well those that swim as those that are fixed to one place, are fed in the places in which they are produced, or in others similar to these; for the nutriment adapted to each is in these places. But carnivorous fishes wander more than others, and all of them nearly are carnivorous, a few excepted; such as the mullet, the stock-fish, the barbel, and the *chalcis*. The fish, however, called the *pholis* spreads round itself a slime which it emits, and this becomes as it were a bed to it.

But of testaceous animals, and which are without feet, the cockle principally moves itself with a flying motion; for the purple, and other animals which resemble it, are moved with the smallest progressive motion.

The fishes, however, which dwell in the Pyrrhæan Euripus, swim out of it in winter, except the gudgeon, on account of the cold, for the Euripus is very cold; but on the return of spring, they again swim into it. The Euripus, however, neither contains the *scarus*, nor the alosa, nor any other fish of the thorny kind, nor the dog, nor the *acanthiæ*, nor the lobster, nor the polypi, nor the *bolitænæ*, nor certain other fishes. But of the fishes contained in the Euripus, the white gudgeon is not found in the sea.

Oviparous fishes, however, flourish in the spring, till the time in which they bring forth; but viviparous fishes flourish in autumn; and besides these, mullets, barbels, and all the rest of a similar kind. But all marine fishes, and those that dwell in the Euripus, bring forth in the Lesbian Euripus. They copulate, therefore, in autumn, but bring forth in the spring. The male and female selache, also, mingling together, copulate in autumn; but in spring they swim apart from each other, till they bring forth. At the time, however, in which they copulate many of them are caught joined together.

But the most crafty of the mollia is the cuttle-fish, which alone employs its ink, or black liquor, for the sake of concealing itself, and not only when it is terrified. The polypus, however, and the small calamary discharge their ink through fear. But none of these animals ever discharge the whole of their ink; and when they discharge it, it is reproduced. The cuttle-fish, however, as we have just observed, frequently uses its ink for the sake of concealment, and when it has
622 proceeded out of the ink so as to be visible, it returns into it. Farther still, it not only hunts after little fishes with its long appendices, but, also, after mullets. But the polypus is a stupid animal; for if a man puts his hand into the water, it walks towards it. It is, however, economic; but it collects every thing into the retreat in which it dwells. But when it has consumed whatever is most useful to it, it throws away the shells and crusts of crabs and shell-fish, and the thorns of little fishes. It, likewise, ensnares fishes by changing its colour, and becoming similar to

the stones to which it approaches. It, also, does this when it is terrified. Some persons, likewise, assert that the cuttle-fish does the very same thing; for they say that it causes its colour to resemble the place in which it dwells. The file-fish, however, is the only fish that does this; for it changes its colour in the same manner as the polypus.

The greater part, therefore, of the genera of polypi do not live two years; for they naturally waste away; of which this is an indication, that their flesh when pressed always emits a certain moisture, and at length is wholly consumed. The female polypi, however, after delivery suffer this in a greater degree, and become so stupid as to be agitated by the waves without perceiving it, and suffer themselves to be caught with facility by the hand of one who is swimming. They, also, become mucous, and so heavy with filth that they can no longer hunt the fishes on which they feed. But the males become leathery and viscous. And that they do not live longer than two years seems to be indicated by this, that after the birth of the polypi, which is in summer and in autumn, it is no longer easy to see a large polypus; but a little before this time the polypi are the largest. It is, likewise, said that when they have brought forth their eggs, both the male and female polypi become so old and feeble as to be devoured by little fishes, and easily divulsed from their retreats; but that prior to this they suffer nothing of this kind. They also say, that small and young polypi suffer nothing of this kind after their birth, but are stronger than the larger polypi. The cuttle-fish, likewise, does not live two years. But the polypus is the only one of the mollia that comes forth from the water on dry land. It also walks through rough places, and avoids such as are smooth. In other respects, likewise, it is a strong animal; but its neck is weak when it is 622 forcibly pressed.

Such, therefore, are the particulars respecting the mollia. But they say that the thin and rough conchæ form about themselves, as it were, a hard cuirass; that this becomes larger as the conchæ increase, and that from this they come forth as from a place of retreat or a house.

The polypus, also, which is called nautilus differs from other polypi, both in its nature and actions. For it sails on the sea, from the profundities of which it raises itself to its surface, and raises itself by inverting its shell [so that the concave part of it may be turned towards the bottom of the sea], in order that it may ascend more easily, and sail in the concavity of the shell. But after it has emerged to the superficies of the sea, it inverts its shell [so that the concavity may be turned towards the heavens]. Between its arms, also, there is a certain membrane, resembling the membranes which are between the toes of the palmipedes birds; except that in them it is thick, but in the polypi nautili it is thin, and similar to the web of a spider. But the nautilus uses this membrane as a sail when there is any wind; and instead of a rudder it uses its arms, which descend from each side into the water. When, also, it is terrified it merges into the sea, and fills its shell with water. The manner, however, in which its shell is formed, and increases in conjunction with the nautilus, has not yet been accurately ascertained. But it seems that it is not produced from copulation, but is formed spontaneously, in the same manner as other shells. It, likewise, is not yet manifest whether it is able to live when divested of its shell.

38 Of insects, however, the genus of ants and the genus of bees are nearly the most industrious, when compared with all other animals. And farther still, drones, wasps, and all such insects, as I may say, that are allied to these. Of spiders, also, some are more elegant, and agile, and are more skilful with respect to the purposes of life than others. The operations, therefore, of ants may easily be seen by every one; such as that all of them perpetually walk in one path, and collect and hoard their nutriment; for they work by night when the moon is at the full.

39 But of spiders, and the phalangia, there are many genera. Of the phalangia that bite, indeed, there are two genera; one resembling the spiders called wolves, which are small, variegated, quick, and adapted to leap, and these spiders are called *psylla*; but the other genus is large, of a black colour, has its fore-legs long, is slow in its motions, walks leisurely, is not strong, and does not leap. All the other phalangia,
623 however, which the vendors of medicine exhibit, either do not bite at all, or their bite is feeble. But there is another genus of spiders which are called wolves. Of this genus, therefore, the small spider does not weave, but the large spider weaves a rough and bad web near the ground, and in ruins; but it always forms its web in holes, and holding within them the threads which are the principles of the web, observes when any thing falling on the web moves it, and then comes forth from its hole. But the variegated spider forms a small and bad web under trees.

The third genus, however, of these is most wise and elegant. For it weaves a web first extending its threads on all sides to their boundaries [*i.e.* to the middle], and afterwards extending them from the middle; but it fixes on the middle with sufficient sagacity. In the next place, it adds to these as it were a woof, and perfects the web. It forms its nest, therefore, where it deposits its prey, in a place apart from the web; but catches its prey in the middle of the web. In the next place, when any thing falls on the web, the middle of it being moved, this spider first surveys on all sides, and enfolds its prey with its threads, till it is rendered incapable of resistance. After this, it raises its prey to its nest, and if it happens to be hungry sucks it; for in this it delights; but if it is not hungry, it again betakes itself to catching its prey, first repairing the web where it was torn. If, also, any thing falls on the web between the middle of it and the extremities, it first proceeds to the middle as from a principle, and from thence to the substance which fell on the web. But if any one injures the web, it again begins to weave it about the rising or setting of the sun, because at these times especially its prey fall into the web. The female, likewise, weaves and hunts after prey; but the male partakes of the booty.

Of elegant spiders, however, and which weave a thick web, there are two genera; the one, indeed, being larger, but the other less. The larger spider, therefore, which has longer legs, hanging under the web, observes the little animals that are his prey, in order that they may not through fear escape him, but may fall on the superior part of the web; for this spider, on account of its magnitude, cannot well conceal itself. But the less spider, whose body possesses more symmetry, conceals itself in a small hole in the superior part of the web.

Spiders, likewise, as soon as they are born are able to emit a thread, not from the interior part of their body, as if it were an excrement, as Democritus says it is, but from the external part, as from a rind, or after the manner of the porcupine, who hurls forth his hairs. Spiders, also, circumvent and enfold larger animals. For they attack small lizards, and surround them with their threads, so as at length to close their mouth; after which they approach to and bite them. And such are the particulars respecting these animals.

40 There is, however, a certain genus of insects which is anonymous by one name, but all the individuals of which have a congeneric form. And these are such insects as construct waxen cells; such as bees, and the insects which resemble them in form. Of these there are nine genera; of which, indeed, six are gregarious bees, the kings of bees, the drone, which lives among the bees, the annual wasp, and besides these, the hornet, and the *tenthredon* [an insect resembling a wasp]. But there are three that live alone; the small *siren*, which is of a brown colour; the

larger *siren*, which is black and variegated; and a third, which is called *bombylius*, and is the largest of these. Ants, therefore, do not hunt any thing, but collect what is made by others; but spiders neither make any thing, nor collect what is made by others, but only hunt after food. Of the before-mentioned nine genera, however, we shall afterwards speak about the rest; but bees hunt after no prey, but make and deposit honey, which is their food. That it is their food, indeed, is evident from this, that when those that keep bees endeavour to take the honey from the hives by smoking them, the bees who suffer very much from the smoke, at that time eat more honey than usual, but at another time they are not seen to eat any great quantity of it, as reserving and depositing it for the sake of nutriment. They have, also, another food, which some call *cerinthus*. But this food is inferior to honey, and possesses the sweetness of figs. They, likewise, bring the cerinthus with their legs, in the same manner as they bring their wax.

In the operations of bees, also, and in their mode of living, there is an abundant variety. For where a pure hive is delivered to them, they construct their cells from different flowers, and various trees; such as the willow, the elm, and others that abound with rosin. With these, also, they anoint the bottom of their hives, in order to prevent other animals from entering into them; but the keepers of bees call this work incrustation. If the avenues of the hives, likewise, are broad the bees build near it [so as to make it narrower]. They, also, first fabricate the cells in which the [common] bees are born; and afterwards the cells in which those that are called the kings and the drones are born. Some of
624 the cells, therefore, they always fabricate; but they construct the cells of the kings when there is a numerous progeny of bees, and the cells of the drones when there are indications of a great abundance of honey. They, likewise, construct the cells of the kings near the cells of the [common] bees; and these are small; but they construct the cells of the drones near the cells of the kings; and these are less than those of the [common] bees. But they begin the series of their cells from the roof of the hive, and form many series or webs of these connected with each other in a descending order till they reach the bottom of the hive. The cells, also, in which they deposit their honey, and those in which they bring forth their young, have a double entrance; for about one basis there are two avenues, after the manner of cups with two handles; the one being internal, and the other external. But the first cells of the honeycomb, which are joined to the hive, are narrow, and void of honey, and consist of two or three circular ranges; but those cells are fuller of honey which

are constructed with an abundance of wax. The first part of the entrance into the hive, likewise, is covered with an ointment called *mitys*; but this is sufficiently black, being as it were a certain sediment of wax, and of an unpleasant smell. This mitys, also, is used as a remedy for wounds, and such like suppurations. But the matter next to this with which the hive is plastered is called *pitch-wax*, which emits a less powerful odour, and is a less efficacious remedy than mitys.

Some persons, however, say that the drones fashion the cells by themselves, and in the same hive and in the same honeycomb divide their employment with the bees, yet they do not make any honey, but both themselves and their offspring are nourished with the honey made by the bees. The drones, also, for a great part of the time remain within the hive; but when they fly out of it, they impetuously raise themselves on high towards the heavens, moving in an orb, and as it were exercising themselves. And when they have done this, they again enter into the hive, and feast.

The kings, however, do not fly out of the hive, except in conjunction with the whole swarm, neither for food, nor for any other purpose. It is, also, said that if the king of a hive wanders from it, the whole swarm seeks and pursues him, till it discovers him by the smell. And it is farther said, that he is carried by the swarm, when he is unable to fly; and if he perishes, the swarm becomes dispersed. If, therefore, he continues absent for some time, the bees make honeycombs but no honey, and they themselves in a short time after perish.

Bees, also, climbing along flowers gather wax with great agility with their forefeet. Afterwards they wipe the wax off these feet, and transfer
it to the two middle feet, and from these to the curvatures of the two 624
hind feet, and thus laden fly away. Hence when they return to their hives, they are seen to be heavy laden with wax. The bee, however, at each flight does not proceed to flowers of a different species [but from one flower to another of the same species]; as, for instance, from violet to violet, and does not meddle with any other flower till it has returned to the hive. But when they arrive at the hive they shake themselves, and three or four bees follow each of them. It is not, however, easy to see what they have brought from the flowers, as neither has it yet been seen how they perform their operations; but they have been seen to take their wax from olives, and they remain in these trees for a long time, in consequence of the thickness of the leaves.

After this they direct their attention to the propagation of offspring. And nothing hinders but that in the same honeycomb there may be young bees, honey, and drones. It is said, therefore, that if the leader [*i.e.* the king] is living, the drones are generated separate from the bees; but if he is not living, the drones are generated in the cells of the bees, than whom they, also, become more irascible. Hence they are called stingers, though they have no sting, because they wish though they are unable to sting. The cells of the drones, also, are larger than those of the others. But the bees sometimes fashion their cells separate from those of the drones, yet for the most part the cells of the latter are in those of the former; on which account, they are to be separated from each other.

There are, however, as has been before observed, many genera of bees; two, indeed, of the leaders, the more excellent of which is of a red colour, but the other is more black and more variegated. The magnitude, also, of this genus is double that of the useful bee. But the most excellent bee is small, round, and variegated; another is long, and resembles the wasp; and a third, which is called a thief, is black, and has a broad belly. Besides these, likewise, there is the drone; this is the largest of all of them, is without a sting, and is slothful. Bees, also, which inhabit cultivated places differ from those that are the inhabitants of mountains. For the bees which dwell in forests are more hairy, are less, more industrious, and more irascible. The good bees, therefore, fashion their hives equable, and render the whole of the surface that covers them smooth. They, also, assign different cells to different uses; as, for instance, the whole of one cell, is appropriated to honey, another to the young bees, and another to the drones; and if it should happen that all these are in the same cell, the cell immediately following is found empty. The longer bees, however, construct anomalous hives, and the surface that covers them is tuberous, and resembles the covering which hornets place over their hives. Besides this, also, they do not dispose in an orderly but in a casual manner either their offspring, or any thing else. From these, likewise, leaders of a bad species are produced, many drones, and those bees that are called thieves; and they either make a very small quantity of honey, or none at all.

Bees, also, after they have constructed their honeycombs sit on them, and concoct them; and if this is not done by them, it is said that the honeycombs are corrupted, and become filled with the webs of spiders. If, however, they are able to continue sitting on their honeycombs during the whole of the necessary time, the whole offspring does not perish, but they produce many abortions; and if they are unable to do this, the whole progeny perish. Little worms, also, are generated in the corrupted honeycombs, which acquire wings, and fly away. Bees, likewise, are careful to place in an upright position those honeycombs that have fallen, and they support them with piles, in order that they may be able to enter into them. For when they have not a passage by which they can enter into their honeycombs, they do not sit on them; and afterwards they are filled with the webs of spiders.

There is no work, however, of the thief and the drone, and they injure the works of the other bees. But when they are caught they are put to death by the bees of the good species. These good bees, also, particularly destroy the leaders when they are numerous, and above all those that are bad, lest the multitude of them should be the cause of dispersing the swarm. Then, however, they especially destroy the leaders when the swarm is not prolific, and they are not able to produce a new swarm; for at these times they destroy the honey-combs of the kings, if they happen to be constructed, because a new progeny is wanting of which these kings might be the leaders. They, also, destroy the honeycombs of the drones, if they foresee there will be a scarcity of honey, and the swarms do not possess it in abundance. They, likewise, strenuously oppose those that take away their honey, and eject the drones that are then in the hive; on which account exiled drones are frequently seen sitting before the hives. The small bees, also, fight strenuously with the long species of bees, and endeavour to eject them from the swarm; and if they are victorious, it is considered as an indication that the swarm will be a good one; but if the long bees conquer, and, therefore remain alone in the hives, they lead an indolent life, and effect nothing good. These, however, also, perish before autumn.

With respect, also, to such as the good bees destroy, they endeavour to do this out of the swarm; but if any one dies within the hive, they, in a similar manner, carry the dead body out of it. But those bees that are called thieves injure their own honeycombs, and if they are able latently 625
enter into other hives; but if they are caught they are put to death. It is not, however, difficult to take them, since there are guards at the entrance of each hive. But the thief, if he can latently enter any hive, in consequence of immoderate repletion is unable to fly, and rolls himself before the hive, so that it is difficult for him to escape. But the kings themselves are never seen out of the hives, except when they migrate with the swarm, and then the other bees are seen collected about the king. When, however, the bees are about to migrate, for some days prior to their emigration a certain peculiar voice is heard, and for two

or three days a few bees fly about the hive; but whether the king is among these is not yet ascertained, because it is difficult to ascertain it. When, however, they are collected together, the other bees fly and arrange themselves about their respective kings. And if a few bees happen to be situated near many, the few betake themselves to the many, and destroy the king they have left if he follows them.

After this manner, therefore, bees quit their hive, and settle in some other place. Different works, also, are assigned to different bees. Thus, for instance, some collect the juice of flowers; others bring water; and others polish and erect the honeycombs. They, likewise, bring water when they nourish their offspring; but they do not sit on any flesh, nor eat any thing that is prepared for the table. There is not, however, any definite time in which they begin to work; but if there is a fit opportunity, and they are well, they principally apply themselves to work in summer; and when the weather is serene, they labour without intermission. Young bees, also, begin to work on the third day after they are born, if they are not in want of food. And when a swarm stops any where, some of the bees go in search of food, and afterwards return to the swarm. If a swarm, also, is in a good condition, the bees cease after the winter solstice from prolifiation for forty days. But when their young are increased in size, they place food before them, and close the entrance of their cell; and the young bee, when he is able, pierces through the covering of the cell, and comes forth. The good bees, also, cleanse their hives from the little animals which are engendered in them, and which injure the honeycombs; but the other bees through the
626 badness of their nature neglect to do this, and thus suffer their works to be destroyed.

When, however, those that keep bees take away the honey from the hives, they leave the bees a quantity of honey to support them during the winter. If, therefore, a sufficient quantity is left, the swarm is preserved, but if not, and the winter [prevents them from leaving the hive] the swarm dies; but if the weather is serene, the swarm quits the hive. Bees, also, feed on honey both in summer and winter; and they, likewise, use for aliment another kind of food, which resembles wax in hardness, and which is called by some persons sandarac, or red arsenic.

The animals, also, which principally injure bees are wasps, and the birds called titmouses, and besides these, the swallow, and the merops. The frogs of the marsh, likewise, hunt bees when they approach the water; on which account the keepers of bees endeavour to catch the frogs in the lakes and marshes which the bees frequent for the purpose of obtaining water. They, also, destroy the wasps and swallows which are near to the swarms, and the nests of the merops. Bees, too, avoid no animal except a bee. But they fight both with each other and with wasps; and when they are at a distance, indeed, from the hive, they neither injure each other, nor any other animal; but when they are near the swarm, they destroy every animal that they vanquish.

The bees, however, that have stung any animal [generally] die, because together with the sting they lose their intestine; but they are frequently saved, if the animal that is stung is careful that the bee does not leave its sting behind. If the bee, however, does not recover its sting it dies. Sometimes, also, they sting large animals to death, and a horse has died from the sting of bees. The kings, however, are not irritated, nor do they sting. Bees, likewise, carry out of their hives those bees that die in them; and the bee is in other respects a most clean animal; on which account they fly away with the excrement which they are frequently obliged to discharge in the hive, on account of its feted odour [in order that they may deposit it out of the hive]. For they abhor, as we have before observed, fetid smells, and, also, the smell of perfumes; and on this account they sting those that use perfumes. They, likewise, perish from other accidental circumstances, and, also, when the leaders or kings being many each of them draws away with him a part of the swarm. The rubeta, also, or red toad, destroys bees; for he comes to the entrance of the hive, blows into it, and observing when the bees fly out of it, seizes and devours them. It is impossible, therefore, for this toad to suffer any injury from the bees; but the person that takes care of the swarm kills him.

Some keepers of bees, also, assert that the irregular honeycombs, which, as we have before observed, are constructed by bees of an inferior kind, are made by young bees, in consequence of their unskilfulness; but these young bees are one year old. Young bees, however, do not sting similarly with the old ones; and on this account swarms are carried away without danger, for they are composed of young bees. But when honey fails, the keepers of bees expel the drones, and supply the bees with figs and other sweet substances. Of bees, however, those that are older work within the hive, and are more hairy, in consequence of remaining within it. But the young bees work out of the hive, are smoother [and less hairy], and kill the drones, when they have no longer sufficient room to work in; for the drones dwell in the innermost part of the hive.

A certain swarm, also, being once diseased, some bees came from another hive, fought with those that were diseased, vanquished them, and took away their honey. The guardian of the hive, however, killed the foreign bees, and was assisted in so doing by the other bees, who did not sting the man [that revenged the injury they had sustained]. Swarms of bees, likewise, that are in a prosperous condition are principally subject to the disease which is called *clerus*. This disease arises from small worms which are formed in the floor of the hive, and from the increase of which the whole swarm is filled with certain webs, as it were of spiders, and the honeycombs become rotten. Bees, likewise, are subject to another disease, which is a kind of lethargy, and then the swarm emits a fetid odour.

Thyme, also, is the food of bees; and the white is better than the red thyme. But the place in which they settle is in summer not fervid, and in winter is warm. They are, likewise, principally diseased when they operate on rusty materials. And when there is much wind, they carry with them a stone, as a support against the violence of it. If, likewise, a river is near, they drink out of it, and in no other place, first laying down their burden; but if a river is not near, they drink elsewhere, vomit their honey, and immediately proceed to their work. For the making of honey, however, there are two seasons, spring and autumn. But the honey which is made in spring is whiter, and, in short, better than the honey which is made in autumn. The honey, also, is better which is made from new wax and new plants. But the red honey is not so good, on account of the waxen cells in which it is deposited; for it is corrupted by these, in the same manner as wine by the vessel in which it is kept. Hence it is necessary to dry this vitiated honey. When
627 thyme, however, is in flower, and the cells are full, the honey does not become hard. But the honey which is of a golden colour is good; and the white honey is not produced from genuine thyme. The white honey, however, is good for the eyes and ulcers. The weak part of honey, also, always ascends to the surface, which part must be separated; but the pure part subsides.

When, however, the trees are in flower, the bees make their wax; on which account the wax must then be taken from the hive, for the bees immediately produce new wax. The plants, also, from which the bees bring their wax are the following: the atractyllis, or wild bastard saffron, the melilote, the asphodelus, or daffodil, the myrtle, the phleus, the willow called agnus, and the sparton. But when the bees work upon thyme, they carry water into their cell, before they anoint them. All bees, likewise, emit their ordure, either flying, as we have before observed, or in one of the cells of the hive. The small bees, also, as we have already observed, are more industrious than the large bees. Their wings, too, are bruised, are of a black colour, and are burnt [by the sun]. But the shining and elegant bees are, like women, indolent.

Bees, also, appear to be delighted with a noise; and hence it is said, that they are collected into the hives by the sound produced from striking earthern vessels. It is, however, perfectly immanifest whether they hear or not, and whether they do this from pleasure or from fear. Bees, likewise, expel from their hives both those that are indolent and those that consume too much honey. Every bee, also, as was before observed, is assigned a certain employment. And some, indeed, build the waxen cells, others make the honey, and others form the glue for the cement of the hives. Some, also, polish the cells; others bring water to the cells, and mingle it with honey; and others betake themselves to work out of the hive. But at day-break they are silent, till one of them by humming twice or thrice rouses the rest. Then all of them fly to their work; and again returning to the hive, at first they are tumultuous, but they gradually become less noisy, till one of them flying round hums, as if signifying that they should go to sleep; after which they are suddenly silent.

The signs, also, by which a swarm is known to be in a good condition are, much noise, and the motion of those that are going out of and entering into the hive; for then the younger bees are at work. But they are especially industrious, when the winter being elapsed they begin their work. And if any one in taking away honey from a hive leaves more than is necessary, the bees become more indolent. It is necessary, therefore, to leave just as much honey in the honeycombs as is sufficient for the multitude of bees. But if less honey is left than is sufficient, they labour with less alacrity. They, also, become more indolent if the hive is very large, as despairing that they shall be able to fill it. One hive, however, furnishes one measure, or one measure and a half of honey; but a very full hive furnishes two measures, or two measures and a half; and rarely three measures.

Sheep and wasps, however, are enemies to bees, as was before observed. But the keepers of bees catch the wasps, by placing a pot and putting flesh into it; and when many wasps have entered into the pot, they cover it and put fire under it. If, likewise, there are but few drones, they benefit the swarm; for they make the bees more industrious.

Bees, also, have a foreknowledge of tempestuous weather and rain; for prior to these taking place, they do not fly far from their hives, but while the weather is yet serene they fly about the hives; by which the keepers of bees know that they expect tempestuous weather. When bees, likewise, in a swarm are suspended from each other, this is a sign that the swarm will leave the hive. The keepers of bees, however, when they perceive this, sprinkle the hive with sweet wine. It is, also, advantageous to plant about a swarm of bees, pear-trees, beans, lucern, the Syrian grass, orobus, myrtles, poppies, wild thyme, and almond trees. Certain keepers of bees, likewise, know their own bees by sprinkling them in the meadow with meal. But if the spring is tardy in its approach, or the season is dry and hot, and the corn has the rust, bees are less prolific. Such, therefore, are the particulars respecting bees.

41 But of wasps there are two genera. For some of these, indeed, are savage and rare; and these dwell in mountains, and bring forth, not under the earth, but in oaks. Their form, also, is larger and longer, and they are blacker than the other wasps. All these, however, are variegated, have stings, and are stronger than the others, and the wound which they inflict is more painful; for the sting of these is, also, proportionally larger than that of the other wasps. These wasps, therefore, live two years, and are seen in winter to fly from oaks which are cut down. But they live in winter in concealment, and pass their time in the trunks of trees.

Of these wasps, also, some are mothers, and others are labourers, in the same manner as among the milder wasps. But what the nature is of the labouring wasp, and of the mother, will be evident in what will be said
628 of the milder wasps. For of mild wasps there are two genera; some of them, indeed, being leaders, which are called mothers, and others labourers. But the leaders are much larger and more mild than the labourers. And the labourers, indeed, do not live two years, but all of them die at the commencement of the winter. This, however, is evident. For at the commencement of the winter, the labouring wasps become stupid, and about the winter solstice they are not at all seen. But the leaders, which are called mothers, are seen during the whole of the winter, and conceal themselves under the earth. For many persons that plough and dig the earth in winter have seen the mothers, but no one in the winter has seen the labourers.

The generation, also, of wasps is as follows: the leaders when they have found an opportune place at the commencement of the summer fashion their honeycombs, and construct what are called small wasp-hives, which have four avenues, or nearly so, and in which the [labouring] wasps and not the mothers are born. When these young labouring wasps are increased in bulk, the leaders again construct other larger wasp-hives, and when the young wasps that are born in them grow larger, they again construct others; so that at the end of autumn many and the largest wasp-hives are formed, in which the leader called the mother no longer begets labouring wasps, but mothers. But these are generated in the superior part of the wasp-hives in the form of large worms, in four or a few more cells connected with each other. And, in a similar manner, the other wasps are produced in the cells of the honeycombs. When, however, the labouring wasps are produced in the honeycombs, the leaders no longer work, but the labouring wasps bring them food. But this is evident from the leaders no longer flying out of the wasp-hives, but remaining quietly within them. Whether, however, the leaders of the preceding year, when they have produced new leaders, die together with the new wasps [at the beginning of the winter], or whether they are able to extend their life through another winter, has not yet been ascertained; as neither have the particulars respecting the old age, the mothers, and such like affections of the savage wasps, been yet explored by any one.

The mother-wasp, however, is broad and heavy, and is thicker and larger than the labouring wasp. It, likewise, does not possess much strength for flying, and is incapable of flying far; hence, the mother-wasps always sit in the wasp-hives engaged in domestic employments. In most wasp-hives, also, there are what are called the mothers; but it is 628
dubious whether they have stings or are without stings. It seems, however, that in this respect they resemble the leaders or kings of bees, who have, indeed, a sting, but who do not employ it in piercing and wounding. But of wasps, some are without stings, in the same manner as the drones, but others have a sting. Those wasps, also, that are without stings are less and more imbecil than those that have, and are not capable of defending themselves; but those that have stings are large and strong. And some, indeed, call those that have stings males, but those that are without stings females. Many, however, of those wasps that have stings appear to lose them in winter; but of this we have not yet had ocular evidence.

Wasps, also, are produced in greater abundance in dry and hot seasons, and in uncultivated places. They are, also, produced under the earth, and they fashion their honeycombs from earth and other materials which they collect together; each honeycomb being constructed from one principle, as from a root. The food, also, which they use is derived from certain flowers and fruits; but, for the most part, they feed on the flesh of animals. Some of the mild wasps, also, are seen to copulate; but whether both the wasps that copulate are without, or both have stings, or the one has but the other has not, has not yet been ascertained. Some of the savage wasps, likewise, have been seen to copulate; and one of them has been found to have a sting, but whether the other had a sting it has not yet been seen.

Wasps, also, do not appear to be produced from conception; because immediately as they are born, they are larger than they could be if they were the offspring of wasps. But if any one taking hold of a wasp by the feet, suffers it to hum with its wings, the other wasps that are without stings fly to it, but those that have stings do not; which some persons employ as an argument that the former are males, and the latter females. Wasps, likewise, are caught in caverns in winter, some of which have stings, and others have not; and some of them, indeed, construct a few small wasp-hives, but others construct many, and such as are large.

But those wasps that are called mothers are caught as the season advances, and many of them are caught about elms; for wasps collect from these a viscous and glutinous matter. Sometimes, also, many mothers are produced, when in the preceding year there has been a great number of wasps, and an abundance of rain. They are, likewise caught in steep places, and in perpendicular chasms of the earth, and all of them are seen to have stings. Such, therefore, are the particulars respecting wasps.

42 Hornets, however, do not live by the nutriment they derive from flowers, like bees, but most of them feed on flesh, and on this account they live near dunghills. For they hunt large flies; and when they have caught them, having taken off their head, they fly away with the rest of
629 the body. They, also, meddle with sweet fruit, which they use for nutriment; and they have a leader, in the same manner as bees and wasps. They, likewise, exceed the other hornets in a greater proportion than the leaders of wasps the other wasps, and the leaders of bees the other bees; and they live within the hive in the same manner as the leader of wasps. Hornets, also, construct their hive under the earth, excavating the earth for this purpose, in the same manner as ants; for neither hornets nor wasps send out swarms after the manner of bees; but the new hornets always remain within the hive, which they enlarge by carrying earth out of it. Hence their hives become very large; for from

swarms that have been in a good condition, three or four baskets of honeycombs have been taken.

Hornets, likewise, do not deposit their food in magazines, in the same manner as bees, but they hide themselves in the winter, and the greater part of them die; but it has not yet been ascertained whether all of them die. There are not, however, in the swarms of hornets, as in the swarms of bees, more leaders than one, which among bees cause a divulsion of the swarm. But when certain hornets wander from the swarm, they assemble about a tree, and construct their hives, which have often been distinctly seen. In each of their hives, also, they produce one leader, who, when he is increased in size, leads the troop from the tree, and dwells together with them in a hive. With respect, however, to the copulation of hornets, nothing has yet been ascertained, nor is it yet known whence their offspring is derived. Among bees, therefore, the drones and the kings are without stings, and this is, also, the case with some wasps, as was before observed; but all hornets are seen to have stings. It remains, however, to be more particularly considered, whether the leaders of hornets have stings or not.

43 With respect to silk-worms, they bring forth under stones in the earth, and construct two or a few more cells, in which the commencement of a certain bad honey is found.

But the tenthredon resembles the hornet, indeed, but is variegated, and in the breadth of its body it is similar to the bee. It is, also, dainty, and flies to kitchens, and fishes, and such like food. It, likewise, brings forth under the earth, in the same manner as wasps. But it is multifarious, and its hive is much larger and longer than the hive of wasps.

And such are the particulars respecting the operations and life of bees, wasps, and other such-like insects.

44 With respect, however, to the manners of animals, a difference in them may be surveyed, as was before observed; and especially as to fortitude and timidity, and, also, as to mildness and ferocity, even among savage animals themselves. For the lion in the act of eating is most ferocious, but when he is not hungry, and has been well fed, he is most mild. He is, likewise, by no means of a suspicious disposition; and towards his associates and familiars he is very playful and bland. When he is hunted, also, he never flies if he is seen, nor hides himself through fear; but if, on account of the multitude of those that hunt him, he is compelled to withdraw himself, he retires slowly step by step, and turns round [and faces his opponents] at every step. And if, indeed, he meets with a thicket, he flies rapidly through it till he comes into a plain; but

then, again, he slowly retires. If, however, in plains and open places, he is at any time compelled to fly, through the multitude of his pursuers, he runs in a continued course, and does not leap; so that the continuity of his course resembles that of a dog. But when he pursues his prey, he leaps upon it when he is near to it. What is asserted, likewise, of the lion is true, that he is particularly afraid of fire, as Homer, also, says:

And glowing torches, which, though fierce, he dreads.

Iliad, II, 553

And, also, that he recognizes the person by whom he is struck, and attacks him. If, however, any one strikes, but does not hurt him, the lion if he catches him does him no injury, nor lacerates him with his claws; but shaking and terrifying, he again dismisses him. Lions, also, principally betake themselves to cities, and injure men, when they become old, and are unable through age to hunt, and their teeth are in a bad condition. They, likewise, live many years; and a lame lion that was caught had a great number of broken teeth; by which circumstance, some persons conjecture, that the life of lions is very extended; for their teeth would not be broken, unless they lived long.

But there are two genera of lions; and of these the one is more round, his hair is more curled, and he is more timid; but the body of the other is longer, his hair is thicker, and he is more courageous. Sometimes, also, lions fly with their tail hanging down, in the same manner as dogs. A lion, likewise, has been seen to run away, when about to attach a hog, as soon as he perceived that the hog prepared to defend himself. With respect to blows, however, the lion is incapable of supporting them in his flank, but in the rest of his body he can receive many blows, and his head is strong. From the wounds, also, of such animals as he bites, or lacerates with his nails, a very yellow lymph flows, and which so tenaciously adheres to bandages and sponges, as not to be washed off. But the remedy for these wounds is the same as the remedy for the bite of a dog.

Again, the thos is friendly to man, and, therefore, neither injures nor is very much afraid of men; but it is hostile to dogs and lions; and hence they are not found in the same place. But the small is better than the large thos; and with respect to the genera of them, some say that there are two, but others that there are three. It does not appear, however, that there are many genera of them, but in the same manner as certain fishes, birds, and quadrupeds, the thos changes at certain times, and its colour is different in summer and winter; and in summer, indeed, its hair is thin, but in winter thick.

45 With respect to the bonassus, it is produced in Pœonia, in the mountain Messapius, which separates Pœonia and Media; and the Pœonians call this animal monapus. In size, also, it is as large as a bull, but is more bulky than an ox; for its body is not long. Its skin, likewise, when distended occupies as much space as would be sufficient for seven guests [at a banquet], but the rest of its form resembles than of an ox; except that it has a mane, which extends to its shoulders, in the same manner as a horse. Its hair, however, is softer, and descends lower than that of the horse; but the colour of its hair is yellow, and its mane extends as far as to its eyes, and is thick. The colour, also, of this animal is a medium between ash and red; not like the colour of the horses which are called sorrel, but is more dirty, and in the inferior part is woolly. None of these animals, however, are black or very red; the voice of the bonassus, likewise, resembles that of an ox, but its horns are inflected towards each other, and are useless for the purposes of defence: their size, also, is that of one palm, or a little more, and their thickness is such that each of them is capable of containing not much less than the half of a congius [or Roman gallon]. The black of the horn, likewise, is beautiful and rich; but the hair which hangs from its forehead extends 630 to its eyes, so that is rather sees obliquely than with direct vision. In the same manner, also, as the ox and other horned animals, it has no teeth in the upper jaw; but its legs are hairy, and its feet are cloven. Moreover, its tail is less than the magnitude of its body requires, and resembles the tail of an ox. It, likewise, scatters the dust and digs into the earth in the same manner as a bull; and its skin is strong with respect to the endurance of blows. Its flesh, also, has a good taste, and on this account the bonassus is hunted; but when it is struck [by the hunters], it flies, and stops when it is incapable of flying any farther. The defence of itself, likewise, consists in kicking, and in discharging its excrement, which it throws from itself to the distance of four paces; but it easily and frequently employs this mode of defence; and its excrement is so hot as to destroy the hair of the dogs that pursue it. The dung, therefore, of the bonassus produces this effect when the animal is agitated and terrified; but when it is not agitated it does not burn. Such, therefore, is the form and nature of the bonassus. At the season of bringing forth, likewise, the females assemble in the mountains, and are there delivered. But before they bring forth, they entrench themselves, as it were, with their excrement, for this animal emits excrement in great abundance.

46 Of all savage animals, however, the elephant is the most gentle and mild; for he may be instructed in, and possesses a knowledge of many

things; since he is taught to prostrate himself before the king. The senses, also, of the elephant are very acute, and he excels [other quadrupeds] in sagacity. But he does not again meddle with the female with which he copulates, and which he causes to be pregnant. Some, likewise, say that the elephant lives two hundred, and others that he lives a hundred and twenty years, and the females live nearly as many years as the males. It is, also, asserted that they arrive at their acme about the sixtieth year of their age; and that they are impatient of the winter and cold. The elephant, also, loves rivers, though he is not a river animal, since he dwells near the banks of rivers. He, likewise, walks through the water, and proceeds into it as long as his proboscis can keep above it; for he blows and respires through this; but he is unable to swim through the weight of his body.

47 Camels, however, do not copulate with their mothers, but are unwilling to do this, though some one should compel them to it. For a camel-driver once who had not a stallion, having veiled a female camel, admitted a young camel, the offspring of this camel, to copulate with her. But when the veil was taken off, the copulation being finished, the
631 young camel shortly after destroyed the camel-driver by biting him to death. It is, also, said, that the king of Scythia had a horse of a good race, from which all the colts that were produced were of a generous breed. When the king was willing that the best of these should procreate from the mother, it was led to the mother that it might copulate, but unwillingly. The mother, however, being veiled, it copulated with her ignorantly. But as soon as it had copulated, the face of the mother being unveiled, on seeing its mother it fled, and hurled itself from a precipice.

48 Among marine animals, also, many indications are related of the gentleness and mildness of the dolphin; and, likewise, of the love and attachment which it has shown to boys, about Tarentum, Caria, and other places. For a dolphin being caught and wounded about Caria, a great multitude of dolphins came to the port, and continued there till the fisherman dismissed the dolphin which he had taken; for then all of them at once departed. Some one of the large, likewise, always follow the small dolphins, as their guardian. A troop, also, of large and small dolphins has been seen; and of these, two leaving the rest, were shortly after seen carrying on their back, and raising to the top of the water, a small dead dolphin which had fallen to the bottom of the sea, as if influenced by pity, lest it should be devoured by some other fish.

Incredible things, also, are related respecting the swiftness of this animal. For it appears to be the swiftest of all animals both aquatic and terrestrial, and to surpass in rapidity the masts of large ships. This, however, particularly happens to be the case with them when they pursue any fish for the sake of food. For then, if the fish flies from them, they follow it to the depths of the sea, impelled by hunger. But when their return from the bottom of the sea is long [through the distance] they hold their breath, as if estimating the distance, and turning themselves, they are carried upward with the velocity of an arrow, in order that by their rapidity they may arrive at the place in which they can respire; and in doing this they surpass in velocity the masts of ships, if they happen to meet with them. Divers, also, do the same thing, when they merge themselves into the depths of the sea; for they raise themselves with all their force to the surface of the sea, in order that they may respire. The male and female dolphins, likewise, live in pairs 631
with each other. But it is dubious with respect to them why they rush into the earth; for it is said that they sometimes do this from no assignable cause.

49 As, however, it happens that the actions of all animals are conformable to their passions, so again their manners change conformably to their actions. This, likewise, is frequently the case with certain parts of the body; as, for instance, in birds. For hens when they vanquish the cocks crow, and imitating the males endeavour to copulate. They likewise, raise their crest and their tail, so that it is not easy to know that they are females; and sometimes hens have certain small spurs. Some males, also, have been seen on the death of the female performing the office of the female in attention to the chickens, by leading them about, and nurturing them, so as no longer either to crow or endeavour to copulate. Certain male birds, likewise, become so effeminate from their birth, as neither any longer to sing, nor endeavour to copulate, but suffer the males that endeavour to copulate with them [as if they were females].

Many birds, likewise, change both their voice and their colour at 632
certain seasons. Thus, for instance, the blackbird, instead of black has a yellow colour and a various voice. For in the summer it sings, but in the winter it makes a stridulous noise, and utters tumultuous sounds. The thrush, also, changes its colour; for in winter it is of an ash colour, but in summer its neck is variegated; yet it suffers no mutation in its voice.

But the nightingale, indeed, sings continually day and night for fifteen days, during the time in which the mountains are thick with foliage.

After this period, likewise, she sings, yet no longer without intermission. But as the summer advances, she emits another and no longer an all-various voice, nor a rapid and modulated, but a simple voice. She, likewise, changes her colour; and in Italy, at this season of the year, is called by a different name. She is not, however, seen for a long time; for she conceals herself.

The robin-redbreast, likewise, and what is called the roach, are changed into each other. But the robin-redbreast is a bird of the winter; and the roach is a summer bird. They do not, however, at all differ from each other, as I may say, except in colour. In a similar manner, also, the becafigo [a bird like a wheat-ear] and the black-head are changed into each other. But the becafigo makes its appearance about the beginning
633 of autumn, and the black-head immediately after autumn. These birds, also, differ in no respect from each other, except in their colour and voice. That the becafigo, however, and the black-head are the same bird is evident from this, that at the time of their change each has something which is the peculiarity of each. Nor is it at all wonderful that these birds change their voice and their colour, since the pigeon, also, it not at all vocal in winter, and, therefore, those who are skilled in the manners of birds are surprised if pigeons are vocal when fair weather succeeds a very severe winter; but when the spring commences, then the pigeon begins to be vocal. In short, birds are especially and most frequently vocal at the time of copulation.

The cuckoo, likewise, changes its colour, and its voice is not clear, when it is about to conceal itself. But it disappears when the dog-star rises. It, also, becomes visible at the commencement of the spring, and continues to be so till the rising of the dog-star. But the bird which some persons call *oenanthe* disappears at the rising of Sirius, and becomes visible when Sirius sets. It, likewise, avoids at one time cold and at another time heat.

The epops, also, changes its colour and its figure, as Eschylus asserts in the following lines: "The epops, the surveyor of its own evils, presents itself to the view variegated and florid, a bold bird of the rock, every way armed. On the first appearance of daylight, this bird shakes the wing of the white buzzard. For it exhibits a twofold form, though it is the offspring of one womb. But in the recent autumn, when the ears of corn are dry, again, [as if it were another bird] it extends about its body speckled plumes. It, likewise, hates cities, and withdraws itself into the deserts or rocks of Pallene."

Of birds, likewise, some roll themselves in the dust; others wash themselves; and others do neither of these. And those birds, indeed, which do not fly high, but near the earth, roll themselves in the dust;

such as the hen, the partridge, the *attagen*, the lark, and the pheasant. 633
But some birds that have straight claws, and such as frequent rivers, or marshes, or the sea, wash themselves. Others, again, both roll themselves in the dust and wash themselves; such as the pigeon and the sparrow. The greater part, however, of birds with crooked claws do neither the one or the other. And thus much concerning the mutations of certain animals.

To some birds, likewise, it is peculiar, as to turtle-doves, to produce a crashing noise. For the ultimate part of their alvus is agitated when they are vocal, and from this a crashing sound is produced.

631

50 Some animals, also, change their forms and manners, not only according to ages and seasons, but, likewise, when they are castrated. But those animals are castrated that have testicles. And some birds, indeed, and oviparous quadrupeds, have their testicles inwardly, near the kidneys. But with respect to viviparous and pedestrious animals, the greater part have their testicles externally, and some have them internally; but all of them have them near the end of the belly. Some birds, however, are castrated near the rump, at the part with which they come into contact with the female when they copulate. For if any one burns this part with two or three hot irons, if the castrated bird is perfect, its crest will become pale, and it will no longer sing nor endeavour to copulate; but if it is castrated while it is yet young, it will never sing, nor copulate, nor have a red crest.

The same thing, also, happens to men. For if they are mutilated in their private parts while they are children, they will neither have hairs of posterior origin, nor will they experience a mutation of voice, but their voice will continue to be acute. But if they are castrated after 632
puberty, the hairs of posterior origin will fall off, except those which are about the private parts; which, though they will be diminished, yet will not entirely fail. The hairs, however, which are born with men do not fail when they are castrated; for no eunuch becomes bald.

The voice, also, of all castrated quadrupeds changes into the female voice. Other quadrupeds, therefore, if they are castrated when they are not young perish; but in the wild-boar alone it is of no consequence at what period he is castrated. But all quadrupeds, if they are castrated when young, become larger and more elegant than those that are not castrated. If, however, they are castrated after the period of adolescence, they no longer receive any increase. But stags, if they are castrated before they are too young to have horns, never have them; and if they are castrated when they have horns, the magnitude of the horns remains the same, and they do not lose them.

Calves, therefore, are castrated when they are a year old; but if they are not then castrated, they become deformed and less. They are, also, castrated after the following manner: The testicles being drawn downward, distended and pressed to the bottom of the scrotum or purse in which they are contained, they are thrust out, and afterwards the roots of the ligaments from which the testicles depend are as much as possible drawn upward. Afterwards, the incisure is filled with hairs, in order that the lymph may not flow forth externally. And if an inflammation takes place the scrotum is burnt, and then anointed. But adult bulls, if they copulate as soon as they are castrated, are able to generate.

The matrix, also, of female swine is cut off, in order that they may no longer require copulation, but may in a short time become fat. But it is cut off after they have fasted for two days; and in the next place, suspending them by their hind-legs, the lower belly is cut in that part in which the testicles of males principally grow. For this is the natural situation of the matrix in females; and a small part of it being cut off, the place where the incisure was made is sewed up. Female camels, likewise, are castrated, when there is occasion to make use of them in war, in order that they may not become pregnant. But some persons possess three thousand camels. Camels, likewise, run much swifter than the Nisæan horses, through the magnitude of their pace. And, in short, castrated animals become longer than those that are not castrated.

Animals, also, that ruminate delight in ruminating; and ruminate as if they were eating. But those animals ruminate which have not teeth both in the upper and lower jaws; such as oxen, sheep, and goats. No savage animal, however, has yet been seen to ruminate, except those that sometimes live with men, such as the hind; for she ruminates. All ruminating animals, likewise, principally ruminate when they are couchant. But they especially ruminate in winter; and animals which are nurtured in a house do this for nearly seven months. Gregarious animals, however, ruminate less, and for a less time, in consequence of deriving their nourishment abroad. Some animals, too, that have teeth in both jaws ruminate, such as the Pontic rats, and fishes; which some, from the operation, call *ruminating*.

Of animals, likewise, that have long legs the bellies are lax [so that they easily suffer a flux of the belly], but those that have a broad breast are subject to vomiting; and this is the case for the most part with quadrupeds, with birds, and with men.

ON PHYSIOGNOMY

ON PHYSIOGNOMY

1 That the soul follows the body, and is not itself by itself impassive[†] 805
to the body, is perfectly manifest in intoxication and sickness. For the soul is seen to suffer a great change from the passions of the body. And, on the contrary, that the body sympathizes with the passions of the soul, becomes evident in love, fear, pain, and pleasure. And farther still, in those things which are produced by nature, it may be more clearly seen that the soul and body are so connascently affected towards each other, as to become to each other the cause of most of their passions. For an animal is never so generated as to have the form of one animal, and the soul of another; but it has always the body and soul of the same animal; so that a particular disposition must necessarily follow a particular body. Farther still, those who are skilled in the nature of other animals are able from the form [of the body] to survey in each [the passions of the soul]. In this way, he who is skilled in the horses surveys horses, and hunters, dogs. But if these things are true (and they are always true), there will be an art of physiognomy.

Prior physiognomists, therefore, endeavoured to physiognomize according to three modes, and each according to one mode. For some, indeed, physiognomize by establishing according to each genus a certain form and character of animal. But others, in the comparison of animals with men, were of opinion that the man whose body resembled the body of some animal had, also, a similar soul. And others did not derive their knowledge from all animals, but from the genus itself of men, dividing according to nations, such of them as differ in sight and manners; such, for instance, as Egyptians, Thracians, and Scythians; in a similar manner making their selection of signs. But others, from apparent manners, inferred that a certain disposition of the body was attended with analogous manners and affections of the soul, in him who is angry or terrified, or has connection with a female, and in each of the other passions.[‡] It is possible, however, to physiognomize according to

† This is only true of the soul so far as it proceeds into body, or rather of those energies of the soul which proceed into body, and which from their union with the body may, also, be called passions. See the Treatise *On the Soul*.

‡ *i.e.* Because paleness follows from the passion of fear, they inferred that those who are naturally pale are naturally timid: and because a redness in the countenance follows from the passion of anger, they inferred that those who naturally have this redness in the countenance are naturally irascible. And again, because those who are engaged in the act of copulation elevate their eyes, as is shown in the fourth section of the *Problems*, they inferred that those whose eyes are naturally drawn upward are naturally libidinous.

all these modes, and, likewise, according to other modes, and that the
805 selection of signs may be dissimilarly† made. These, therefore, alone physiognomizing according to manners erred, in the first place, indeed, because the same disposition in the face may follow from different passions of the soul. Thus, for instance, the brave and the impudent man have the same disposition in the face [*i.e.* the face of each is unchanged], but in mind they are much separated from each other. And they erred in the second place, because at different times the same persons have not the same, but different manners. For it sometimes happens that those who are [for the most part] sad pass through the day pleasantly, and assume the manners of one who is cheerful; and on the contrary, those who are [for the most part] cheerful are [sometimes] sad, so that the manner in the face is changed. Farther still, if any one alone considers the habit of the countenance, he will be able to frame a conjecture of but few passions of the soul [because there are but few which change this habit]. But those that physiognomize from savage animals do not rightly make the selection of signs. For it is not possible by directing our attention to the form of every animal to assert, that the animal whose body resembles that of another animal has, also, a similar soul. For, in the first place, in short, no one will find a man thus similar to a wild beast, but resembling it in a certain respect. And in addition to what has been said, few animals have peculiar, but many have common signs. Hence when any one resembles a brute, not according to a peculiar, but according to a common sign, in what respect will he be more similar to a lion than to a stag? For it is reasonable that peculiar signs should signify something peculiar; and common signs something common. Common signs, therefore, will not manifest any thing to him who physiognomizes. But if any one selects the peculiarities of each savage animal, he will not be able to assign the man of whom these are the signs; for it is reasonable that peculiarity should belong to no other animal than that of which it is the peculiarity. No one, however, can assume in his mind the peculiarity of any one of those animals which are the subject of physiognomy. For neither is a lion alone endued with fortitude, but many other animals are brave; nor is a hare alone timid, but an infinite number of other animals. If, therefore, nothing becomes manifest either to him who selects common, or to him who selects peculiar signs, we must not in physiognomizing

† Thus, if large extremities denote fortitude, on the contrary, small extremities indicate fear, and, as it were, an exility of soul.

direct our attention to one genus of animals, but a selection must be made from all of them that suffer the same passion. Thus, for instance, if any one surveys the indications of a brave animal, he ought, by collecting brave animals into one, to investigate what passions [*i.e.* properties] are inherent in all these, which do not happen to any other animal. For if any one should conclude that there are the same signs of fortitude in the before-mentioned animals, it would be requisite that not only the affection of fortitude should be common to them, but, also, something else.† And thus it will be dubious to him whether the indications are those of fortitude, or of something else. It is necessary, however, to make a selection from most animals, and which have not any other common passion, the signs of which are the objects of speculation. Such signs, therefore, as are stable, also, signify something stable. But with respect to such as accede and depart, and fail, how can the sign be true of that which no longer remains in the mind.‡ For if any one receives a sign which accedes and departs of that which is stable, it will happen, indeed, that it will be true; yet it will not be appropriate, unless it is perpetually consequent to the thing. But with respect to such passions as being ingenerated in the soul do not in any respect change the signs in the body which are employed by the physiognomist, things of this kind will not be the objects of knowledge to art. Thus, for instance, from opinions and sciences§ it is not possible to know a physician or a harper. For he who learns any discipline does not change any of the signs which the physiognomist employs.

2 It is necessary, therefore, to determine what the objects are with which physiognomy is conversant, since it is not conversant with all things; and, also, from what particulars the several signs are assumed.

† For if all brutes, for instance, which have large extremities, not only agree in this that they are naturally brave, but, also, in some other property, as that they are naturally cruel, there will be no reason why large extremities should be assumed as the sign of fortitude rather than as the sign of cruelty.

‡ For a sign is always essentially connected with the thing of which it is the sign; but things which are not permanent, but are at one time present, and at another absent, are not essentially connected with a permanent passion; since the passion is permanent, but the sign is not.

§ *i.e.* From the possession of those opinions and sciences by which a man becomes a physician or a harper, it is not possible to know that he is a physician or a harper, because this possession produces no change in physiognomic signs.

Afterwards, it will be necessary to render manifest, in a consequent order, each of the more apparent signs. Physiognomy, therefore, as the name evinces, is conversant with the natural passions of the soul, and, also, with those adventitious circumstances to which some accidents and passions of the body correspond, and which may be known by physiognomic signs. But what these are will afterwards be rendered manifest. Now, however, I will show from what genera all the signs are assumed. For the decisions of physiognomists are derived from figures and colours, and from the manners apparent in the face; from the hairs, also, from the smoothness of the skin, from the voice and the flesh, and from the parts and figure of the whole body. Such, therefore, in short, are the genera to which physiognomic signs are reduced. If, therefore, a discussion of this kind were immanifest, or not properly significant, what has been said might be deemed sufficient. But now, perhaps, it is
806 better more accurately to explain whatever is manifest among all those particulars which are the subjects of physiognomy; enumerating, also, the several signs, and unfolding what is signified by them, so as to render such of the before-mentioned particulars clear as were not manifest.

With respect to colours, therefore, such of them as are acute [*i.e.* vivid] signify a hot and sanguineous complexion. But white colours, mixed with red, denote a good disposition, when they happen in a smooth body. With respect to the hairs, also, such as are soft signify a timid disposition; but hard hairs denote one that is brave. This sign, however, is derived from all animals. For the stag, the hare, and the sheep, are most timid, and have the softest hair; but the lion and the wild-boar are most brave, and have the hardest hair. This very same thing, likewise, may be seen in birds. For, universally, such birds as have hard feathers are brave; but such as have soft feathers are timid. And this is particularly evident in quails and cocks. In a similar manner, also, this same thing happens in the genera of men. For those who dwell in the north are bold, and have hard hair; but those who dwell in the south are timid, and have soft hair. But hairiness about the belly signifies loquacity; and this is referred to the genus of birds. For of things pertaining to the body, hairiness about the belly is the peculiarity of birds; but of things pertaining to the soul, the peculiarity of these animals is loquacity.

With respect to the flesh, that which is hard, and naturally of a good habit, signifies a man who is dull, and of a rude genius; but that which is smooth signifies one of a naturally good disposition, yet unstable, unless this happens in a strong body, and which has hard extremities.

But with respect to motions, such as are slow signify a soft [and effeminate] mind; but such as are swift denote a hot complexion.

In the voice, also, that which is grave and intense signifies a brave man; but that which is acute and relaxed, a timid man.

The figures, also, and passions which appear in the face, are assumed according to the similitudes of passions. Thus, for instance, a habit of the face resembling that of a man when he is angry signifies a man naturally irascible.

The male, however, is larger and stronger than the female, and the extremities of his body are stronger and thicker. The habit, also, of his body is more excellent, and he is better according to all the virtues.

But of the signs which are in the parts, those are stronger which are assumed in the manners that are manifest; and, also, those which are assumed according to motions and figures.

In short, it is stupid to believe in one sign only. But when many signs 807
accord of one thing, it may be more confidently admitted that the signs are true.

There is, also, another mode of physiognomizing, but which no one has yet adopted. Thus, for instance, if it is necessary [from signs] that some one should be irascible, morose, and pusillanimous, he will, also, be envious, though he has no signs of an envious man; but from those former signs it will be possible for a physiognomist to discover that he is envious; and this will be especially the case when such a mode will be proper to one who philosophizes. For to be able, certain things existing, to discover what is necessary, is, we apprehend, the peculiarity of philosophy. We may, also, argue from a contrary; for if from certain signs certain good things are inferred, from contrary signs we may, also, infer evils, which are their contraries.

With respect to voice, indeed, some one directing his attention to passion may think that an acute voice is to be considered as indicative of an irascible disposition, and this for a twofold cause. For he who is indignant and angry is accustomed to give intention to his voice, and to speak acutely; but he who is indolently disposed remits the tone of his voice, and speaks gravely. Again, of animals, those that are brave have a grave voice; but those that are timid, an acute voice. The lion, indeed, the bull, and the barking dog, and the more animated cocks, emit a grave voice; but the stag and the hare, an acute voice.

Perhaps, however, in these instances, also, it is better not to consider the animals as brave or timid, because their voice is acute or grave, but to conclude that a robust voice is the sign of a brave animal, and that a remiss and feeble voice is the sign of a timid animal. But when, in any one, contrary and not acknowledged signs present themselves to the view, it is then best not to affirm any thing of such a one, unless of the discordant signs some are more credible than others; and signs, indeed, ought especially to be referred to species, but not to whole genera; for those which are referred to species are more similar to the object of investigation. For we do not physiognomize the whole genus of men, but some of those which are in that genus.

3 The signs of fortitude are, hard hair, an erect figure of the body, the bones and ribs and extremities of the body strong and large, and a belly ample and contracted; shoulders broad and distant, neither very much bound together nor entirely loose; the neck robust, and not very fleshy; the breast fleshy and broad; the hollow part of the thigh in which the
807 head of the thigh turns (*ισχιον*) contracted; the fleshy part of the leg not raised above the rest of the leg [as in women], but contracted downward; the eyes of a yellow† colour, not very expanded, nor entirely closed; the colour of the body as it were adust; the forehead acute, straight, and not large; and the cheeks neither smooth nor entirely wrinkled.

The signs of timidity are, soft hairs; the body incurvated, and not erect; the fleshy part of the leg drawn upward [as in women]; a palish colour of the face; the eyes languid, and frequently winking; and the extremities of the body imbecil; the legs small; the hands thin and long; the loins small and weak; the figure of the body agile in motions, not precipitous but supine, and after a manner stupid; and the face easily moved and sorrowful.

The signs of a naturally good disposition‡ are, moister and softer flesh, which is not of a good habit, nor very fat; the parts about the shoulders and neck, and, also, about the face, more lean; the parts about the shoulders bound together, and the inferior parts [*i.e.* the shoulder-blades] remiss; the parts about the ribs disencumbered, and the back not fleshy; the colour of the body white mixed with red, and pure; the skin thin; the hair not very hard nor very black, and the eyes very pleasing and moist.

† This colour of the eyes according to Baptista Porta verges to blackness in conjunction with splendour. It is called in Greek *χαροπον*, and according to Averroes and Rhases is a variegated colour.

‡ *i.e.* A disposition naturally adapted to moral and mental excellence; for this is the proper meaning of the word *ευφυης*.

The signs of an insensate person are these: the parts about the neck and leg are fleshy, complex, and bound together; the cavity of the hip-bone is round; the muscles of the shoulder-blade are drawn upward; the forehead is large, round, and fleshy; the eyes are green, and dull; the tibiæ about the ankle-bones are thick, fleshy, and round; the jaws are large and fleshy; the loins are fleshy; the legs are long; the neck is thick; the face is fleshy, and sufficiently long. But the motions, figures, and manner apparent in the face of such a one, subsist according to similitudes.

The signs of an impudent person are, an expanded and splendid eye; the eyelids sanguineous, and thick, and in a small degree curved; the muscles of the shoulder-blades drawn upward; the figure of the body not erect, but a little prone. Such a one in his motions is quick; his body is of a ruddy and sanguineous colour; his face is round; and his breast is drawn upward.

The signs of a man of elegant manners (*κοσμιος*) are these; he is eloquent, and slow in his motions; his voice is slow, but full of spirit, easy, and bland; his eye is void of splendour, and black, and is neither very open, nor entirely closed, and he opens and shuts his eyelids slowly. For the eyes that open and shut the eyelids rapidly, sometimes 808
denote timidity, and sometimes audacity.

The signs of a man of a cheerful disposition are, a forehead of a good magnitude, fleshy, and smooth; the parts about the eyes [*i.e.* the eyelids and eyebrows] more depressed; the face somnolent, neither viewing any thing intensely nor grave. Such a one in his motions will be slow and remiss; and from the form and manner of his face it will be apparent, that he is not precipitous, but considerate.

The signs of a sorrowful person are, a wrinkled and lean face. Such a one, also, has distorted eyes; which signify two things, softness and effeminacy, anxiety and sadness. The form, likewise, of his body is without dignity, and all his motions indicate despondency.

The signs of a catamite are, the eyes distorted; a crepitation of the knees; an inclination of the head to the right-hand parts; the motion of the hands supine and careless; twofold motions, one being that of a person inclining round, but the other that of one restraining his loins. The eyes, also, of such a one are wandering; and such was Dionysius the sophist.

The signs of acerbity (*πικπος*) are, a remiss face [the mouth of which is open and gaping]; the colour black, and the body lean; the parts about the face, as it were, worn off; the face itself wrinkled and fleshless; and the hair straight and black.

The signs of an irascible person are, the body erect, and the figure of it broad. Such a one is full of spirit, and of a reddish colour. His shoulders are broad, distant from each other, and large. The extremities of his body, also, are large and robust; and he is smooth about the breast and private parts. He is well bearded, and the circuit of his hair is florid, flowing downward.

The signs of a mild person are, gravity of deportment, and a sufficiency of flesh; the flesh humid and abundant; a good and commensurate magnitude of the body; the figure of the body curved, and the circuit of the hair flowing upward.

The signs of a dissembler, or one given to irony are, the parts about the face fat, and about the eyes wrinkled; and the manner of the face somnolent.

The signs of a pusillanimous person are, small members and small joints; leanness; small eyes, and a small face; and such was Leucadius the Corinthian.

The lovers of dice have the cubit shorter than is fit, and delight in dancing.

The lovers of contumely have the upper lip elevated, the form of the body prominent, and the colour of it red.

Compassionate persons are such as have an elegant form of the body, a white colour, fat eyes, the nostrils distant in the inferior part, and who perpetually weep. The very same persons, also, are lovers of the female sex, begetters of females, in their manners amatory, never forgetful, of a naturally good disposition, and of a hot temperament of body. These,
808 therefore, are the signs of a compassionate person. Such a one, also, is wise, and timid, and elegant. But he who is void of compassion is ignorant and impudent.

Those who are voracious have that part which is extended from the navel to the breast larger than that part which is extended from the breast to the neck.

He who is very much addicted to venery is of a white colour, and hairy; his hairs, also, are straight, thick, and black. His temples too are hairy, and his eye is fat and lascivious.

Those who are somnolent have the superior parts of the body [proportionally] larger than the inferior. They are, also, of a hot temperament, they have flesh of a good habit, the form of their body is elegant, and they are hairy about the belly.

Those who remember well have the superior parts [proportionally] less than the inferior, elegant, and more fleshy.

4 It appears, however, to me that the soul and the body sympathize with each other; and when the habit of the soul suffers a mutation in quality, it, also, changes the form† of the body. Again the form (*μορπη*) of the body when changed in quality, changes, also, the habit of the soul. For since it is the province of the soul to be sorrowful and to rejoice, it is evident that the face of the sorrowful is sad, and of those who rejoice cheerful. If, therefore, it were possible that when the soul is liberated from the body the form (*μορφη*) of the body could yet remain, even then the soul and body would sympathize with, though they would no longer contribute any thing to each other. Now, however, it is manifest that each follows each; and this will become especially evident from what follows: For insanity appears to subsist about the soul; and physicians purging the body by medicines, and certain diet which they employ for this purpose, liberate the soul from insanity. But by the remedies which they apply to the body, at one and the same time, the form (*μορφη*) of the body is liberated [from disease], and the soul is freed from insanity. Since, therefore, both are liberated, it is evident that they contribute to each other.

At the same time, likewise, it is manifest that the forms (*αι μορπαι*) of bodies become similar to the powers of the soul; so that all such particulars as are similar in animals are indicative of the same thing. With respect, also, to such things as are performed by animals, some, indeed, are the peculiar, but others the common passions of each genus of animals. In the proper works, therefore, of the soul, the passions, according to the body, are peculiar; but in the common works of the soul, these passions are common. The common passions, therefore, are wantonness and venereal ecstasy. Hence both of animals that have a tuft of hair on the forehead, and of those that have not, wantonness is a common passion; but of asses and swine, the common passion is venereal ecstasy. But the peculiar passion in dogs, indeed, is reviling by barking; and in asses freedom from sorrow. And thus we have shown how the common and the peculiar ought to be distinguished. 809

Much exercise, however, and the habit acquired by exercise, are requisite in order to be sufficiently qualified to discuss each of these

† The word here used by Aristotle is *μορφη*, which, as we have frequently observed, is that which pertains to the colour, figure, and magnitude of superficies.

particulars. For since the things which are seen in bodies are said to be referred to the similitudes produced from animals and their actions, certain other peculiarities, also, arise from heat and cold. But some of these appearances are in bodies which employ small differences, and are called by the same name; such, for instance, as the paleness arising from terror and from labour. For these have the same appellation, and differ but little from each other. When, however, the difference is small, it is not easy to know [what passion is signified]; but from experience and custom, the fitness of the form (*μορφη*) must be assumed. The swiftest and best mode, therefore, [of knowing the passions and manners of the soul by physiognomy] is that which is derived from fitness [through which a man resembles this or that animal]. And by thus employing this mode, many particulars may be known. It is, also, not only universally useful, but is, likewise, useful in the selection of signs. For it is requisite, that each of the selected signs should have a fitness of such a kind as is conformable to that which is selected. Farther still, [we may, also, investigate the natures of men] both by the selection of signs, and by the syllogism which it is requisite to use wherever it may occur, adding things appropriate to such as are appropriate. Thus, for instance, if some one is impudent, and pays attention to trifles, he will, also, be a thief and illiberal. And a thievish disposition is, indeed, consequent to impudence; and illiberality to an attention to trifles. By this method, therefore, from the signs of the body, we may investigate natural propensities.

5 Now, however, I shall endeavour in the first place to discuss the genera of animals, and to show what those particulars in them are which it is fit should be changed, in order that they may be brave or timid, just or unjust. But the genus of animals must be divided into two forms (*μορφαι*), *i.e.* into the male and female, at the same time adapting to each form that which is fit. For these forms are similar. With respect to those milder animals, therefore, which we endeavour to nurture, the females are of a softer nature than the males, and less robust; and they are, also, more easily accustomed to the hand of man, and more readily receive nutriment from it. Hence such being the nature of the females, they are in a certain respect less irascible than the males. And this is manifest from what we experience in ourselves. For when we are vanquished by anger, we yield to persuasion with difficulty, and are less tractable. We are, likewise, hurried on to violent conduct, and to the perpetration of whatever anger may impel us to do. It, also, appears to

me that females are more malignant than males, more precipitate, illiberal, and imbecil. This, therefore, is very evident in women, and in 809
the females of those animals which are nourished by us. And the same thing is testified by all shepherds and hunters of animals of the forest.

This, likewise, is manifest, that the female in each genus of animals has a less head, a narrower face, a more slender neck, a more imbecil breast, and ribs less firm and robust than the male. The hips, also, and the thighs of females are more fleshy than those of males. But the knees of females are flexible, and their legs are slender. Their feet however, are more elegant than the feet of males, and the form (*μορφη*) of their body is more pleasing than generous (*γενναιος*). They are, also, less nervous, softer, and have moister flesh. But males, possessing the contraries to all these, are naturally braver, and more just in their kind; and the female is naturally more timid and unjust.

This being the case, the lion appears to participate the idea [*i.e.* form] of the male the most perfectly of all animals. For he has a very large mouth; his face is more square, and not very bony; his upper does not project beyond but equally corresponds to the lower jaw; his nostrils are more thick than thin; his eyes are yellow and concave, neither very round nor very oblong, but of a moderate magnitude. His eyebrows, also, are large, and his forehead is square, but more concave in the middle, and about the eyebrows and the nose rises up like a cloud. But in the superior part of the forehead, near the nose, he has hairs inclining [to the nose]. His head is of a moderate size; his neck is sufficiently long, thick, and commensurate; his hair is yellow, neither straight nor very much curled; and the parts about the collar-bone are more detached than conjoined. His shoulders are strong, his breast is juvenile [*i.e.* is robust], his metaphrenon [*i.e.* the part opposite to the breast] is broad, and his ribs and his back are very strong. He is, also, a less fleshy animal in his hips and his thighs. His legs are robust and nervous, neither very hard, nor very moist. But he walks slowly, takes large steps, and vehemently agitates his shoulders as he marches. And such are the particulars respecting his body. But with respect to his soul, he is disposed to give, is liberal, magnanimous, and aspires after victory; is mild and just, and possesses an affection for those with whom he associates resembling parental and conjugal love.

The panther, however, among animals which appear to be brave, is of a more feminine form, since by reason of its legs it is inferior in its energies to all other brave animals. For what robust work does it
perform? It has, indeed, a small face, a large mouth, small white and 810

concave eyes, and which are more superficially situated. Its forehead is more oblong; its ears are more round than plain; its neck is very long and thin; its breast is not defended with sides; and its back is long. Its hips, also, and its thighs are fleshy; but the parts about the flank and the belly are more equable. Its colour is variegated; and the whole of its body is without articulation, and incommensurate. Such, therefore, is the form of its body. But with respect to its soul, it is little, given to theft, and, in short, fraudulent. It is evident, therefore, how among animals which are considered as brave the lion subsists, who possesses most perfectly the form of the male, and the panther which possesses in equal perfection the form of the female. How the males and females subsist in other animals it is now easy to determine. But with respect to such particulars as it belongs to physiognomy to consider, these will be enumerated in the selection of signs.

6 The selection, therefore, of signs by which we may detect the nature and manners of men must be made as follows:

Those whose feet are well formed and large, have articulation, and are nervous, are robust in what pertains to the soul, and are referred to the male genus. But such as have small and narrow feet, without articulation, are more pleasing to the view than robust, and are effeminate in what pertains to the soul. They are, also, referred to the female genus.

Those whose toes and the nails of their toes are curved are impudent, and are referred to birds with crooked claws. Those whose toes are conjoined are timid, and are referred to quails that frequent lakes, whose feet are narrow [and their toes conjoined].

Those who have the parts about the ankle-bone nervous, and possessing articulation, have a robust soul, and are referred to the male genus. But such as have fleshy ankle-bones, and without articulation, have effeminate souls, and are referred to the female genus.

Those whose legs have articulation, and are nervous and robust, have a robust soul, and are referred to the male genus. But such as have thin and nervous legs are lascivious, and are referred to birds. Those whose legs are very thick, and but little distant from each other, are odious and impudent persons, and are referred to the fitness of their legs to their character.

Those whose knees are so closed so that they nearly knock against each other, are catamites, and are referred to the fitness of their knees to their character.

Those who have bony and nervous thighs are robust; and are referred to the male genus. But those whose thighs are bony and full are soft, and are referred to the feminine genus.

Those whose buttocks are sharp and bony are robust; but those whose buttocks are fleshy and fat are effeminate. And those whose buttocks have but little flesh, and as it were wiped dry, are of depraved manners, and are referred to apes.

Those whose loins are firm are lovers of hunting, and are referred to lions and dogs. For any one may see that dogs who are lovers of hunting have strong loins.

Those who have the parts about the belly slender are robust, and are referred to the male genus. But those that have these parts not slender are effeminate, and are referred to the fitness of these parts to their character.

Those whose back is of a good size and strong are of a robust soul, and are referred to the male genus. But those whose back is narrow and weak are soft, and are referred to the female genus.

Those whose sides are strong have a robust soul, and are referred to the male genus. But those whose sides are weak have an effeminate soul, and are referred to the female genus. And those who have their sides timid, and as it were inflated, are loquacious, and talk foolishly. These persons are referred to oxen and frogs.

Those who have the part which is extended from the navel to the bottom of the breast larger than the part which is extended from the bottom of the breast to the neck, are voracious and insensate. They are voracious, indeed, because they have a great ventricle by which they receive their nutriment; and they are insensate, because their senses have a narrow receptacle,† and which is contracted by the cavity that receives the food, so that the senses are oppressed either through the abundance or the want of food.

Those who have a large breast, and which possesses articulation, have robust souls, and are referred to the male genus.

Those who have the metaphrenon [or that part of the back which is opposite to the breast] large and fleshy, and possessing articulation, have robust souls, and are referred to the male genus. But such as have this

† The reason why such are insensate is because the senses originate from the heart, which resides in the breast. But on account of the magnitude of the ventricle in these persons, the breast is narrow and contracted. Hence the sense is dulled from the narrowness of the place, and the abundance or vacuity of food, since each has a power of blunting the senses.

part weak, not fleshy, and without articulation, have effeminate souls, and are referred to the female genus. And such as have the metaphrenon very much curved, and the muscles at the breast contracted, are of depraved manners, and are referred to the fitness of such a metaphrenon to their character, because the anterior parts do not appear to be appropriate. But such as have the metaphrenon supine are proud and stupid, and are referred to horses. Since, however, it is necessary that the metaphrenon should neither be convex nor concave, the medium between these must be investigated in him who is well born.

Those who have a thick neck and shoulders have robust souls, and are referred to the male genus. But those whose neck is imbecil, and without articulation, have effeminate souls, and are referred to the female genus. And thus much concerning the feet and the thighs.

811 Those whose shoulders are properly detached have liberal souls, and are referred to their apparent structure, because liberality is appropriate to the apparent form (*μορφη*). But those whose shoulders are not well detached, and are contracted, are illiberal; and they are referred to the fitness of such shoulders to their character.

Those whose clavicles or collar-bones are properly detached possess acute senses. For the collar-bones being properly detached, they easily receive the motion of the senses. But those whose collar-bones are moved with difficulty are insensate. For the collar-bones, not being properly detached, are incapable of receiving the motion of the senses.

Those that have a thick neck have robust souls, and are referred to the male genus. But such as have a thin neck are imbecil, and are referred to the female genus. Those who have a thick and fleshy neck are irascible, and are referred to irascible bulls. But those whose neck is of a proper size, and not very thick, are magnanimous, and are referred to lions. Those whose neck is thin and long are timid, and are referred to stags. But those whose neck is very short are insidious, and are referred to wolves.

Those whose lips are thin, and loose at the corners, so that the superior fall upon the inferior parts, are magnanimous, and are referred to lions. This, also, may be seen in large and robust dogs. Those whose lips are thin and hard, and rising up about the canine teeth, are ignoble, and are referred to swine. But those whose lips are thick, and so formed that the upper bends downward to the lower are foolish, and are referred to asses and apes. And such as have the upper lip and the gums prominent are lovers of strife, and are referred to dogs.

Those whose nose is thick in the superior part are indolent, and are referred to oxen. But those whose nose is thick at the extremity are insensate, and are referred to swine. Those whose nose is acute in the extremity are wrathful with severity, and are referred to dogs. Those the extremity of whose nose is round, but obtuse, are magnanimous, and are referred to lions. But those whose nose is thin in the extremity participate of the nature of birds. Those whose nose is curved near the forehead are impudent, and are referred to crows. But those that have an aquiline nose, and which possesses articulation from the forehead, are magnanimous, and are referred to eagles. Those whose nose is concave and round from the region of the forehead, but the circumference of it tends to the superior parts, are lascivious, and are referred to cocks. 811 And those that have a flat[†] nose are, likewise, lascivious, and are referred to stags. Those whose nostrils are expanded are irascible, and are referred to the passion which takes place in anger.

Those whose face is fleshy are indolent, and are referred to oxen. Those whose face is lean are active and diligent; but those whose face is fleshy are timid, and are referred to asses and stags. Those that have a very little face are pusillanimous, and are referred to the cat and the ape. Those whose face is large are sluggish, and are referred to asses and oxen. Since, however, the face ought neither to be small nor large, the middle habit of it [*i.e.* the face which possesses mediocrity] will be adapted [both to genius and good manners]. Those whose face appears to be sordid and squalid are illiberal, and are referred to the fitness of such a face to their character.

Those from whose eyes vesicles as it were depend are lovers of wine, and are referred to ebriety. For those that have the parts which appear before the eyes very much dependent, and resembling vesicles, are somnolent, and are referred to the passion [*i.e.* to sleep]; since those that rise from sleep have the parts about the eyes dependent. Those that have small eyes are pusillanimous, and are referred to the fitness of such eyes to their character, and, also, to the ape. But those that have large eyes are sluggish, and are referred to oxen. He that is well born, therefore, ought neither to have small nor large eyes. Those that have hollow[‡] eyes are pernicious characters, and are referred to the ape. Those that have prominent eyes are stupid, and are referred to the

† This fault of the nose is perceived in the compression of the superior part of the nostrils, and is called *a defect of the nose.*

‡ *i.e.* Eyes hollow in the extreme.

fitness of such eyes to their character, and to asses. Since, however, the eyes ought neither to be prominent nor hollow, the middle habit will be the best. Those whose eyes are in a small degree hollow are magnanimous, and are referred to lions. But those whose eyes are more [*i.e.* moderately] hollow are mild, and are referred to oxen.

Those that have a small forehead are ignorant persons, and are referred to swine. But those that have a very large forehead are sluggish, and are referred to oxen. Those that have a round forehead are insensate, and are referred to asses. Those that have the superficies of the forehead less round are sagacious, and are referred to dogs. But those that have a square and commensurate forehead are magnanimous, and are referred to lions. Those that have a wrinkled forehead are arrogant, and are referred to the bull and the lion; but those that have an expanded forehead are flatterers, and are referred to the passion produced. And
812 this may be seen in dogs, who when they fawn have a serene forehead. Since, therefore, a wrinkled forehead signifies arrogance, and a serene forehead flattery, the habit of the forehead which is the medium between these will subsist in a becoming manner. Those that have an austere forehead are sorrowful, and are referred to the passion, because those that are sorrowful have an austere face. Those that have a depressed forehead are querulous, and are referred to the passion, because those who lament have a depressed forehead.

Those that have a large head† are sagacious, and are referred to dogs. But those that have a small head are insensate, and are referred to asses. Those whose head is as it were turbinated are impudent, and are referred to birds that have crooked claws.

Those that have small ears resemble apes [in their nature and manners]; but those that have large ears resemble asses. It may, also, be seen that the best dogs have moderate ears.

Those that are very black are timid, and are referred to Egyptians. Those, also, that are very white are timid, and are referred to women. It is necessary, however, that the colour which contributes to fortitude should subsist between these. Those that are yellow are magnanimous, and are referred to lions. Those that are very red are very crafty, and are referred to foxes. But those that are pale, and whose colour is confused, are timid, and are referred to the passion which is produced from fear. Those who have a colour resembling that of honey are frigid. But frigid substances are moved with difficulty; and things pertaining to

† *i.e.* A head, which does not exceed the proportion of the body.

the body being moved with difficulty, those that have this colour will be slow. Those that have a red colour are acute; because every thing pertaining to the body being heated by motion becomes red.† Those whose colour resembles that of flame are maniacs; because things pertaining to the body when vehemently heated have a colour similar to that of flame. Those, also, who are heated in the extreme will be maniacs. Those who have a colour resembling that of flame about the breast are very irascible, and are referred to the passion of anger; for the parts about the breast are inflamed in those that are enraged. Those, likewise, whose veins about the neck and temples are distended are very irascible; and are referred to the passion, because these things happen to those that are enraged. Those who have a blushing face are bashful, and are referred to the passion; because the face blushes of those that are bashful. Those who have blushy cheeks are lovers of wine, and are referred to the passion, because the cheeks of those that are intoxicated have a blushing colour. But those that have eyes of a blushy colour are furious through anger, and are referred to the passion of anger, because those who are transported by anger have eyes of a blushy colour. Those whose eyes are very black are timid. For a very black colour appears to signify timidity. But those whose eyes are not very black, but inclining to a yellow colour, are magnanimous. Those whose eyes are of an azure or white colour are timid; for it seems that a white colour denotes timidity. But those whose eyes are not of an azure, but of a yellow colour, are magnanimous; and are referred to the lion and the eagle. Those whose eyes are of a vinous colour are foolish, and are referred to she-goats. But those that have fiery eyes are impudent, and are referred to dogs. Those that have discoloured and disturbed eyes are timid and are referred to the passion of fear; because those who are terrified are pale with a colour which is not equable. But those that have shining eyes are lascivious, and are referred to cocks and crows.

Those that have hairy legs are lascivious, and are referred to goats. But those that have the breast and belly very hairy are unstable, and are referred to birds, because these animals have a most hairy breast and belly. Those whose breast is very smooth [*i.e.* free from hairs] are impudent, and are referred to women. Since, therefore, the breast ought neither to be very hairy nor very smooth, the middle habit is the best. Those whose shoulders are hairy never persevere in the same things, and are referred to birds. Those who have a very hairy back are impudent,

† Because a red colour arises from heat; and, therefore, bodies that are heated become red.

and are referred to savage animals. But those that have the posterior part of the neck hairy are liberal, and are referred to lions. Those that have a long chin are full of spirit, and are referred to dogs. But those that have the eyebrows conjoined are sad, and are referred to the similitude of the passion.[†] Those who have the eyebrows divulsed and extended towards the temples are stupid, and are referred to swine. Those who have the hairs of the head erect are timid, and are referred to the passion of fear; because the hairs of those that are terrified become erect. Those whose hair is very much curled are timid, and are referred to the Ethiopians. Since, therefore, both hair that is erect and that is very much curled signify timidity, the hair that is curled at the extremity will be that which is referred to fortitude, and which is, also, referred to the lion. Those that have the superior part of the forehead, which is near the vertex of the head, elevated, are liberal, and are referred to lions. Those that have the hairs of the head growing on the forehead, near the nose, are illiberal, and are referred to the adaptation of the hairs
813 thus growing, to illiberality; because it has a servile appearance, or such as becomes a slave.

Those whose step is long and slow are efficacious in acting, and bring to perfection whatever they attempt to do; because a long step is efficacious, but a slow step denotes a thoughtful[‡] person. Those whose step is short and slow, and attended with delay, do not bring to perfection what they undertake; because a short and slow step is not efficacious. Those whose step is long, rapid, and unmolested, bring what they undertake to perfection; because celerity is perfective, and length is not inefficacious. Those whose step is short, rapid, and unmolested, are not perfective of what they undertake.

Similar conclusions may, also, be deduced concerning the motions of the hand, the cubit, and the arm.

[But the signs which are derived from the motion of the members are these:] Those who are moved with erect and expanded shoulders are proud and arrogant, and are referred to horses. Those whose shoulders when they walk move upward, and are curved, are magnanimous, and are referred to lions.

Those who walk with their feet and legs inverted, [*i.e.* so that with one foot they verge to the right, and with the other to the left] are effeminate, and are referred to women.

† Because when we are in sorrow we are accustomed to conjoin the eyebrows.

‡ For those who in action are efficacious, and at the same time thoughtful, accomplish what they undertake.

Those who in walking turn themselves round, and rub against those they walk with, are flatterers, and are referred to the passion. But those who is walking incline to the right hand, are catamites, and are referred to the fitness of such inclination to such characters.

Those who have voluble eyes are acute and rapacious, and are referred to hawks. Those who frequently wink are timid; because the eyes of those that are terrified are the first parts of the body that are agitated. Those that have a depressed aspect, and who have one of their eyelids depressed, and, therefore, fix their eyes with depressed eyelids, and, in short, those who look with a soft and diffused vision, are effeminate, and are referred to the fitness of such an aspect to such characters, and to women. Those who move their eyes slowly, that is, through great intervals, and have, also, a tincture of whiteness in the eye, so that their eyes are as it were fixed, are intellective persons. For when the soul intensely applies itself to the intellection of any thing, the sight, also, is fixed.

Those who speak with a loud and grave tone of voice are petulant and injurious, and are referred to asses. But such as speak beginning from a grave and ending in an acute tone of voice, are irascible and querulous, and are referred to oxen, and to the fitness of their character to their voice. Those who speak with an acute, soft, and frequently interrupted voice, are catamites, and are referred to the fitness of such a voice to their character. Those who speak with a tone of voice grave, loud, and not sufficiently distinct, are referred to robust dogs, and to the fitness of such a voice to their character. Those whose voice is soft and languid are mild, and are referred to sheep. Those who speak with a tone of voice acute and strong are insane and petulant, and are referred to she-goats.

Those whose body is very small are quick in their motions and acute. For the motions of the blood occupying but a small place, the motions [of the animal spirits] very rapidly arrive at that part which is the seat of intelligence. But those whose body is very large are slow. For the motion of blood occupying a large place, the motions [of the animal spirits] slowly arrive at the seat of intelligence. Those whose body is small, and whose flesh is dry, or who have a colour which arises from heat in the body, never bring any thing to perfection which they undertake. For since the animal spirits on account of the fiery complexion are rapidly moved in a small space, intelligence is never in the same state, but at different times is in a different condition, before that which is super-eminent is able to produce perfection. Those large bodies that have moist flesh, or that have a colour ingenerated from frigidity, do not accomplish any thing. For the motion [of the animal spirits] being slowly effected in a large space, on account of frigidity, does not permit any thing to arrive at the seat of intelligence. But such small bodies as have humid flesh, and a colour arising from frigidity, effectively accomplish what they intend. For the motion [of the animal spirits] existing in a small space, the slowness is compensated by the smallness of the space, and, therefore, they are moved in such a manner as is requisite to the accomplishment of what is intended. Such large bodies as have dry flesh, and a dry colour arising from heat, accomplish what they intend. For the spirits are moved rapidly in a large space, and, therefore, the velocity is compensated by the magnitude of the space, and the motion is commensurate to the accomplishment of what is intended. And thus we have shown how bodies which surpass in magnitude, or are deficient in smallness, are either efficacious or imperfect. But the nature which is between these is the best and most perfect for the purposes of sensation in those bodies in which it is supposed to be inherent. For the motions [of the animal spirits] existing in a space neither too ample nor too narrow, easily arrive at the seat of intelligence. Hence to the perfect accomplishment of what is intended, and for the purpose of energizing well, according to the senses, it is necessary that the body should be of a moderate magnitude.

814 Those whose bodies are without symmetry are very crafty, and are referred to the passion, and, also, to the female genus. For if those whose bodies are without symmetry are very crafty, those whose bodies possess symmetry will be just and brave. The relation, however, of symmetry must be referred to the proper discipline and naturally good condition of bodies, and not to the form of the male, as was, also, asserted by us in the beginning. But it will be well to refer all such signs as have been before-mentioned to the fitness of them to the characters to which they belong, and, also, to the male and female. For these are divided according to the most perfect divisions; and it has been demonstrated that the male is more just and brave, and, in short, is
814 better than the female. In every selection of signs, however, one sign more clearly manifests the subject than another. But the most manifest signs are those which exist in the most principal places. And the most principal place is that which is about the eyes, the forehead, the head, and the face. The place, also, next to this is that which is about the breast and shoulders. The third place in rank is that which is about the legs and feet; and the least principal place is that which is about the belly. In short, those places afford the most manifest signs in which the ornament of wisdom is most abundant.

An

ALPHABETICAL TABLE

of the

NAMES OF THE ANIMALS

EMPLOYED BY

ARISTOTLE

in his

HISTORY OF ANIMALS

(Key to type: A = Aquatic Animal, B = Bird, C = Crustaceous, F = Fish, fa = Flying Animal, fq = Flying Quaduped, I = Insect, M = Rational Mortal Animal, Q = Quadruped, R = Reptile, T = Testaceous, ta = Terrestrial Animal, W = Aquatic Bird, 2 = Amphibious.

Type	Terms used By Aristotle	Ancient Translation	Translation of Gaza	Translation of Scaliger	English Translation
B	Αηδων, η	Filomena	Luscinia	Luscinia	Nightingale
F	Αθερινη, η	Atherina	Atherina quam Aristam appello	Antherina	Ear-of-corn
B	Αιγιθαλος, ο	Egitalus	Parus	Ægithalus	Titmouse
B	Αιγιθος, ο	Ægithus	Salus	Ægiothus	Œgithus
B	Αιγοθηλας, ο	Igothilas	Caprimulgus	Caprimulgus	Goat-teat
B	Αιγοκεφαλος, ο	Egocephalus	Capriceps	Ægocephalus et Capriceps	Ægocephalus
B	Αιγοπιος, ο	Egiptius	Vultur	Vultur	Ægypius
B	Αιγωλιος, ο	Egolius	Ulula	Capriceps et Ægolius	Owl
B	Αιετος, ο	Aquila	Aquila	Aquila	Eagle
F	Αιετος, ο	Aquila	Aquila	Aquila	Eagle
B	Αιθυια, ο	Ethya	Mergus	Mergus	Plungeon or Diver

Q	*Αιλουρος, η*	Fissam, Marturina et Martirio	Feles et Felis	Feles	Cat
Q	*Αιξ, ο & η*	Caper et Capra	Caper et Capra	Hircus et Capra	He-goat and She-goat
B	*Αιξ*	Aix	Capella	Capella	She-goat
B	*Αισαλων, ο*	Esalon	Œsalo	Æsalo	Merlin
A	*Ακαληφη, η*	Acalefe et Akalifa	Urtica	Urtica	Nettle of the sea
F	*ακανθιας*	Acanthia	Spinax	Acanthias	Acanthias
B	*Ακανθις, η*	Acanthis	Spinus sive Legurinus	Spinula et Acanthis	Thorn
B	*Ακανθυλλις, η*	Achanthyllis	Argathylis	Acantylis	Thistle-finch
I	*Ακαρι*	Akare	Acari	Acari	Acari
I	*Ακρις, η*	Locusta	Locusta	Locusta	Cricket
B	*Αλεκτορις, η*	Gallina	Gallina	Gallina	Hen
B	*Αλεκτρυων*	Gallus	Gallinaceus	Gallus	Cock
B	*Αλιαιετος*	Alictus	Aquila marina et haliætus	Aquila marina, nisus haliætus, haliætus	Eagle of the sea
B	*Αλκυων, η*	Alkion	Alcedo	Alcedo	Halcyon
Q	*Αλωπηξ, ο*	Vulpis	Vulpes	Vulpes	Fox

fq	Αλωπηξ	Alopex	Vulpecula	Vulpes	Flying fox
F	Αλωπηξ, ο	Vulpes	Vulpecula	Vulpes	Fox
F	Αμια, η	Amia	Amia	Amia	Amia
F	Ανθιας, ο	Anthias	Sacer et Anthias	Anthias	Anthias
B	Ανθος, ο	Anthus	Florus	Florus	Flower
I	Ανθρινη, η	Anthrena	Crabro	Crabro	Drone
M	Ανθρωπος, ο	Homo	Homo	Homo	Man
T	Απορραις, η	Emorois	Murex	Aporrais et Defluctia	Aporrhais
B	Απους, ο	Apodes, pl.	Apodes	Apodes	Apodes, or without feet
F	Απας, η	Membrates, pl.	Membrades	Arades	Arades
I	Απαχνης, ο	Aranea	Araneus	Araneus	Spider
Q	Αρκτος, η	Ursa	Ursus et Ursa	Ursus et Ursa	Bear
C	Αρκτος, η	Ursa	Ursa	Ursa	She-bear
Q	Αρουραιος, ο	Mus Campestris	Sorex	Sorex	Field-rat
B	Αρπη, η	Arpa	Harpa	Harpe	Harpa
F	Αρχανος, ο	Acharinus	Archanus	Archanus	Archanus

ta	Ασκαλαβωτης	Ascalabotes	Stellio	Stellio	Stellio
B	Ασκαλαφος, ο	Ascalafus	Ascalaphus	Ascalaphus	Aschalaphus
B	Ασκαλωπας, ο	Ascalopas	Gallinago	Ascalopas	Ascalopas
I	Ασκαρις, η	Askaris	Tipula	Ascaris	Ascaris
Q	Ασπαλαξ, ο	Talpa	Talpa	Talpa	Mole
R	Ασπις, η	Aspis	Aspis	Aspis	Asp
C	Αστακος, ο	Astacus	Gammarus	Gammarus	Lobster
B	Αστεριας, ο	Asterias vel Astur	Stellaris	Asterias	Asterias, or Dog Star
T	Αστηρ, ο	Astrum	Stella	Stella	Star
B	Ατταγην, ο	Attagis	Attagena	Attagen	Attagas
I	Αττελαβος, ο	Attalabus	Bruchus	Bruchus	Attelabus
F	Αυλωπιας, ο	Aulopias	-	Aulopias	Aulopias
F	Αυξις, η	Auxis	Auxuma	Auxis	Auxis
F	Αφυη, η	Afia	Apua	Apua	Aphye
F	Αχαρνας, ο	Archanas	Acarnas	Acarnas	Acharnas
I	Αχετης, ο	Acheta	Acheta	Canora	Achetes, or the Singing Grasshopper

T	*Βαλανος, ο*	Balanum	Glans	Glans	Acorn
F	*Βαλλερος, ο*	Ballirus	Ballerus	Ballerus	Ballerus
B	*Βαθις*	Sacios	Rubetra	Rubicola	Bathis
F	*Βαρινος, ο*	Balagrus	Balagri, pl.	Barinus	Barinus
B	*Βασιλευς, ο*	Rex	Regulus	Regulus	Wren
F	*Βατος, ο*	Raia	Raia	Raia	Ray, or Thornback
A	*Βατραχος, ο*	Rana	Rana	Rana	Frog
F	*Βατραχος, ο*	Rana	Rana	Rana	Frog of the sea
F	*Βελονη, η*	Acus	Acus	Acus	Needle
F	*Βολιταινα, η*	Bolitæna	Bolitæna	Bolitæna	Bolitæna
I	*Βομβυλιος, ο*	Bombilius	Bombyx	Bombyx	Bombylius
I	*ΒομΒυξ, ο*	Bombyzans	Bombyx	Bombyx	Silk-worm
Q	*Βονασσος, ο*	Bonassus	Bonasus	Bonassus	Bonassus
B	*Βοσκας, ο*	Baskas	Bosca	Pascula, *i.e.* Sarcella	Teal
I	*Βοστρυχος, ο*	Bostrichus	Cirrus	Cirrus	Buckle, or Hair-curl
Q	*Βυβαλις, η*	Bubala	Bubalus	Bubalus	Bubalus

Q	*Βους, ο & η*	Bos et Vacca	Bos et Vacca	Bos et Vacca	Ox and Cow
F	*Βους, ο*	Bos	Bos	Bos	Marine Ox
B	*Βρενθος, ο*	Brentalus	Anas	Brenthus	Brenthus
B	*Βρενθος, ο*	Brenthus	Brenthus	Brinthus	Brinthus
A	*Βροσος, ο*	Virsorus	Brissus	Brissus	Bryssus
B	*Βρας, ο*	Bryas	Bubo	Bubo	Great Screechowl
F	*Βοκες, οι*	Bokes	Vocæ	Boces	Bocæ
T	*Γαλαδες αι*	Galakes	Galades	Galades	Galades
F	*Γαλεος, ο*	Galeus	Mustelus	Mustelus	Sea-dog
Q	*Γαλη, η*	Catus	Cattus et Mustela	Mustela	Weasel
B	*Γερανος, ο*	Grus	Grus	Grus	Crane
Q	*Γιννος, ο*	Rippa	Ginnus	Ginnus	Colt
F	*Γλανις, ο*	Glanis	Silurus	Glanis et Glanius	Glanis
Q	*Γλανος, ο*	Glanus	Glanus	Glanus	Glanus
F	*Γλαυκος, ο*	Glaucus	Glaucus	Glaucus	Glaucus
B	*Γλαυξ, η*	Bubo	Noctua	Noctua	Horned Owl

B	Γλωττες, η	Glottis	Lingulaca	Ligula	Glottis
B	Γναφαλυς, ο	-		Gnaphalus	Gnaphalus
F	Γογγρος, ο	Conger	Conger	Conger	Conger, or the Sea-eel
B	Γυψ, ο	Vultur	Vultur	Vultur	Vulture
F	Δασκιλλος, ο	Dalkillus	Dasquillus	Dascillus	Dascillus
Q	Δασυπους, ο	Lepus et Lepurculus	Lepus	Lepus et Cuniculus	Hare
F	Δελφις, ο	Delphis	Delphinus	Delphinus	Dolphin
Q	Δορκας, η	Hinnulus	Capra	Caprea	Roe-buck
R	Δρακων, ο	Dracon	Draco	Draco	Dragon
F	Δρακων, ο	Dracon	Draco	Draco	Dragon
B	Δρεπανις, η	Drepanis	Falcula sive Riparia	Riparia	Martinet
B	Δρυοκολαπτης	Diryocolaptes	Picus Martius	Picus	Woodpecker
B	Εγκρασιχολοι	Encrasicoli	Encrauli	Encrasicholi	Encrasicholi
F	Εγχελυς, η	Anguilla	Anguilla	Anguilla	Eel
Q	Ελαφος, ο & η	Cervus et Cerva	Cervus et Cerva	Cervus et Cerva	Stag and Hind
B	Ελεα, η	Elea	Velia	Helea	Elea

F	Ελεγινος, ο	Eleginus	Eleginus	Eleginus	Eleginus
F	Ελεδωνη, η	Eloanis	Eledona	Eledona	Eledone
Q	Ελειος, ο	Elius	Glis	Glis	Dormouse
B	Ελεος, ο	Eleus	Aluco	Ulula	Eleus
Q	Ελεφας, ο	Elefas	Elephantus	Elephantus	Elephant
F	Ελλοψ, ο	Elops	Acipenser	Elops	Ellops
I	Ελμινς, η	Lumbricus	Tinea Sive Lumbricus	Lumbricus	Worm
I	Εμπις, η	Empys	Culex Mulio	Culex Mulio	Empis, a kind of Gnat
A	Εμυς, η	Emis, Mis	Lutaria et Mus Aquatilis	Mys et Lutaria	Emys, or the fresh water Tortoise
I	Εντερα γης, τα	Terræ Intestina	Terræ Intestina	Terræ Intestina	Entrails of the Earth
2	Ενυδρις, η	Enidris	Lutris	Lutra	Otter
B	Επιλαις, η	Epilis	Curuca	Curuca	Epilais
B	Εποψ, ο	Epops	Upupa	Upupa et Epops	Epops
B	Εριθακος	Erithatus	Silvia et Rubecula	Erithacus	Robin-Redbreast
F	Εριθρινος, ο	Erutrinus	Rubellio	Rubellus	Roach
B	Ερωδιος, ο	Erodius	Ardeola	Ardeola	Heron

F	Ετελις	Entelis	Etelis	Etelis	Etelis
I	Ευλαι, αι	Vimmiculus	Vermes qui carnes putres edunt	Vermes qui purtribus in carnibus enascuntur	Worms that are produced in putrid flesh
I	Εφημερον, το	Efimerum	Ephemerum	Ephemerum	Ephemeron-worm
F	Εχενηις	Echeineda	Remora	Echeneis	Sea-lamprey
R	Εχιδνα, η	Vipera	Vipera	Vipera	Viper
T	Εχινομητρα, η	Echinomitra	Echinometra	Echinometra	Urchin-mother
T	Εχινος, ο	Ericius	Echinus	Echinus	Sea-urchin
Q	Εχινος, ο	Ericius	Erinaceus	Erinaceus	Hedgehog, or Urchin
F	Εψητος, ο	-	Naricæ	Elixarii pisces	Fishes from which sauce is made
F	Ζυγαινα, η	Ygena	Libellæ	Zygæna	Zygæna
ta	Ζυγνις, η	Zignis	Dygnis	Zygnis	Zygnis
Q	Ημιονος, ο	Mulus	Mulus	Mulus	Mule
F	Ηπατος, ο	Epatus	Jecorinus	Jecorinus	Hepatus
I	Ηπιολος, ο	Ipiliotis	Hepiolus	Hepliolus	Butterfly
C	Ηρακλεοτικος	Eracheoticus	Heracleoticus [Cancer]	Heracleoticus	Heracleotic Crab

B	Θραυπιθ, η	Thlipis	Carduelus	Thraupis	Thraupis
F	Θρισσς, η	Thrissa	Alosa	Thrissa	Alosa
F	Θυννος, ο, Θυννις, η	Thynus, et Tinnus	Thunnus et Lunaria	Thunnus, Thunilla, Thunna	Tunny, male and female
Q	Θως, ο	Lupus Cervarius	Lupus Cervarious	Thos et Lupus Cervarius	Thos
B	Ιβις, η	Ibis	Ibis	Ibis	Ibis
B	Ιεραξ, ο	Accipiter	Accipiter	Accipiter	Sparrow-hawk
R	Ιεραθ, ο	Sacer	Sacer	Sacer	Sacred
B	Ικτινος, ο	Milvus	Milvus	Milvus	Kite
Q	Ικτις, η	Ictis	Ictis quam Viverram interpretor	Viverra	Pole-cat
Q	Ιννος, ο	Ippa	Innus	Ginnus	Little Nag, or Dwarf Horse
F	Ιουλις, η	Eulis	Julia	Julis	Julis
I	Ιουλος, ο	Iulus	Iulus	Iulus	Iulus
Q	Ιππαρδιον, το	Pardium	Hippardium	Pardium	Hippardium
Q	Ιππελαφος, ο	Equicervus	Hippelaphus	Equicervus	Horse-stag
C	Ιππεας, ο	Ippes	Hippea hoc est eques	Eques	Crab-fish-trooper

I	Ιππομορμηξ, ο	Equestris Formica	Formica quæ eques appellatur	Hippomyrmex	Ant-horse
Q	Ιππος, ο	Equus	Equus	Equus	Horse
2	Ιππος ποταμιος	Equus Fluvialis	Equus Fluviatilis	Equus Fluviatilis	Hippopotamus
F	Ιππουρος, ο	Ippurus	Equiselis	Hippurus	Hippurus
B	Ιυγξ, ο	Linx	Iynx	Iynx	Wry-neck
Q	Ιχνευμων, ο	Ichneumon	Ichneumo	Ichneumo	Ichneumon
I	Ιχνευμων, ο	Ichneumon	Ichneumo	Ichneumo	Ichneumon
B	Καλιδρις, ο	Skandris	Calidris	Calidris	Calidris
F	Καλλιονιμυς, ο	Kallionimus	Pulcher	Callionymus	Callionymus
Q	Καμηλος	Camelus	Camelus	Camelus	Camel
I	Καμπη, η	Kampa	Eruca	Eruca	Caterpillar
I	Κανθαρις, η	Kantharis	Cantharida	Cantharis	Cantharida
I	Κανθαρος, ο	Chanthurus	Scarabæus pillurarius	Scarabeus pillularius	Beetle
F	Λανθαρος, ο	Kantharus	Scarabæus	Cantharus	Beetle pillurarius
Q	Καπρος, ο	Caper, Porca	Aper	Aper	Wild-Boar
F	Καπρος, ο	Capros	Aper	Aper	Wild-Boar

I	*Καραβος, ο*	Carabus et Karambius	Fullo et Scarabæus taurus	Fullo et scarabæus	Locust, or Grasshopper
C	*Καραβος, ο*	Carabus	Locust	Locusta	Kind of Lobster
C	*καρις, η*	Karis	Squilla	Squilla	Squill
C	*Καρκινιον, το*	Karkinum	Cancellus	Cancellus	Small Crab
C	*Καρκινος, ο*	Cancer	Cancer	Cancer	Crab
Q	*Καστωρ, ο*	Castor	Fiber	Fiber	Castor
B	*Καταρρακτης, ο*	Cataractes	Cataracta	Mergula	Catarractes
B	*Κεγχρις, η*	Cenkris	Tinnunculus	Tinnunculus	Kestrel
B	*Κελεος, ο*	Eleus	Galgulus	Galgulus	Runner
B	*Κεπφος, ο*	Keifos et Kepsus	Fulica	Fulica	Coot
F	*Κεστρευς, ο*	Kestreus	Mugilis	Mugil	Mullet
F	*Κεφαλος, ο*	Kefalus	Capito	Cephalus	Cephalus
ta	*Κηβος*	Kebus	Cebus	Cebus	Kebus
B	*Κηρυλος, ο*	Ceralus	Carulus	Cerulus	Cerylus
T	*Κηρυξ, ο*	Kerix	Buccinum	Buccinus	Buccinum
F	*Κητωδη, τα*	Cetosa	Cetarium genus	Cetacea	Cetaceous

I	*Κηφην, ο*	Cefenus	Fucus	Fucus	Drone
B	*Κιγκλος, ο*	Kikhlus	Cinchlus	Cinclus	Cinclus
F	*Κιθαρος, ο*	Cythara	Fidicula	Citharus	Citharus
B	*Κινναμωνμον, το*	Cinamomus	Cinnamomus	Cinnamomus	Cinnamomus
B	*Κιρκος, ο*	Kircus	Circus	Circus	Circus, or gentle Falcon
B	*Κιττα, η*	Pica et Kitta	Pica	Pica	Magpy, or Pie
B	*Κιχλη, η*	Kikla et Maninx	Turdus	Turdus	Thrush
F	*Κιχλη, η*	Kikhla	Turdus	Turdus	Thrush of the Sea
I	*Κληρος, ο*	Cleros	Clerus	Clerus	Clerus
A	*Κνιδη, η*	Knida	Urtica	Urtica	Nettle of the Sea
B	*Κνιπολογος, ο*	Knidolegas	Culicilega	Culicilega	Wagtail
I	*Κνιψ, ο*	Cnips	Mulio	Mulio	Cnips
T	*Κογχη, η*	Concha	Concha	Concha	Shell-Fish
F	*κοιτος, ο*	Cottus	Cottus	Boetus	Gull
B	*Κοκκυξ, ο*	Coccux	Cuculus	Coculus	Cuckow
F	*Κοκκυξ, ο*	Cockyx	Cuculus	Cuculus	Cuckow

B	*Κολαρις, ο*	Calapis	Colaris	Colaris	Colaris
F	*Κολιας, ο*	Kolias	Monedula	Gracculus	Colias
B	*Κολιος, ο*	Skeleus	Galgulus	Galgulus	Green-Woodpecker
B	*Κολλυριων, ο*	Killuron	Collyrion	Collyrion	Collyrion
B	*Κολοιος, ο*	Kolonus	Monedula	Monedula et Gracculus	Jackdaw
B	*Κολυμβις, η*	Colibis et Kalimbus	Urinatrix	Urinatrix	Didapper
-	*Κονις, η*	Lens	Lens	Lens	Nit
B	*Κορακιας, ο*	Corakias	Graculus	Coracias	Coracias
F	*Κορακινος, ο*	Korakinus	Corvulus sive Graculus	Coracinus	Coracinus
B	*Κοραξ, ο*	Corvus	Corvus	Corvus	Raven
W	*Κοραξ, ο*	Corvus	Corvus	Corvus	Raven
A	*Κορδυλος, ο*	Cordilus	Cordulus	Cordulus	Cordulus
I	*Κορις, ο & η*	Coreis	Cimex	Cimex	Bug
B	*Κορυδος, ο*	Coridus	Alauda	Alauda	Lark
B	*Κορωνη, η*	Cornicula	Cornix	Cornix	Crow
B	*Κοττυφος, ο*	Merula	Merula	Merula	Blackbird

F	*Κοττυφες, ο*	Cottifus	Merula	Merula	Blackbird
T	*Κοχλιας, ο*	Coklia	Limax, Umbilicus, Cochlea	Limax, Umbilicus, Cochlea	Snail
T	*Κοκλος, ο*	Cochlus	Cochlea	Umbilicus	Snail
C	*Κραγγων, η*	Cransgo	Crango	Crango	Crango, a kind of Squill
B	*Κρεξ, η*	Krex	Crex	Crex	Crex
Q	*Κριος, ο*	Aries	Aries	Aries	Ram
2	*Κροκοδειλος, ο*	Cocodrillus	Crocodilus	Crocodilus	Crocodile
I	*Κροτων, ο*	Kroton	Redivius	Ricinus	Tick
T	*Κτεις, ο*	Kteis	Pectunculus	Pecten	Cockle
B	*Κυανος, ο*	Kianos	Cæruleus	Cyaneus	Bluebird
B	*Κυκιος, ο*	Cygnus	Olor	Olor	Swan
C	*Κυλλαρος, ο*	-	Parva Squilla	Scyllarus	Small Squill
B	*Κυμινδις, η*	Kymindis	Cymindis	Cymindis	Cymindis
ta	*Κυνοκεφαλος, ο*	Cynocephali	Caniceps	Caniceps	Cynocephalus
I	*Κυνοραιστη*	Musca Canina	Cynoraista	Cynoraista	Flail of the Dog
F	*Κυπρινος, ο*	Carpra	Cyprinus	Cyprinus	Carp

B	*Κυχραμος, ο*	Kekharmis	Cynchramus	Cychramus	Cychramus
B	*Κυψελλος, ο*	Kipselus	Cypsellus	Cypsellus	Cyosellus, or Martlett
Q	*Κυων, ο & η*	Canis	Canis	Canis	Dog
F	*Κυων, ο*	Canis	Canis	Canicula	Dog
F	*Κωβιος, ο*	Kobius	Gobio	Gobio et Gobius	Gudgeon
T	*Κωκαλια, τα*	Cocalia	Cocalia	Cocalia	Cocalia
I	*Κωλωτης, ο*	Colotes	Colota	Colota	Colotes
I	*Κωνωψ, ο*	Konops	Culex Vinarius	Culex	Conops
F	*Λαβραξ, ο*	Spinula	Lupus	Lupus	Wolf of the Sea
Q	*Λαγωνος, ο*	Lepus	Lepus	Lepus	Hare
B	*Λαεδυς, ο*	Lædus	Lædus	-	Laedus
B	*Λαιος, ο*	Laios	-	-	Laius
F	*Λαμια, η*	Lamia	Lamia	Lamia	Lamia
B	*Λαρος, ο*	Larus	Gavia	Gavia	Seagull
2	*Λαταξ, η*	Latax	Latax	Latax	Latax
F	*Λειοβατος, ο*	Liobatus	Leviraia	Leviraia	Smooth-Ray

T	Λεπας, η	Lapas	Patella	Patella	Lepas
Q	Λεων, ο	Leo	Leo	Leo	Lion
B	λιβυος, ο	Kibius	Libyus	Libyus	Libyus
T	Λιμνοστεα, τα	Lemnostrea	Ostrea	Ostrea Vulgaris	Marshy (or Common) Oysters
B	Λοκαλος, ο	-	-	Cocalus	Localus
Q	Λυγξ, η	Lynx	Lynx	Lynx	Lynx
Q	Λυκος, ο	Lupus	Lupus	Lupus	Wolf
I	Λυκος, ο	Lupus	Lupus	Lupus	Wolf
F	Λυρα, η	Lyra	Lyra	Lyra	Lyre
C	Μαια, η	Mea	Maia	Maia	Crab Maia
F	Μαινις, η	Menis	Alec sive Mœna	Mœnula	Mœnis
F	Μαλακια, τα	Malakia	Mollia	Mollusca	Mollia
B	Μαλακοκρανευς, ο	Megalocraneus	Molliceps	Malococraneus	Soft-Skull
A	Μαλακαστρακα	Malacostraca	Crustata	Crustata Intecta	Crustaceous
Q	Μαρτιχορα, η	Manthichora	Mantichora	Mantichora	Martichora

B	*Μελαγκορυφος, ο*	Melanokorifus	Atricapilla	Atriceps et Melancoryphus	Blackhead
F	*Μελανουρος, ο*	Melanurus	Oculata	Melanurus	Melanurus
I	*Μελιττα, η*	Apis	Apis	Apis	Bee
B	*Μελεαγρις, η*	Meliagrida	Meleagris	Meleagris	Meleagris
B	*Μεροψ, ο*	Merops	Merops qui Apiaster	Merops	Merops
I	*Μηλολονθη, η*	Melolontha	Scarabæus et Scarabæus Viridis	Melolontha et Scarabæus Pillurlarius	Scarabæus Beetle, or Chafer
F	*Μηρυξ, ο*	Myrix	Ruminalis	Ruminax	Ruminator
F	*Μορμυρος, ο*	Osmius	Mormur	Murmurus	Mormurus
B	*Μορφνος, ο*	-	Morphna quasi Nævia	Morphnus	Morphnus
Q	*Μυγαλη, η*	Mugis	Mus Araneus	Mus Araneus	Shrew Mouse
I	*Μυια, η*	Musca	Musca	Musca	Fly
F	*Μυξων, ο*	Mixon	Muco	Muco	Muco, or Snotty
F	*Μυραινα, η*	Muræna	Muræna	Muræna	Lamprey
F	*Μυρινος, ο*	Marinus	Murinus	Myxinus	Murinus
I	*Μυρμηξ, ο*	Formica	Formica	Formica	Ant

Q	*Μυς, ο*	Mus	Mus	Mus	Rat
T	*Μυς, ο*	Mys	Mitulus	Musculus	Muscle
-	*Μυστικητος*	Impistocetus	Musculus Piscis	Mysticetus	Mysticetus
I	*Μυωψ, ο*	Myops	Tabanus	Tabanus	Ox-Fly
F	*Ναρκη, η*	Stupor	Torpedo	Torpedo	Cramp-Fish
F	*Ναυτιλος, ο*	Nautilus	Nauta	Nautilus	Nautilus
Q	*Νεβρος, ο & η*	Cervulus	Hinnulus	Hinnulus	Fawn
I	*Νεκυδαλος, ο*	Nekydalus	Necydalus	Necydalus	Necydalus
T	*Νηριτης, ο*	Nerites	Natex	Nerites	Nerites
B	*Νηττα, η*	Anas	Anas	Anas	Duck
fa	*Νυκτερις, η*	Vespertilio	Vespertilio	Vespertilio	Bat
B	*Νυκτικοραξ, ο*	Nicticorax	Cicunia	Nicticorax et Nocticorvus	Raven of the Night
F	*Ξιφιας, ο*	Gladiatus	Gladius	Gladius	Sword
I	*Ξυλοφθορον, το*	Ytyloforum	Xylophtorum Ligniperda	Xylophtorum	Wood-Piercer
B	*Οινανθη, η*	Ynanthe	Oenantha, idem ac si Vitiflora	Oenanthe	Oenanthe

B	*Οινας, ο*	Inas et Ynas	Vinago	Rupilia	Vinous Pigeon
Q	*Οις, η*	Ovis	Ovis	Ovis	Sheep
I	*Οιστρος, ο*	Ystrus	Asilus	Asilus	Flying Asilus
A	*Οιστρος, ο*	Ystrus	Asilus	Asilus	Oistrus
B	*Οιστρος, ο*	Istrus	Asilus	Asilus	Asilus
T	*Ολοθουριον, το*	Olothurium	Vertibula et Calli et Tubera, pl.	Aelothuria	Olothurion
Q	*Ονος, ο & η*	Asinus	Asinus	Asinus	Ass
F	*Ονος, ο*	Asinus	Asellus	Asellus	Ass
I	*Ονος ο πολυπους*	Omux Multiplex	Asellus Multipes	Asellus Multis Pedibus Præditus	Wood-Louse
Q	*Ορευς, ο & η*	Mulus	Mulus	Mulus	Mule
F	*Ορκυνες, οι*	Orkynes	Orcinæ	Orcynes	Orcynes
B	*Οροσπιζης, ο*	Oreospizos	Montifringilla	Monfringilla	Chaffinch of the Mountain
I	*Ορσαδακνη, η*	Orcodagna	Orsodacna quam Mordellam appellem	Orsodacna	Mordella, or Blister-Fly
B	*Ορτυγομητρα*	Ortigometra	Matrix	Ortygometra	Quail Mother

B	*Ορτυξ, ο*	Coturnix	Coturnix	Coturnix	Quail
Q	*Ορυξ*	Oryx	Oryx	Oryx	Oryx
F	*Ορφος, ο*	Orfus, et fortasse Tenka	Cernua	Orphus et Cernua	Orphus
B	*Ορχιλος, ο*	Orchilus	Orchilus	Orchilus	Orchilus
A	*Οστρακοδερμα*	Astracoderma	Silicea testa Conclusa	Testacea	Testaceous
A	*Οστρεα, τα*	Ostrea	Ostrea et Conchæ	Ostrea et Conchæ	Oysters, and Shell-Fish
T	*Ουνθαλαττιον, το*	Avis Marina	Auris Marina	Auricula Maris	Ear of the Sea
R	*Οφις, ο*	Serpens	Serpens	Serpens	Serpent
C	*Παγουρος, ο*	Pagurus	Pagurus	Pagurus	Crab Pagurus
Q	*Πανθηρ, ο*	Panthera	Lupus Canarius	Panthera	Panther
Q	*Παρδαλις, η*	Pardus et Pardalis	Panthera	Panthera	Leopard
B	*Παρδαλος, ο*	Pardales	Pardalus	Pardalus	Pardalus
B	*Πελαργος, ο*	Ciconia	Ciconia	Ciconia	Stork
B	*Πελειας, η*	Pelis	Livia	Turrilia	Kind of Stock-Dove, or Wood-Pigeon
B	*Πελεκαν, ο*	Pellicanus	Platea	Platea	Pelican

B	Περαιας, ο	Pareas	-	-	Peræas
B	Περδιξ, ο	Perdix	Perdix	Perdix	Partridge
B	Περιστερα, η	Columba	Columba	Columba	Pigeon
F	Περκη, η	Perka	Perca	Perca	Perch
B	Περκνοπτερος, ο	Percopteros	Percnopterus	Percnopterus	Percnopterus
B	Περνης, ο	Pternis	Pernix	Pernes	Pernes
F	Πηλαμυς, η	Palamis	Limaria	Pelamis	Pelmys, Pilcher, or Sprat
B	Πηνελοψ	Penelops	Penelops	Penelops	Penelops
I	Πηνια, τα	Ypenia	Aciæ	Peniæ	Penia
ta	Πιθηκος, ο	Symia	Simia	Simia	Ape
T	Πιννα, η	Pina	Pinna	Pinna	Pinna
C	Πιννοτηρας, ο	Prædo Pinnarum	Pinnother	Pinnothera	Pinnoteras
C	Πιννοφυλαξ, ο	Custos Pinnarum	Custos Pinnæ	Custos Pinnarum	Guardian of the Pinna
B	Πιπος, η	Pipo	Pipo	Hippus	Pipus
B	Πιπρα, η	Pipo	Pipo	-	Wood-Pecker
B	Πιφηξ	Tifunx	Pifex	Pifex	Pifex

B	*Πλαγγος, ο*	Plangos	Planga et Clanga	Plancus	Plangus, or the Mourner
A	*Πνευμων, ο*	Pulmo	Pulmo	Pulmo	Lungs of the Sea
B	*Ποικιλις*	Pikilis	Varia	Variola	Speckled Bird
F	*Πολυπους, ο*	Polipes	Polypus	Polypus	Polypus
T	*Πορφυρα*	Purpura	Purpura	Purpura	Purple
B	*Πορφυριων, ο*	Porfirion	Porphyrio	Porphyrio	Porphyrion
I	*Πρασοκουρις, η*	Prassicaris	Blatta	Porricida	Lettuce-Eater
B	*Πρεσβυς, ο*	Senior	Senator	Senator	The Elder, or King of Birds
F	*Πριμαδιαι, αι*	Primades	Primadæ	Primadiæ	Primadiæ
F	*Πριστης, ο*	Prisus	Pristes	Pristes	Pristes
Q	*Προβατον, το*	Ovis	Ovis	Ovis	Sheep
Q	*Προξ, ο*	Prokis	Dama	Dama	Deer
B	*Πτυγξ, ο*	Pox	Ptynx	Ptynx	Ptynx
B	*Πυγαργος*	Pygartus	Pygargus	Pygargus	Pygargus
I	*Πυγολαμπις, η*	Pykokpis	Cicindela	Cicindela	Glow-worm
B	*Πυραλις*	Pirallis	Ignaria	Pyralis	Pyralis

I	*Πυραυστης, ο*	Piraustis	Pyrausta	Pyrausta	Pyraustes
B	*Πυρρουλας, ο*	Pirrus	Rubecula	Byrriola	Robin-Redbreast
F	*Ρινη, η*	Rine	Squatina	Squatina	File
F	*Ρινοβατος, ο*	Rinobatus	Squatino-Raia	Squatino-Raia	File-Ray
Q	*Σαθεριον, το*	Sathrium	Satherium	Satherion	Satherion
Q	*Σαλαμανδρα, η*	Salamandra	Salamandra	Salamandra	Salamander
F	*Σαλπη, η*	Salpa	Salpa	Salpa	Stock-Fish
F	*Σαπερδις, ο*	Saperdis	Saperdis	Saperdis	Saperdis
F	*Σαργινος, ο*	Sarginus	Sargiacus	Sarginus	Sarginus
F	*Σαργος, ο*	Sarga	Sargus	Sargus	Sargus
Q	*Σατυριον, το*	Satirium	Satyrium	Satyrion	Satyrion
Q	*Σαυρα, η*	Lacerta	Lacertus	Lacertus	Lizard
F	*Σαυρος, ο*	Lacerta	Lacertus	Lacertus	Lizard
I	*Σειρην, η*	Eiren	Sirenis	Siren	Siren
A	*Σελαχη, τα*	Selachea	Cartilaginea	Cartilaginea	Selache
F	*Σηπια*	Sepia	Sepia	Sepia	Cuttle-Fish

I	Σητις, ο	Tinea	Tinea	Tinea	Moth-Worm, or Tiny
I	Σιλφη, η	Silpha	Silpha	Silpha	Kind of Moth
F	Σινοδων, ο	Sinodon	Dentex	Dentex	Dentex
B	Σιππη, η	Speghta	Sitta	Sitta	Sippe
B	Σιττη, η	Sitta	Sitta	Sitta	Nut-Jobber
F	Σκαρος, ο	Scarus	Scarus	Scarus	Scarus
F	Σκιαινα, η	Skiena	Umbra	Umbra	Umbra, or Shadow
I	Σκνιψ, ο	Sknipus	Culex	Culex	Gnat
B	Σκολοπαξ, ο	Scolopax	Gallinago	Scolopax	Woodcock
I	Σκολοπενδρα, η	Scolopendra	Centipeda	Centipeda et Scolopendra	Scolopendra
F	Σκομβρος, ο	Scombria	Scomber	Scomber	Mackerel
F	Σκορδυλα	Scordila	Scordula	Scordula	Scordyla
I	Σκορπιος, ο	Scorpius	Scorpio	Scorpius	Scorpion
F	Σκορπιος, ο	Scorpius	Scorpio	Scorpius	Scorpion of the Sea
F	Σκορπις, η	-	Scorpio	Scorpio	Scorpion
I	Σκορπιωδες, το	Scorpiale	Bestiola similis Scorpioni	Scorpiunculus	Species of Small Scorpion

F	*Σκυλιον, το*	Scyllium	Canicula	Canicula	Little Dog
I	*Σκωληξ, ο*	Vermis	Vermis et Cossus	Vermis et Cossus	Worm
B	*Σκωψ, ο*	Scops	Asio	Asio	Small Owl
B	*Σπιζα, η*	Spiza	Fringilla	Fringilla	Chaffinch
F	*Σμαρις, η*	Karis	Cerrus	Smaris	Smaris
F	*Σμυρος, ο*	Smurus	Murus	Smurus	Murus
F	*Σπαρος, ο*	Skarus	Scarus	Scarus	Scarus
T	*Σπαταγος, ο*	Spatagius	Spatagus	Spatagus	Spatagus
A	*Σπογγος, ο*	Sfongus	Spongia	Spongia	Sponge
T	*Στρομβος, ο*	Strombus	Turbo	Turbo	Trumpet
I	*Σταφυλινος, ο*	Stafilion	Pastinaca	Pastinaca	A Fish resembling a Ray
B	*Στρουθιον, το*	Passer	Passer	Passer	Sparrow
B	*Στρουθος, ο*	Passer, et Passer qui in Lybia	Passer et Struthio Africa	Passer et Struthio Afer	Sparrow and Ostrich
B	*Συκαλις, η*	Sicallis	Ficedula	Ficedula	Becafigo, a Bird like a Wheat-Ear
F	*Συναγρις, η*	Synagris	Dentex	Synagris	Synagris

Q	*Συς, ο & η*	Porcus	Sus	Sus	Hog
I	*Σφηξ, η*	Vespa	Vespa	Vespa	Wasp
I	*Σφονδυλη, η*	Spondila	Verticullum	Sphondyla	Sphondyle
F	*Σφυραινα, η*	Firena	Malleolus	Sphyrena	Sphræna
B	*Σχοινικλος, ο*	Skinilus	Junco	Schœniclus	Schoeniclus
B	*Σχοινιων, ο*	Skinium	Junco	Schœnio vel Juncarius	Junco
T	*Σωλην, ο*	Solin	Unguis	Digitellus	Solen
F	*Ταινια, η*	Tainia	Vitta	Vitta	Tænia
Q	*Ταυρος, ο*	Taurus	Taurus	Taurus	Bull
B	*Ταων, ο*	Pavo	Pavo	Pavo	Peacock
I	*Τενθρηδων, ο*	Tenthridon	Teredo	Tenthredo	Tenthredon
I	*Τερηδων, η*	-	Teredo	Harde	Augur Caterpillar
B	*Τετριξ*	Tetrix	Tetrix	Tetrix	Tetrix
I	*Τεττιγομητρα, η*	Cicadarum Matrix	Tettigometra	Tettigometra	Mother of the Cicida, a kind of Grasshopper
I	*Τεττιγονιον, το*	Tettigonium	Cicadaster	Cicadaster	Small Cicada
I	*τεττιξ, ο*	Cicada	Cicada	Cicada	Cicada

F	Τευθις, η	Teuthis	Lolligo	Lolliguncula	Small Calamary
F	Τευθος, ο	Teuthos	Lolius	Lolius	Large Calamary
T	Τηθυα, τα	Tethya	Tubera, Vertibula	Tethya	Tethya
Q	Τιγρις, ο	Tygris	Tigris	Tigris	Tiger
F	Τιλλων, ο	Thron	Tillon	Tillon	Tillon
Q	Τραγος, ο	Hyrcus	Hircus	Hircus	He-Goat
F	Τριγλη, η	Trema, Trelia	Mullus	Mullus	Barbel
B	Τριορχης, ο	Triochis	Buteo	Buteo	Triorches, or Buzzard
F	Τριχια, η	-	-	-	Trichia
F	Τριχας, ο	Sardella	Sarda	Trichia	Trichias
F	Τριχις, η	Trikis	Sardina	Trichis	Trichis
B	Τροχιλος, ο	Trochilus	Trochilus	Trochilus	Wren
B	Τρυγγας, ο	Pigatus	Tringa	Trynga	Tryngas
B	Τρυγων, η	Turtur	Turtur	Turtur	Turtle-Dove
F	Τρυγων, η	Turtur	Pastinaca	Pastinaca	Pastinaca
Q	Τρυγων, η	Turtur	Pastinaca	Pastinaca	Turtle-Dove

B	Τυπανος, ο	Tympanius	Tympanus	Tapynus	Typanus
B	Τυραννος, ο	Tyrannus	Tyrannus	Tyrannus	Tyrant, or Tufted Wren
Q	Υαινα, η	Yena	Hyæna	Hyæna	Hyæna
B	Υβρις, η	Ydris	-	-	Hybris
A	Υδρος, ο	Ydros	Natrix	Natrix	Water-Serpent
I	Υπερα, τα	Ypera	Superæ	Hyperia	Hypera
B	Υπολαις, η	Ypolais	Curuca	Curuca	Hen-Linnet
A	Υποτριορχης, ο	Ypotriorcha	Subbuteo	Subbuteo	Water-Snake
Q	Υς, ο & η	Sus	Sus	Sus	Hog
Q	Υς, αγριος, ο	Porcus Sylvestris	Sus Ferus	Sus Ferus	Wild-Boar
Q	Υστριξ, η	Ystrix	Hystrix	Hystrix, Porcus-Spinus	Porcupine
F	Φαγρος, ο	Fagrus	Phager, Pagrus	Pagrus	Pagrus
I	Φαλαγγιον, το	Falangium, Aranea	Phalangium	Phalangium	Small Phalangium
I	Φαλαγξ, ο	Falanx	Phalanx	Phalangium	Phalangium
F	Φαλαινα, η	Balæna	Balæna	Balæna	Whale
B	Φαλαρις, η	Falaris	Phalaris	Phalaris	Phalaris

F	*Φαλερικη, η*	Falerica	Phalerica	Phalerica	Phalerica
B	*Φασιανος, ο*	Fasianus	Phasianus	Phasianus	Pheasant
B	*Φαττα, η*	Fatga, Fassa	Palumbes	Palumbes	Large Ring-Dove
B	*Φαψ, η*	-		-	Small Ring-Dove
B	*Φηνη, η*	Fena	Ossifraga	Ossifraga	Osprey
I	*Φθειρες, οι*	Pediculi	Pediculi	Pediculi	Lice
B	*Φοινικουρος, ο*	Finikurus	Ruticilla	Phænicurgus	Roach
F	*Φοξινος, ο*	Foxinus	Phoxinus	Phoxinus	Phoxinus
Q	*Φρυνη & Φρυνος*	Fyrnus, Frinus	Rubeta	Rubeta	Red Toad
F	*Φυκις, η*	Fykys	Phuca	Fucilla	Phycis
B	*Φωιξ, η*	Foix	Phoix	Phoix	Phoix
F	*Φωκαινα, η*	Fokena	Phocæna, *i.e.* Tirsio	Phocæna	Phocæna
2	*Φωκη, η*	Bos Marinus	Vitulus Marinus	Vitulus Marinus	Sea-Calf
F	*Φωλις, η*	Folida	Pholis	Pholis	Pholis
I	*Φωρ, ο*	For	Fur	Fur	Bee-Thief
Q	*Χαλκις, η*	Khalkis	Chalcis	Chalcis	Chalcis

B	*Χαλκις, η*	Chalcis	Chalcis	Chalcis	Chalcis
F	*Χαλκις, η*	Khalkis	Chalcis, Ærica	Chalcis	Chalcis
Q	*Χαμαιλεων, ο*	Chameleon	Chamælco	Chamælco	Chameleon
F	*Χανη, η*	Khanna	Hiatula	Channa	Cockle
B	*Χαραδριος, ο*	Kalandra	Rupex	Charadrius	Bird of the Rock
B	*Χελιδων, η*	Hirundo	Hirundo	Hirundo	Swallow
F	*Χελιδων, η*	Yrundo	Hirundo	Hirundo	Swallow of the Sea
F	*Χελων, ο*	Khelo	Labeo	Labeo	Thick-Lip
2	*Χελωνη, η*	Tortuca	Testudo	Testudo	Tortoise
T	*Χημαι, αι*	Chemæ	Chamæ	Chamæ	Muscles
B	*Χην, ο*	Anser	Anser	Anser	Goose
B	*Χηναλωπηξ, ο*	Cinalopex	Vulpanser	Vulpanser	Goose-Fox
B	*Χλωρευς, ο*	Khloreus	Luteus, et Lutca	Chloreus	Chloreus
	Χλωρευς, ο	Khlaris	Lutcola	Vireola, Chloris	Greenfinch
B	*κλωρις, η*	Khlorio	Vireo	Vireo, Virido	Chlorion
Q	*Χοιροπιθηκος, ο*	Choriopithikus	Simia-Porcaria	Simia quæ rostrum præ se fert porcinum	Hog-Ape

F	*Χρεμψ, ο*	Kiremis	-	Chremps	Chremps
F	*Χρομις, ο*	Chromis	Chromis	Chromis	Chromis
B	*Χρυσομιτρις, η*	Krisomettris	Aurivittis	Rhysometris	Gold-Cap
F	*Χρυσοφρυς, ο*	Aurata	Aurata	Aurata	Gold-Fish
B	*Ψαρος, ο*	Sparus	Sturnus	Sturnus	Starling
I	*Ψην, ο*	Judicium	Ficarius Culex	Psen	Psen
F	*Ψηττα, η*	Pecten	Passer	Passer	Plaice
B	*Ψιττακη, η*	Itacus	Psittaca	Psittacus	Parrot
I	*Ψυλλον, το*	Psilla	Pulex	Pulex	Psyllon
I	*Ψυλλη, η*	Pulex	Pulex	Pulex	Flea
F	*Ψυλων, ο*	Tylo	Psyllo	Fullo	Psylon
I	*Ψυχη, η*	Psycha	Papilio	Papilio	Butterfly
B	*Ωτις, η*	Bistarda	Otis, *i.e.* Avis-Tarda	Otis	Bustard
B	*Ωπος, ο*	Otus	Otus	Nocturnus	Great Owl